石油教材出版基金资助项目

石油高等院校特色规划教材

Mechanics of Oil and Gas Flow in Porous Media

# 油气渗流力学

（第二版·英汉对照·富媒体）

李 璗　陈军斌　编著

石油工业出版社

## 内 容 提 要

本书以英汉对照加富媒体展示的方式,将"油气渗流力学"课程所涉及的知识立体地展现在读者面前。全书从驱动力和驱动方式出发,在对达西定律分析的基础上,遵循由浅入深的认识规律,详细介绍了单相不可压缩液体的稳定渗流理论、刚性水压驱动下的油井干扰理论、微可压缩流体的不稳定渗流理论、天然气的渗流规律、水驱油理论基础、油气两相渗流理论(溶解气驱)、双重介质中的渗流理论、非牛顿液体的渗流理论等。

本书可作为石油工程、石油地质、地下水工程、油田化学等专业本科生国际学生的专业教材,也可作为相关专业研究生的参考书,还可供从事油气田勘探与开发的科研技术人员参考。

**图书在版编目(CIP)数据**

油气渗流力学:富媒体:英汉对照 / 李璗,陈军斌编著. -- 2版. -- 北京:石油工业出版社,2025.1. (石油高等院校特色规划教材). -- ISBN 978-7-5183-7367-3

Ⅰ. TE312

中国国家版本馆 CIP 数据核字第2025W3W434号

出版发行:石油工业出版社
    (北京朝阳区安华里二区1号楼  100011)
    网    址:www.petropub.com
    编辑部:(010)64523733  图书营销中心:(010)64523633
经    销:全国新华书店
排    版:北京密东文创科技有限公司
印    刷:北京中石油彩色印刷有限责任公司

2025年1月第2版    2025年1月第1次印刷
787毫米×1092毫米    开本:1/16    印张:22.25
字数:539千字

定价:52.00元
(如出现印装质量问题,我社图书营销中心负责调换)
版权所有,翻印必究

# 第二版前言

当今世界,数字技术深刻改变了人类的生产生活方式、改变了知识传递方式和人类学习方式。千百年来形成的基本教学模式、教学理念面临改变。充分利用现代信息技术、大数据、人工智能技术,加强泛在学习环境建设与个性化学习手段开发,加快发展面向每个人、适合每个人、更加开放灵活的教育体系,抓住数字化带来的新机遇,加强不同层面的系统谋划,丰富不同维度的交流模式等,是我国开辟教育发展新赛道和塑造教育发展新优势的重要突破口。

基于此,西安石油大学渗流力学课程组在原"十二五"国家级规划教材《油气渗流力学》《油气渗流力学(英文版)》《油气渗流力学学习指南》的基础上,对获得陕西省优秀教材特等奖的石油行业特色规划教材《油气渗流力学(英汉对照·富媒体)》开展修订和完善。修订以后的教材具有如下特色:

(1)能始终坚持马克思主义指导地位,坚持以习近平新时代中国特色社会主义思想为指导,能够将马克思主义立场、观点、方法等课程思政内容贯穿教材建设始终,能够遵循教育教学规律和人才培养规律,体现先进教育理念,将价值塑造、知识传授和能力培养三者融为一体。

(2)教材内容涵盖了油气渗流力学的主要内容和知识点,能够准确、科学地把握油气渗流力学的基本概念、基本理论、基本方法。内容涵盖面广、简明扼要,脉络清晰、由浅入深,符合学生的学习规律。在对各章节中的有重大实用价值的公式、方法等方面讲解时,能将启发式思维贯穿教材建设始终,注重发挥学生学习的能动性和培养学生解决实际问题的能力。

(3)将石油与天然气工程学科在低渗(特低渗)油田和非常规油气开发方面的新理论、新技术、新方法以及教师的最新科研成果作为案例适时引入教材,理论联系实际,增强教材的实用性,使学生在学习基本理论、基础知识的同时,能够更好地了解渗流力学理论发展的前沿,开阔学生视野。

(4)练习题目充足,难易适中,便于学生课后对所学知识理解掌握。教材内容编排科学合理,文字准确流畅,符合规范化要求,插图质量高,图文配合得当,版式设计专业,装帧印刷美观,整本教材文字生动简洁,图表清晰规范。

修订以后的教材具有如下创新点:

(1)以英汉对照的形式呈现内容,学习者在学习专业知识的同时,可以了解专业词汇的表达方法,有利于提高英语阅读能力。作为富媒体教材,对应课程建立了成熟的网络教学平台(平台1至平台4),有利于读者利用碎片化时间自由、自主、自如地展开学习,学习方式更加灵活,学习效率可大幅提高。所提供的富媒体资源,形象生动,可以使学习者非常直观地了解油气水在地下流动规律,加深对课程内容的理解。

(2)补充完善了全书及各章知识图谱，在每一章最后都以二维码形式，补充各章学习指南、习题参考答案、课程思政案例，丰富了学习体验。

全书由陈军斌教授统稿，张益教授、袁士宝教授、谢青副教授重点负责线上教学以及课程思政内容的挖掘整理，谢青副教授提供了第九章习题的答案，王晓明博士参与了全书知识图谱的归纳整理和全书校稿。

本书在编写过程中得到石油工业出版社的大力支持和帮助，并融入了教师的科研成果，特此向有关单位和个人致以诚挚的谢意。

本书力求思路清晰、重点突出、题目典型，但由于编著者水平有限，教材中难免有不当甚至错误之处，恳请读者批评指正。

仅以本书奉献给我最敬爱的老师李璆教授！李老师，我们永远怀念您！

陈军斌<br>2024 年 9 月

| 平台1 智慧树网 | 平台2 学银在线 | 平台3 学堂在线 | 平台4 数字石油学院 | 思政案例 |

# 第一版前言

　　1985年，西安石油大学石油工程专业（原采油工程专业）恢复招生，当时我国在"渗流力学"课程教学方面可供选择使用的教材极少。鉴于此，结合行业特色和学校实际，着手进行"渗流力学"课程教材编写工作。1986年，"渗流力学"课程讲义编撰完成，根据学生使用后的反馈意见修订并改进，于1991年"渗流力学"课程内部讲义问世，一直使用到2000年。

　　随着课程教学研究与改革的不断深入，特别是在教育部所实施的以进一步提高人才培养质量为目的的"高等学校教学质量与教学改革工程"建设工作带动下，任课教师以提高教学水平和教学质量为目的持续进行课程建材建设工作。从2001—2015年先后在陕西科学技术出版社和石油工业出版社出版发行了《油气渗流力学基础》《渗流力学与渗流物理》等5部教材。其中，成套教材——《油气渗流力学》《油气渗流力学（英文版）》《油气渗流力学学习指南》于2014年获评"十二五"普通高等教育国家级规划教材，于2015年荣获陕西省高等学校优秀教材一等奖。

　　为方便读者学习，本教材采取中英文对照编排方式，读者在学习专业知识的同时，可以了解专业词汇的表达方法，有利于读者提高英语阅读能力。本教材为富媒体教材，对应课程建立了成熟的网络教学平台（平台1～平台4），为陕西省高等教育MOOC平台和石油工业出版社数字学院平台首批上线的石油类专业基础课程之一，有利于读者利用碎片化时间自由、自主、自如地展开学习，学习方式更加灵活，学习效率可大幅提高。

　　本书在编写过程中得到石油工业出版社的大力支持和帮助，并融入了教师的科研成果，特此向有关单位和个人致以诚挚的谢意。

　　本书力求思路清晰、重点突出、题目典型，但由于作者水平有限，教材中难免有不当甚至错误之处，恳请读者批评指正。

陈军斌

2018年10月

平台1　智慧树网　　　平台2　学银在线　　　平台3　学堂在线　　　平台4　数字石油学院

# Contents/目录

**1 Basic law of seepage/渗流的基本规律**

1.1 Static state in reservoirs ……………… 15
1.2 Driving forces and driving modes in reservoir ……………………………… 20
1.3 Basic laws of Darcy flow ………… 26
1.4 The limitations of Darcy flow and non-Darcy flow ……………………………… 32
Exercises …………………………… 37

1.1 油气藏静态状况 ……… 15
1.2 油气藏中的驱动力和驱动方式 ……………… 20
1.3 达西渗流的基本规律 …… 26
1.4 达西渗流的局限性与非达西渗流 …………… 32
习 题 …………………… 37

**2 Steady-state seepage in porous media of single-phase incompressible fluid/单相不可压缩液体的稳定渗流**

2.1 Three basic forms of seepage in porous media ……………………………… 42
2.2 Planar one-dimensional steady seepage of single-phase incompressible fluid …… 44
2.3 Planar radial steady seepage of single-phase incompressible fluid ………… 52
2.4 Imperfect well ……………………… 59
2.5 Steady well testing ………………… 64
2.6 Differential equation of single-phase fluid seepage in porous media ……… 67
Exercises …………………………… 80

2.1 三种基本渗流方式 …… 42
2.2 单相不可压缩液体的平面一维稳定渗流 ……… 44
2.3 单相不可压缩液体的平面径向稳定渗流 ……… 52
2.4 油井的不完善性 ……… 59
2.5 稳定试井 ……………… 64
2.6 单相流体渗流的微分方程 ……………………… 67
习 题 …………………… 80

**3 Interference theory of wells under rigid water drive/刚性水压驱动下的油井干扰理论**

3.1 Phenomenon of interference between wells ……………………………… 87
3.2 Superposition principle of potential …… 89
3.3 Method of mirror ………………… 96
3.4 Application of complex function theory in the planar seepage field ………… 118
Exercises …………………………… 147

3.1 井间干扰现象 ………… 87
3.2 势的叠加原理 ………… 89
3.3 镜像反映法 …………… 96
3.4 复变函数理论在平面渗流场中的应用 ………… 118
习 题 …………………… 147

— *1* —

# 4 Unsteady seepage of slightly compressible fluid/微可压缩液体的不稳定渗流

- 4.1 The physical process of elastic fluid seepage towards wells ⋯⋯⋯⋯⋯ 153
- 4.2 The planar one-dimensional unsteady seepage in semi-infinite formation ⋯⋯⋯ 155
- 4.3 Pressure transient law of elastic fluid unsteady seepage towards wellbore ⋯⋯⋯ 163
- 4.4 Approximate solution of pressure change of fluid seepage towards well in finite closed elastic formation ⋯⋯⋯⋯⋯⋯ 172
- 4.5 Multiple-well interference of elastic unsteady seepage ⋯⋯⋯⋯⋯⋯⋯ 177
- 4.6 Unsteady well-test analysis ⋯⋯⋯⋯ 184
- Exercises ⋯⋯⋯⋯⋯⋯⋯⋯⋯⋯⋯⋯ 192

- 4.1 弹性液体向井渗流的物理过程 ⋯⋯⋯⋯⋯⋯⋯⋯⋯⋯ 153
- 4.2 半无限大地层平面一维不稳定渗流 ⋯⋯⋯⋯⋯⋯ 155
- 4.3 弹性液体向一口井不稳定渗流的压力传导规律 ⋯⋯ 163
- 4.4 有限封闭弹性地层内液体向井渗流压力变化的近似解 ⋯⋯⋯⋯⋯⋯⋯⋯ 172
- 4.5 弹性不稳定渗流的多井干扰 ⋯⋯⋯⋯⋯⋯⋯⋯ 177
- 4.6 不稳定试井分析 ⋯⋯⋯⋯ 184
- 习 题 ⋯⋯⋯⋯⋯⋯⋯⋯ 192

# 5 Seepage law of natural gas/天然气的渗流规律

- 5.1 Properties of natural gas and its basic differential equation of seepage ⋯⋯⋯⋯ 199
- 5.2 Steady seepage of gas ⋯⋯⋯⋯⋯ 207
- 5.3 Unsteady seepage of gas ⋯⋯⋯⋯ 213
- Exercises ⋯⋯⋯⋯⋯⋯⋯⋯⋯⋯⋯⋯ 218

- 5.1 天然气的性质及其渗流的基本微分方程 ⋯⋯⋯⋯⋯⋯ 199
- 5.2 气体的稳定渗流 ⋯⋯⋯⋯ 207
- 5.3 气体的不稳定渗流 ⋯ 213
- 习 题 ⋯⋯⋯⋯⋯⋯⋯⋯ 218

# 6 Foundation of water/oil displacement theory/水驱油理论基础

- 6.1 Piston like displacement of oil by water ⋯⋯ 224
- 6.2 Bottom water coning ⋯⋯⋯⋯⋯⋯ 233
- 6.3 Theoretical foundation of non-piston like displacement of oil by water ⋯⋯⋯⋯ 236
- Exercises ⋯⋯⋯⋯⋯⋯⋯⋯⋯⋯⋯⋯ 255

- 6.1 活塞式水驱油 ⋯⋯⋯⋯ 224
- 6.2 底水锥进 ⋯⋯⋯⋯⋯⋯ 233
- 6.3 非活塞式水驱油理论基础 ⋯⋯⋯⋯⋯⋯⋯⋯⋯⋯ 236
- 习 题 ⋯⋯⋯⋯⋯⋯⋯⋯ 255

## 7 Seepage theory of oil-gas two phases (dissolved gas drive) / 油气两相渗流理论(溶解气驱)

7.1 Basic differential equation of oil-gas two-phases seepage ·················· 261
7.2 Oil-gas two-phase steady seepage ············ 265
7.3 Unsteady seepage of gassy fluid ······ 273
Exercises ············································· 279

7.1 油气两相渗流的基本微分方程 ················· 261
7.2 油气两相稳定渗流 ······ 265
7.3 混气液体的不稳定渗流 ································· 273
习题 ····················································· 279

## 8 Seepage in dual-medium / 双重介质中的渗流

8.1 Seepage characteristics in dual-medium ······ 283
8.2 Basic seepage equation of single-phase slightly compressible fluid in dual-medium ·········· 285
8.3 Pressure distribution in infinite dual-medium ·············································· 290
8.4 Oil-water two-phase seepage in dual-medium ··· 293
8.5 Seepage of fluid in pure fracture formation ·············································· 301
Exercises ············································· 304

8.1 双重介质中的渗流特征 ··· 283
8.2 双重介质中单相微可压缩液体的基本渗流方程 ·········· 285
8.3 无限大双重介质地层中的压力分布 ················ 290
8.4 双重介质中的油水两相渗流 ································· 293
8.5 在纯裂缝地层中液体的渗流 ································· 301
习题 ····················································· 304

## 9 Seepage of non-Newtonian liquid / 非牛顿液体的渗流

9.1 Mechanical behavior and type of non-Newtonian liquid ·················· 308
9.2 Seepage of non-Newtonian power law liquid ·············································· 315
Exercises ············································· 324

9.1 非牛顿液体的力学特性与类型 ·················· 308
9.2 非牛顿幂律液体的渗流 ································· 315
习题 ····················································· 324

**References/参考文献** ······································································· 326

**Appendix/附录** ··································································· 327

    Appendix A    Seepage equation in cylindrical coordinate/柱坐标形式的渗流方程 ········ 328

    Appendix B    To obtain the solution of one-dimensional unsteady seepage by the Laplace transformation/用拉氏变换求液体作平面一维不稳定渗流的解 ··················································································· 331

    Appendix C    Calculation procedure of error function erf($x$)/误差函数 erf($x$)计算程序 ··················································································· 333

    Appendix D    Calculation procedure of exponential integral function $-\mathrm{Ei}(-x)$/指数积分函数 $-\mathrm{Ei}(-x)$计算程序 ····································· 338

    Appendix E    Calculation procedure of function $\Gamma(x)$ /$\Gamma(x)$函数计算程序 ············ 339

    Appendix F    Conversion relationship between the commonly used units in reservoir engineering/油气藏工程常用单位之间的换算关系································ 341

# 富媒体资源目录

| 序号 | 名称 | 页码 |
|---|---|---|
| 1 | Dynamic graph 1-1 Porosity/动态图1-1 孔隙 | 15 |
| 2 | Dynamic graph 1-2 Oil,gas and water in pores/动态图1-2 孔隙中的油气水 | 15 |
| 3 | Dynamic graph 1-3 Open reservoir/动态图1-3 敞开式油气藏 | 16 |
| 4 | Dynamic graph 1-4 Deformation of rock frame/动态图1-4 岩石骨架的变形 | 23 |
| 5 | Dynamic graph 1-5 Darcy flow/动态图1-5 达西渗流 | 29 |
| 6 | Chapter 1 Study Guide/第一章学习指南 | 38 |
| 7 | Dynamic graph 2-1 Planar one-dimensional seepage/动态图2-1 平面一维渗流 | 44 |
| 8 | Dynamic graph 2-2 Planar radial fluid flow/动态图2-2 平面径向渗流 | 44 |
| 9 | Dynamic graph 2-3 (a) Partial penetration well/动态图2-3 (a)打开程度不完善井 | 60 |
| 10 | Dynamic graph 2-3 (b) Imperfect characteristic well/动态图2-3 (b)打开性质不完善井 | 60 |
| 11 | Dynamic graph 2-3 (c) Double imperfect well/动态图2-3 (c)双重不完善井 | 60 |
| 12 | Chapter 2 Study Guide/第二章学习指南 | 83 |
| 13 | Dynamic graph 3-1 The interference of two production wells/动态图3-1 两口生产井的干扰 | 89 |
| 14 | Dynamic graph 3-2 The interference of a production well and an injection well/动态图3-2 一口生产和一口注水井的干扰 | 89 |
| 15 | Dynamic graph 3-3 Streamline diagram of uniform-intensity one source and one sink in infinite formation/动态图3-3 无穷大地层等强度一源一汇流线图 | 99 |
| 16 | Dynamic graph 3-4 Streamline diagram of the interference of two sinks in infinite formation/动态图3-4 无穷大地层两汇干扰流线图 | 106 |
| 17 | Chapter 3 Study Guide/第三章学习指南 | 149 |
| 18 | Dynamic graph 4-1 The curve of pressure distribution producing by single constant rate of well in constant pressure boundary/动态图4-1 定压边界,井以定产量生产的压力分布曲线 | 154 |
| 19 | Dynamic graph 4-2 The curve of pressure distribution producing by constant pressure of well with constant pressure boundary/动态图4-2 定压边界,井以恒定压力生产的压力分布曲线 | 154 |

— 1 —

续上表

| 序号 | 名称 | 页码 |
|---|---|---|
| 20 | Dynamic graph 4-3 The curve of pressure distribution producing with constant rate in closed boundary/动态图4-3 封闭边界,井以定产量生产的压力分布曲线 | 156 |
| 21 | Dynamic graph 4-4 The curve of pressure distribution producing with constant pressure in closed boundary/动态图4-4 封闭边界,井以恒定压力生产的压力分布曲线 | 156 |
| 22 | Dynamic graph 4-5 The pressure distribution curve of unstable seepage flow with constant bottom hole pressure/动态图4-5 井底压力为定值的不稳定渗流压力分布曲线 | 159 |
| 23 | Dynamic graph 4-6 The production curve of unsteady seepage/动态图4-6 不稳定渗流产量分布曲线 | 162 |
| 24 | Dynamic graph 4-7 The pressure distribution curve of unstable seepage with constant production/动态图4-7 产量恒定时的不稳定渗流压力分布曲线 | 167 |
| 25 | Dynamic graph 4-8 The pressure distribution curve of injection well/动态图4-8 注水井的压力分布曲线 | 169 |
| 26 | Chapter 4 Study Guide/第四章学习指南 | 195 |
| 27 | Chapter 5 Study Guide/第五章学习指南 | 219 |
| 28 | Dynamic graph 6-1 Planar radial piston like displacement of oil by water/动态图6-1 平面径向活塞式水驱油 | 229 |
| 29 | Dynamic graph 6-2 Non-piston like displacement of oil by water/动态图6-2 非活塞式水驱油 | 238 |
| 30 | Chapter 6 Study Guide/第六章学习指南 | 257 |
| 31 | Chapter 7 Study Guide/第七章学习指南 | 279 |
| 32 | Chapter 8 Study Guide/第八章学习指南 | 304 |
| 33 | Chapter 9 Study Guide/第九章学习指南 | 325 |

# Basic law of seepage

# 渗流的基本规律

```
┌──────────────┐
│石油、天然气、  │
│煤层气、地下水  │      ┌────────┐      ┌────────┐      ┌────────┐
│和地热等在储层  │─────▶│ 地下渗流 │─────▶│ 渗流力学 │◀─────│ 工程渗流 │
│中的流动       │      └────────┘      └────────┘      └────────┘
└──────────────┘                          ▲                ▲
                                          │                │
                                    ┌────────┐      ┌──────────────┐
                                    │ 生物渗流 │      │流体在人造多孔 │
                                    └────────┘      │介质中的流动，  │
                                          ▲         │比如加工业上各 │
                                          │         │种过滤器中的流 │
                                  ┌──────────────┐  │动，流化床的流 │
                                  │人、动物和植物 │  │化床通过流体等 │
                                  │体内流体的流动，│  └──────────────┘
                                  │如血液、淋巴液、│
                                  │水分和养料在人 │
                                  │体内的运动，在 │
                                  │植物体内的输运 │
                                  │等            │
                                  └──────────────┘
```

渗流力学分类

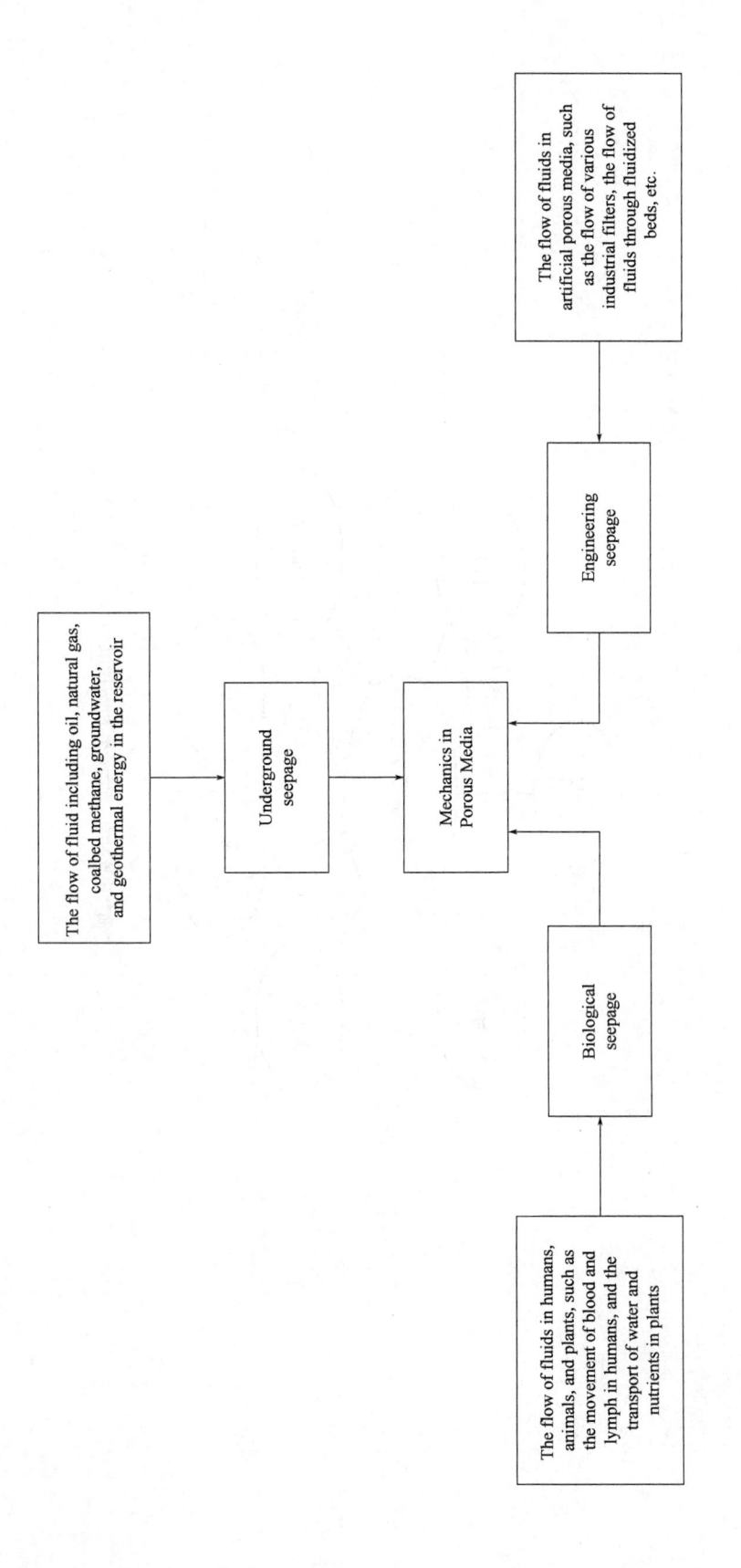

```
┌─────────────────┐
│ The flow of     │
│ fluid including │         ┌──────────┐      ┌──────────┐
│ oil, natural    │         │Underground│     │Mechanics │
│ gas, coalbed    │────────▶│ seepage  │────▶│in Porous │
│ methane,        │         │          │     │  Media   │
│ groundwater,    │         └──────────┘     └──────────┘
│ and geothermal  │
│ energy in the   │
│ reservoir       │
└─────────────────┘
```

Classification of Seepage Mechanics

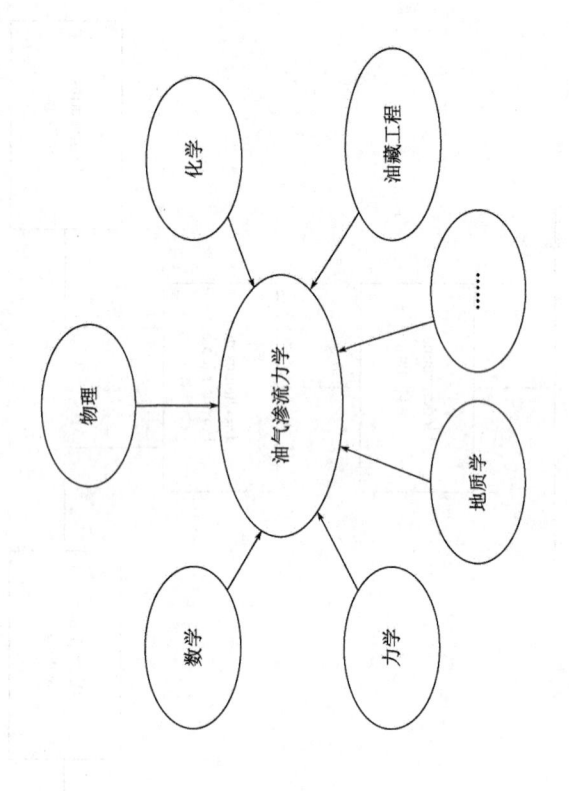

"油气渗流力学"课程内容的学科构成

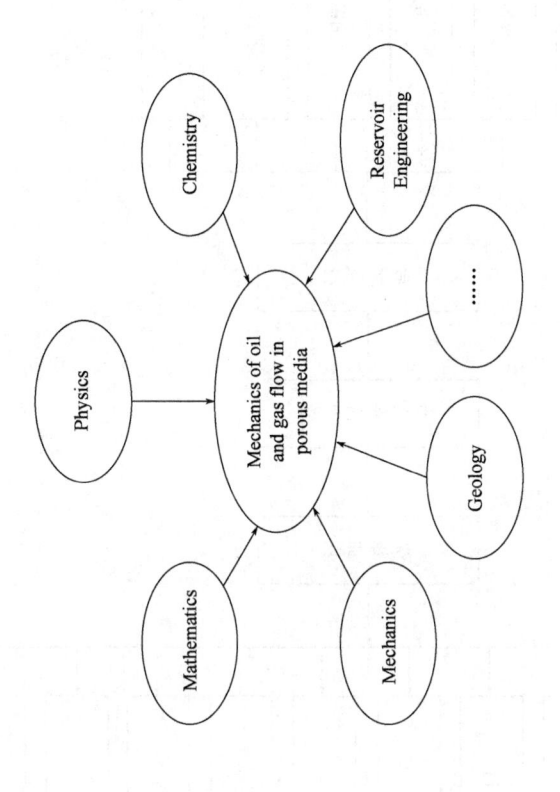

The subject composition of the course content of "Mechanics of oil and gas flow in porous media"

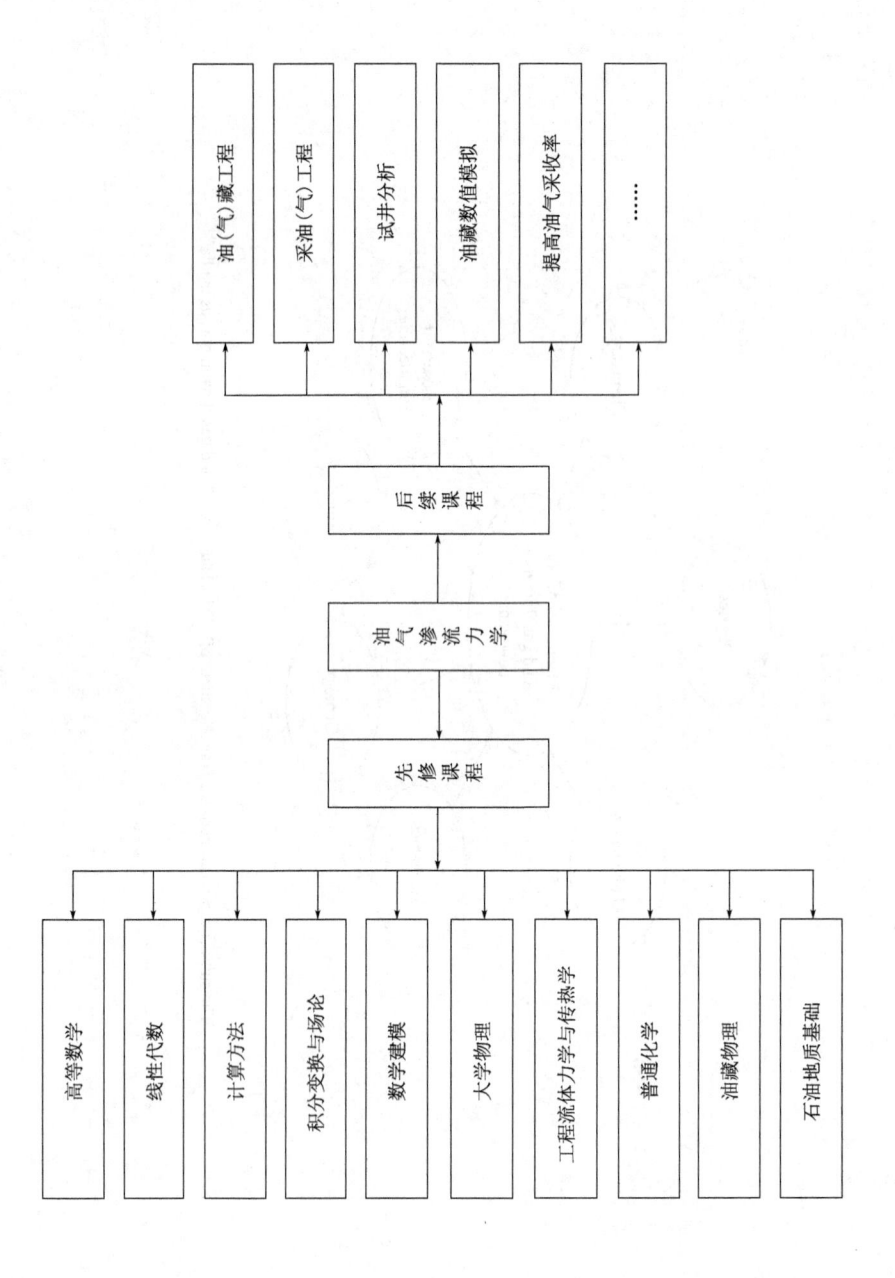

"油气渗流力学"的先修与后续课程

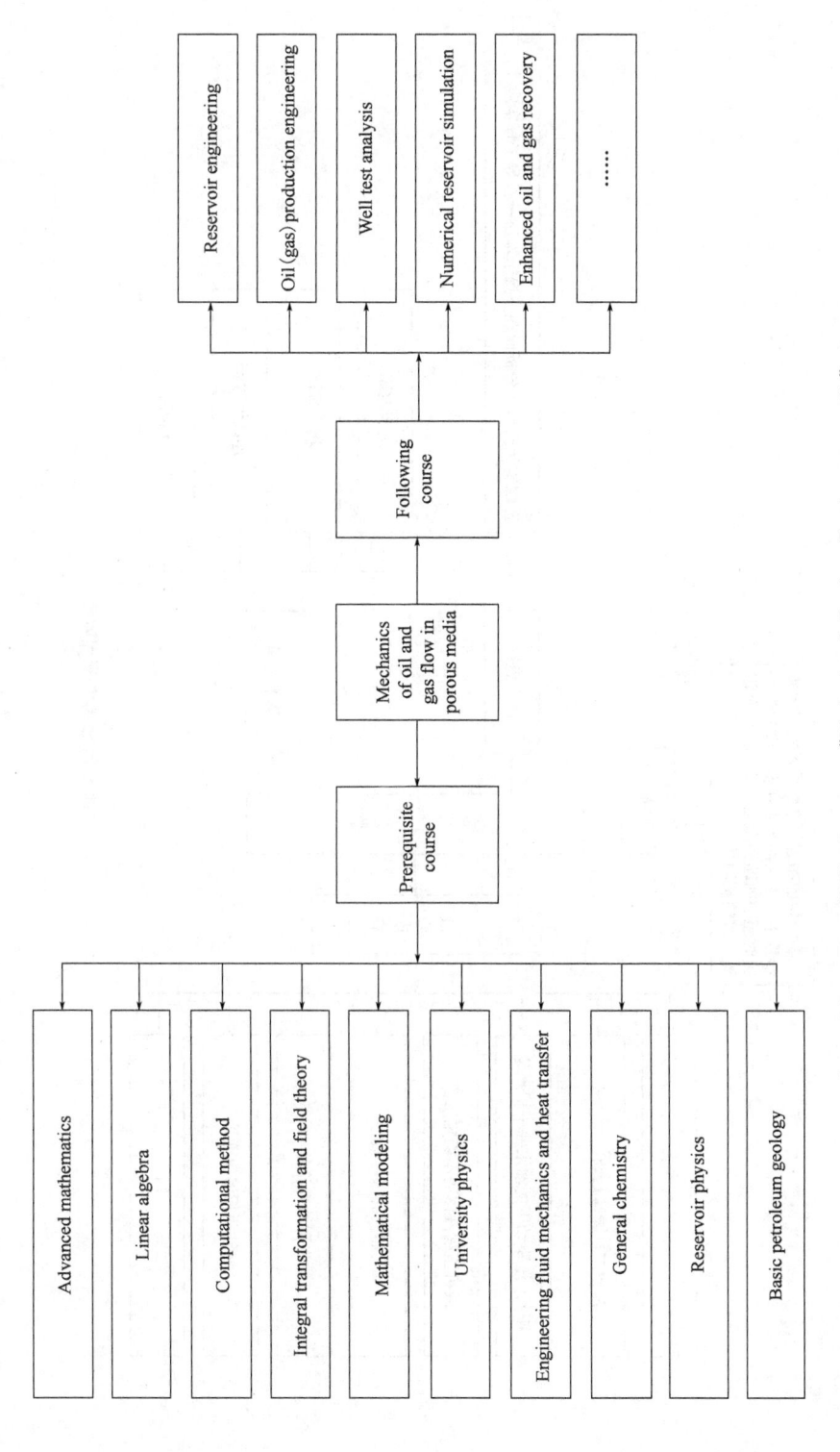

```
                    ┌─ Reservoir engineering
                    ├─ Oil (gas) production engineering
                    ├─ Well test analysis
  Following course ─┤
                    ├─ Numerical reservoir simulation
                    ├─ Enhanced oil and gas recovery
                    └─ ……

  Mechanics of oil and gas flow in porous media

                     ┌─ Advanced mathematics
                     ├─ Linear algebra
                     ├─ Computational method
                     ├─ Integral transformation and field theory
                     ├─ Mathematical modeling
  Prerequisite course┤─ University physics
                     ├─ Engineering fluid mechanics and heat transfer
                     ├─ General chemistry
                     ├─ Reservoir physics
                     └─ Basic petroleum geology
```

Prerequisite course and Following course "Mechanics of oil and gas flow in porous media"

"油气渗流力学"课程体系

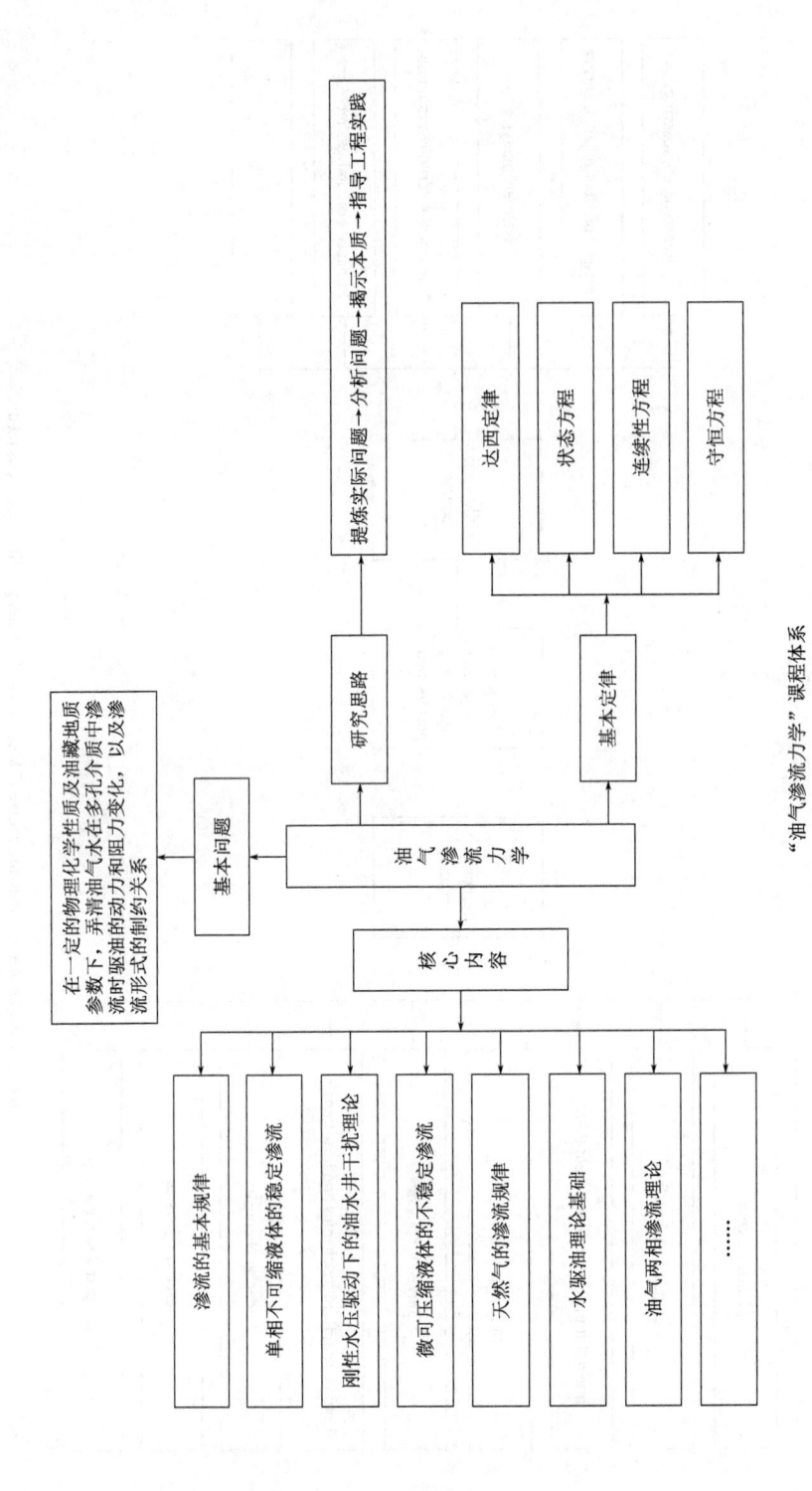

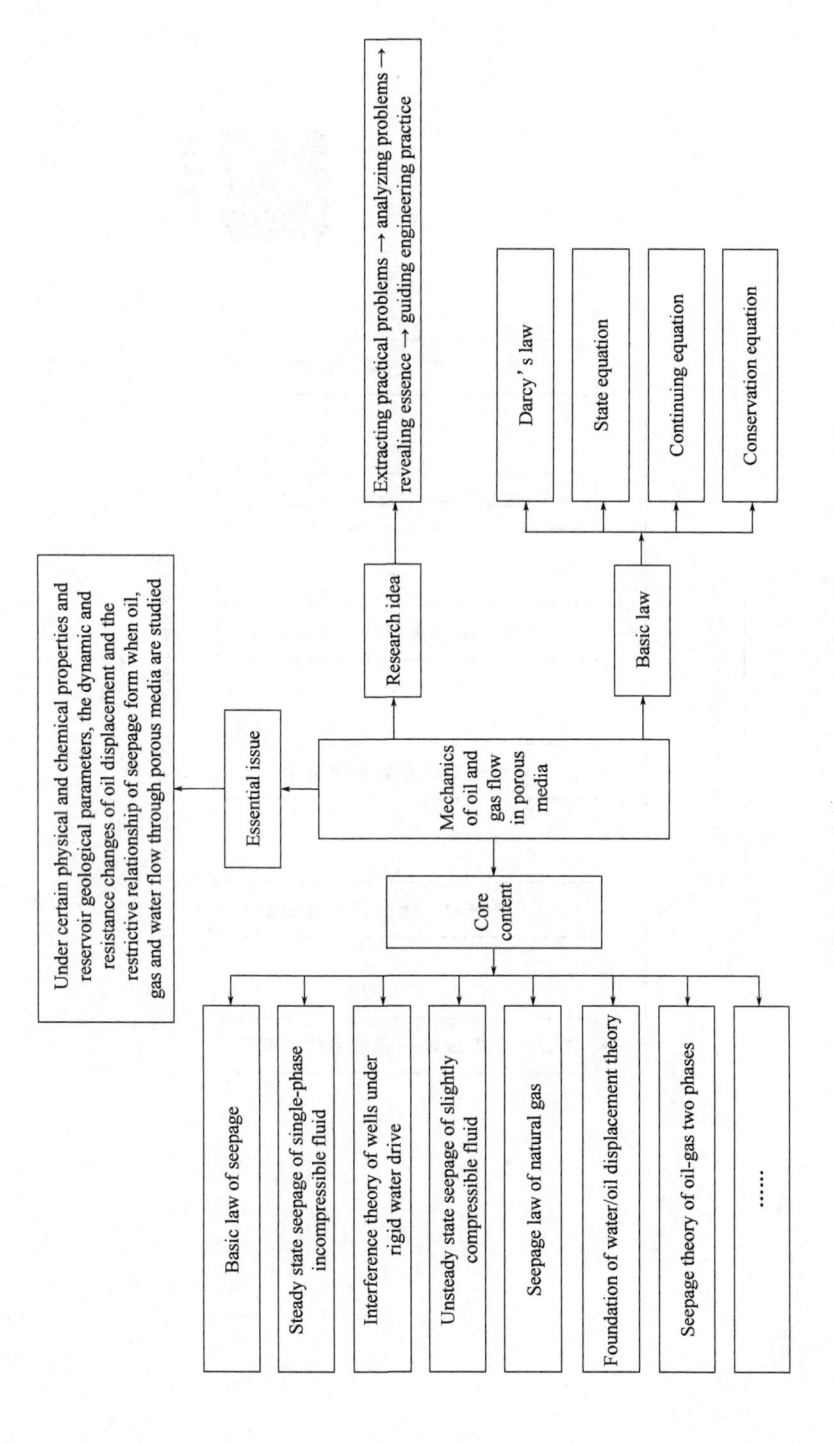

Curriculum system of "Mechanics of oil and gas flow in porous media"

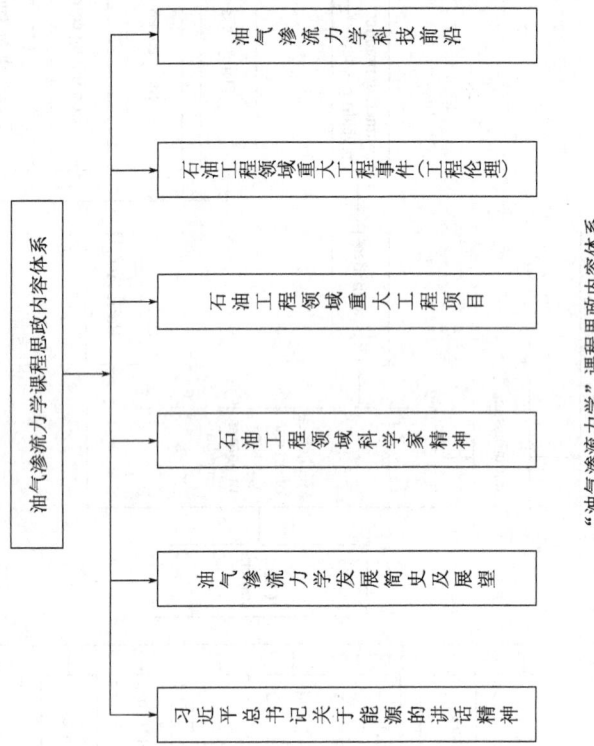

思政案例

```
                        ┌─── 油气渗流力学科技前沿
                        │
                        ├─── 石油工程领域重大工程事件（工程伦理）
                        │
油气渗流力学课程思政内容体系 ──┼─── 石油工程领域重大工程项目
                        │
                        ├─── 石油工程领域科学家精神
                        │
                        ├─── 油气渗流力学发展简史及展望
                        │
                        └─── 习近平总书记关于能源的讲话精神
```

"油气渗流力学"课程思政内容体系

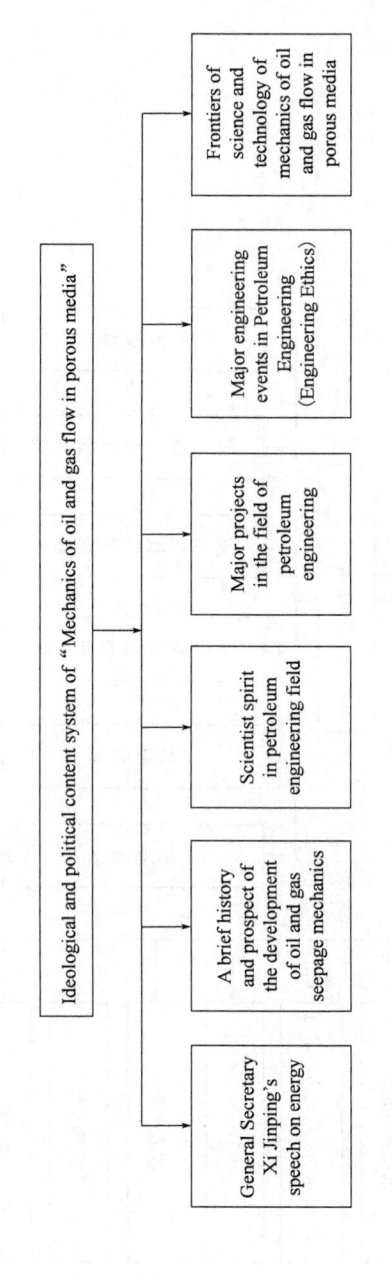

Ideological and political content system of "Mechanics of oil and gas flow in porous media"

- General Secretary Xi Jinping's speech on energy
- A brief history and prospect of the development of oil and gas seepage mechanics
- Scientist spirit in petroleum engineering field
- Major projects in the field of petroleum engineering
- Major engineering events in Petroleum Engineering (Engineering Ethics)
- Frontiers of science and technology of mechanics of oil and gas flow in porous media

Ideological and political content system of "Mechanics of oil and gas flow in porous media"

# 第一章知识图谱

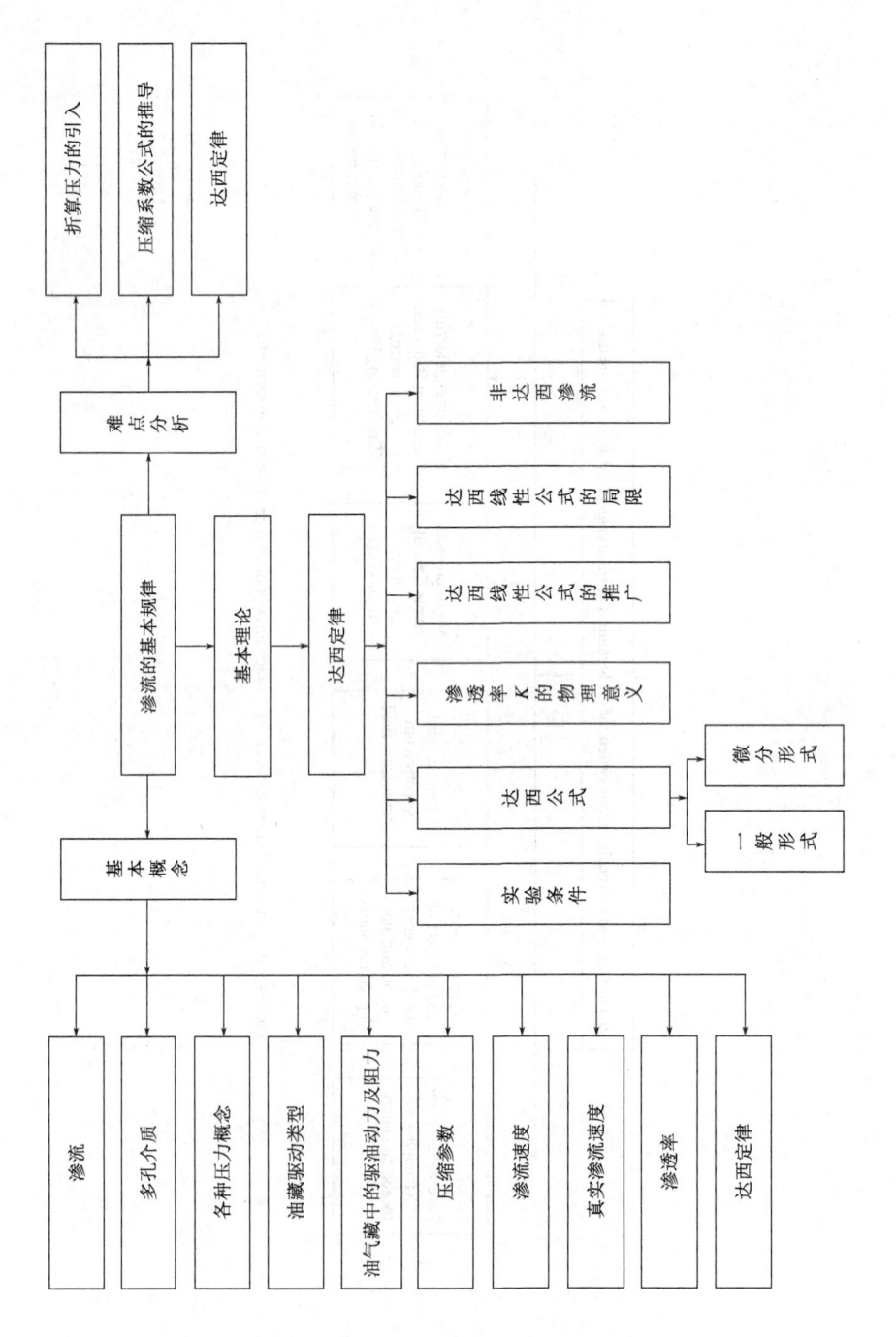

```
                                    ┌─ 折算压力的引入
                                    │
                          难点分析 ──┼─ 压缩系数公式的推导
                          ↑         │
                          │         └─ 达西定律
                          │
                                              ┌─ 非达西渗流
                                              │
                                              ├─ 达西线性公式的局限
                                              │
                          渗流的基本规律       ├─ 达西线性公式的推广
                          → 基本理论 → 达西定律┤
                          │                   ├─ 渗透率K的物理意义
                          │                   │
                          │                   │              ┌─ 微分形式
                          │                   ├─ 达西公式 ───┤
                          │                   │              └─ 一般形式
                          │                   │
                          │                   └─ 实验条件
                          ↓
                        基本概念
                          │
      ┌──────┬──────┬──────┬──────┬────────────┬──────┬──────┬──────────┬──────┬──────┐
      │      │      │      │      │            │      │      │          │      │
    渗流  多孔介质 各种  油藏  油气藏中的    压缩  渗流  真实渗      渗透率 达西
              压力  驱动  驱油动力       参数  速度  流速度          定律
              概念  类型  及阻力
```

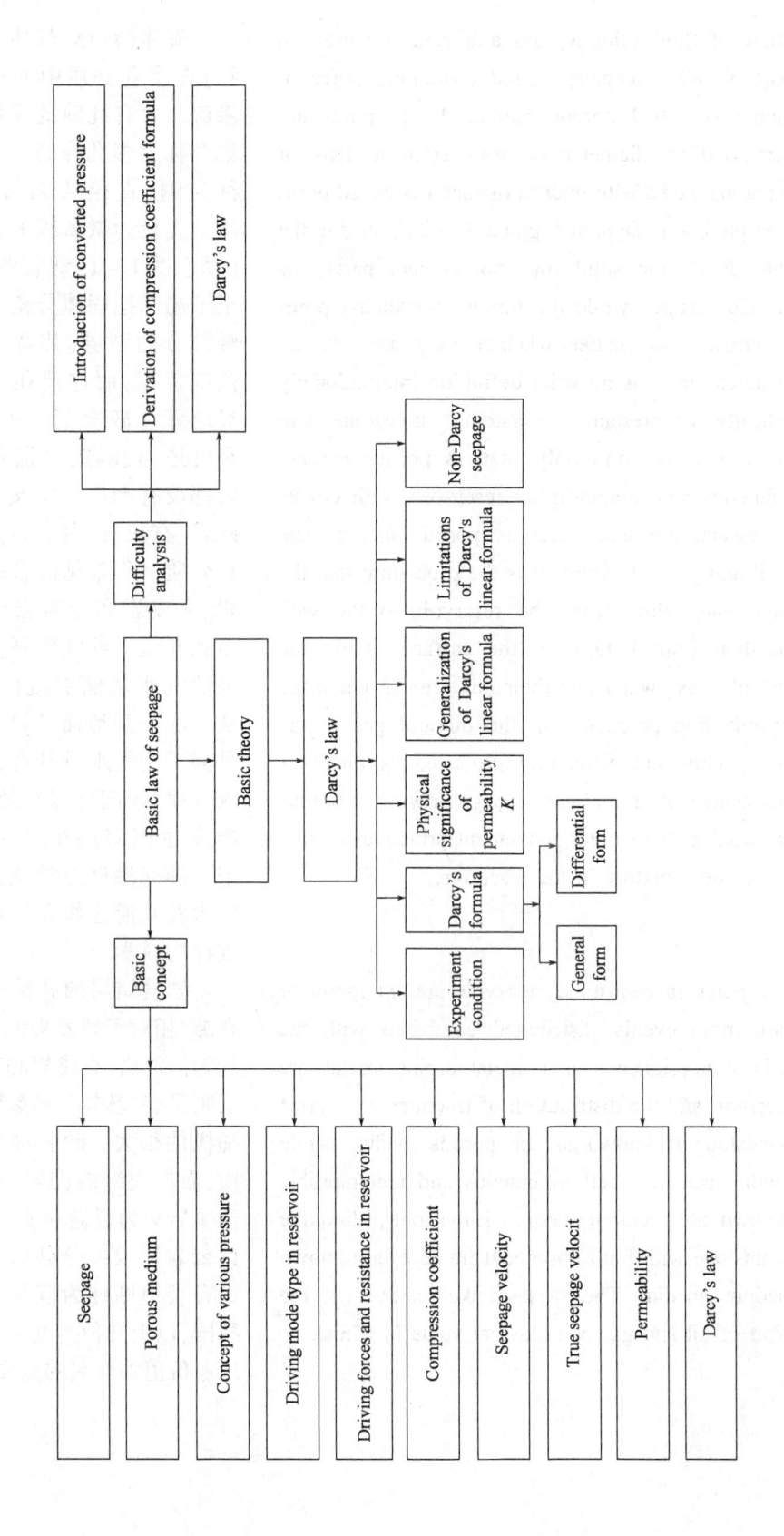

# Chapter 1 Knowledge Graph

- Basic law of seepage
  - Basic concept
    - Seepage
    - Porous medium
    - Concept of various pressure
    - Driving mode type in reservoir
    - Driving forces and resistance in reservoir
    - Compression coefficient
    - Seepage velocity
    - True seepage velocit
    - Permeability
    - Darcy's law
  - Basic theory
    - Darcy's law
      - Experiment condition
      - Darcy's formula
        - General form
        - Differential form
      - Physical significance of permeability $K$
      - Generalization of Darcy's linear formula
      - Limitations of Darcy's linear formula
      - Non-Darcy seepage
  - Difficulty analysis
    - Introduction of converted pressure
    - Derivation of compression coefficient formula
    - Darcy's law

— 13 —

The flow of fluid (liquid, gas and their mixture) in porous media is called seepage. Solid containing pores or solid collection is called porous media. Large pores are called cavern, and its diameter is from 2mm to tens of meters. the porous media with micron diameter is called pores (Dynamic graph 1 – 1、Dynamic graph 1 – 2), as for the fractures that divide the solid into non-contact parts, its width has a wide range, while the throats connecting pores are rather narrow. No matter whether they are cavern, fractures or pores, there is no strict definition internationally and domestically at present. Sandstone, limestone and dolomite reservoirs are commonly seen as porous media. mudstone, igneous and metamorphic reservoirs with cracks which are unusual are also seen as porous media. The process of oil and gas development is the procedure that the oil, gas and water flow from the reservoir to the well bottom and then from bottom to the surface. Only the principles of oil, gas, water and their mixture can guide the petroleum production practice, can the oil and gas in the reservoir be quickly and efficiently produced as much as possible. Mechanics of oil and gas seepage flowing in porous media is the science that studies movement principles of oil, gas, water and their mixture in the reservoir.

Most of pores in sandstone reservoir are intergranular and they are more evenly distributed compared with the fractures. Mudstone, igneous and metamorphic rocks are fractured reservoir and the distribution of fractures is of great random. Sandstone is known as the porous media; while mudstone with cracks, as well as igneous and metamorphic rocks are known as fractured media. Limestone, dolomite and some sandstone with both pores and fractures are known as dual- medium media. The stones talked above are the main reservoir of oil and gas of industrial value in China.

流体(液体、气体及其混合物)在多孔介质中的流动称为渗流。含有孔隙的固体或固体集合称为多孔介质。大的孔隙称为洞,直径从2mm到几十米。直径为微米级的称为孔隙(动态图1-1、动态图1-2),至于局部把固体分割成互不接触部分的裂缝,其宽度的变化范围很宽,而连接孔隙空间的喉道就比较窄了。不管是洞、裂缝还是孔隙,目前在国际国内都没有严格的定义。常见的砂岩、石灰岩、白云岩储层是多孔介质。不常见的含裂缝的泥岩、火成岩和变质岩储层也是多孔介质。油气田开发过程就是油气水从储层流向井底,再从井底采到地面的过程。只有掌握了油气水及其混合物的渗流规律,才能迅速有效地、尽可能多地把储层中的油气采出来。油气渗流力学就是研究油气水及其混合物在储层中运动规律的科学。

砂岩储层的孔隙多是粒间孔隙,相对于裂缝来说分布比较均匀。泥岩、火成岩和变质岩储层则属裂缝储层,而裂缝分布的随机性很大。砂岩称为孔隙介质;而带裂缝的泥岩、火成岩和变质岩称为裂缝介质。石灰岩、白云岩以及部分砂岩既有孔隙又有裂缝则称为双重介质。在我国,以上几种类型的岩石是有工业价值油气藏的主要储层。

Dynamic graph 1-1 Porosity
动态图1-1 孔隙

Dynamic graph 1-2 Oil,gas and water in pores
动态图1-2 孔隙中的油气水

According to the current teaching program of relevant specialty, fluid statics in porous media and character of fluid and porous media are introduced in the course of oil and gas reservoir physics. For closer integration with the needs of petroleum and without loss of generality, the static distribution of oil, gas and water in reservoir is introduced first in this chapter, and several oil displacement forces are introduced focusing on teaching the concepts. On this basis, the fundamental basic law of seepage—Darcy's law is introduced and then used to analyze a variety of practical problems.

## 1.1 Static state in reservoirs

### 1.1.1 Distribution of oil, gas and water in reservoirs

Most of oil and gas reservoirs remain a state of relative equilibrium before development, the distribution of oil, gas and water in reservoir is associated with the properties of rocks and fluid in it. If the oil, gas and water exist together in resevoir, as the lightest component, gas will occupy the top of the structure, so it is known as a gas cap; oil will gather in the lower wings, and water with higher density will assemble under oil, as shown in figure 1 - 1.

In reservoir, the oil-water contact surface is called oil-water interface, and it shall be the oil-bearing boundary when projected onto the horizontal plane. Strictly speaking, the oil-bearing boundary should be classified as inner boundary and outer boundary. In fact, an average of the two boundaries is generally taken as the oil-bearing boundary. The intersecting line of oil-gas interface and reservoir top is called gas boundary. If the outer-ring of reservoir connects to a natural water source that supplies for reservoir, then the

按照我国现行的有关专业的教学计划,多孔介质中的流体静力学、流体及多孔介质的性质是在油气藏物理学中讲授的。为了更紧密地结合石油工程专业的需要,而又不失一般性,本章先从油气水在油气藏中的静态分布讲起,然后介绍各种驱油动力,着重讲授有关的概念。在此基础上介绍渗流的基本定律——达西定律,然后用它来分析各种各样的实际问题。

## 1.1 油气藏静态状况

### 1.1.1 油气水的分布

大多数油气藏在未开发以前,处于相对平衡状态,储层中的油气水分布与储层的岩石性质、流体性质有关。如果在储层中同时存在油气水,由于气体密度最小,将占据构造的顶部,称为气顶;石油聚集在稍低的翼部;而密度更大的水则位于油的下部,如图1-1所示。

油气藏中油和水的接触面称为油水界面,投影到水平面上即为含油边缘。严格来说,含油边缘应划分为含油内缘和含油外缘。实际上,一般取两者的平均值作为含油边缘。油气界面与储层顶面的交线称为含气边缘。如果油气藏外围与天然水源相连通,可向油气藏供液,

reservoir is called open-reservoir, the projection of overall profile is called supply boundary (Figure 1 – 1、Dynamic graph 1 – 3). If the outer-ring of reservoir is closed and its altitude is the same as oil-water interface, then the reservoir is called closed reservoir, and its overall profile is called closed boundary(Figure 1 – 2).

则称其为敞开式油气藏,外廓的投影称为供给边缘(图1-1、动态图1-3)。如果外围封闭且边缘高程与油水界面高程一致,则称其为封闭式油气藏,其外廓称为封闭边缘(图1-2)。

Dynamic graph 1-3
Open reservoir
动态图1-3 敞开油气藏

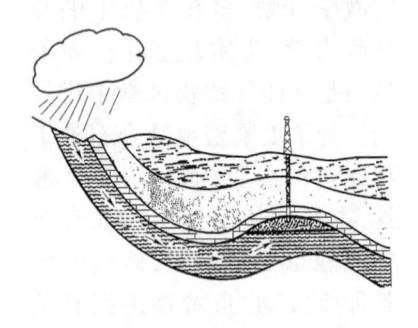

Figure 1 – 1  Reservoir with recharge area
图 1 – 1  有供水区的油气藏

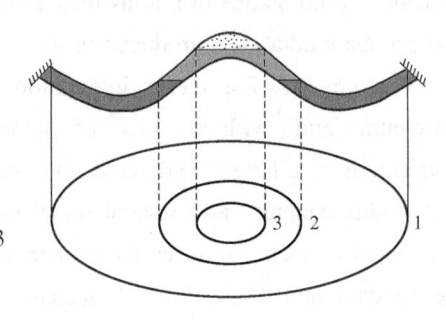

Figure 1 – 2  Oil and gas boundary
图 1 – 2  油气、边缘

1—closed boundary;2—oil bearing boundary;3—gas boundary

According to the distribution of oil, gas and water, water lies outside the oil-bearing boundary is called edge water. If the reservoir is of large thickness or relatively flat structure that makes the water under the oil, then the water is called bottom water.

In the actual oil field, it is rare to see reservoir with only a single oil layer, most of reservoirs are of sandwich-type oil layers. The lithologic character, oil-water interface and oil-gas interface are not always the same from layer to layer. Therefore, the distribution and features of oil, gas and water should be understood first in the process of oil fields development.

### 1.1.2 Concept of various pressures

Reservoir pressure is the response of formation energy. If the pressure field of the reservoir is obtained, this is equivalent to know the flow state of fluid in oil layers. In the oil field development process, concepts of different pressures are often encountered, some of which will be introduced as follows.

根据油气水分布状况,把位于含油边缘外部的水称为边水。若油层厚度大或构造比较平缓,致使水位于原油之下,这样的水称为底水。

实际油田中,很少有单一油层的油气藏,而往往是多层油气藏,层与层之间岩性也经常不一致,油水、油气界面可能各层都不一样。因此,在开发油田时,首先应了解油气水的分布状况及特点。

### 1.1.2 各种压力概念

压力是地层能量大小的反映,知道了油气藏的压力场就等于知道了油层中的流体流动状态。在油田开发过程中,经常会遇到不同的压力概念,以下介绍其中的主要几种。

1. Initial formation pressure $p_i$

The reservoir pressure is the fluid pressure in the pores where are in the middle of oil formation. The reservoirs remain a state of relative equilibrium before development. Generally, the fluid pressure at this point is called initial formation pressure. When the dip angle of formation is relatively big, the mid-depth of every well is usually not the same. Wells lie on the top of the structure have a shallower mid-depth while the wells lying in the wings have a deeper mid-depth. The total energy (total water head) $H = Z + p/(\rho g)$ of an arbitrary point $M$ is a constant before oil field development, as shown in Figure 1 – 3. In this figure, $OO'$ is base level, and the initial oil-water interface is always chosen as the base level. $Z$ is the height from the base level. $\rho$ is the density of the fluid. $p$ is the pressure of point $M$. $g$ is an acceleration of gravity.

1. 原始地层压力 $p_i$

油层压力是指油层中部的孔隙内的流体压力。油气藏在开发以前,一般处于平衡状态,此时油层中的流体所承受的压力称为原始地层压力。当油层倾角较大时,各井的油层中部深度往往各不相同,处于构造顶部的井,油层中部深度小;而处于翼部的井,油层中部深度大。油气藏在投入开发以前,油层中任一点 $M$ 的总能量(或称总水头)$H = Z + p/(\rho g)$ 为一常数,如图 1 – 3 所示。其中,$OO'$ 为基准面,一般选原始油水界面为基准面;$Z$ 为 $M$ 点离基准面的高度;$\rho$ 为流体密度;$p$ 为 $M$ 点的压力;$g$ 为重力加速度。

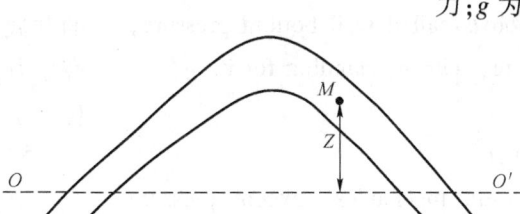

Figure 1 – 3  Converted pressure figure
图 1 – 3  折算压力图

The original formation pressure of every well is not of equality. Shutting down every well and testing its steady state formation pressure after the first exploration well sees the commercial oil and gas flow, and pressure should be tested at the beginning of every exploration well, the measured pressure of mid-depth of the reservoir is the initial formation pressure of this well.

Only one well has been produced in an oil and gas reservoir, will the initial equilibrium be destroyed and the initial formation pressure cannot be obtained anymore. If the pressure manometer cannot be put in the mid-depth of the reservoir, the initial formation pressure can be calculated from tested pressure gradient. Clearly, to improve the measurement accuracy, the pressure manometer should be put in the mid-depth of the reservoir as far as possible.

各井原始地层压力是不相等的。应当在第一口探井见到工业油流后,关井测其稳定的地层压力,以后每一口探井在试油开始就要测压,所测得的各油井油层中部的压力就是该井原始地层压力。

严格地讲,油气藏中只要有一口井生产过,原始的平衡状态就遭到破坏,以后就不可能再测得原始地层压力。如果压力计不能下到油层中部,则需根据所测得的压力梯度进行折算。显然,压力计应尽可能下到油层中部,以提高测量精度。

2. Current formation pressure p

In the process of oil and gas reservoir development, if a well is off production and others are still producing stably, the bottom-hole pressure of the shut-down well will gradually increase. After a long time, the pressure will no longer grow and stabilize. The measured pressure of mid-depth of reservoir of this well is called current formation pressure, and it is also known as static pressure.

3. Supply pressure $p_e$

When there is a fluid supply zone, the pressure of supply boundary is called supply pressure, with $p_e$ standing for it.

4. Bottom hole pressure $p_w$

The pressure measured in the mid-depth of well bottom in the process of production is called well bottom pressure, also known as flow pressure, with $p_w$ standing for it.

5. Converted pressure $p_r$

As is known from fluid mechanics, except pressure energy, the fluid in the reservoir also has potential energy, if it is in the kinematical state, it also has kinetic energy. As a point M in reservoir shown in Figure 1-3, the total energy of unit mass fluid is:

$$H = Z + \frac{p}{\rho g} + \frac{v^2}{2g}$$

In the formula above, $v$ is the movement velocity of point M. Of course, as for the original state, $v = 0$.

Because flow rate is very small in the reservoir and the order is $10^{-7}$ m/s, and its square is smaller than hydrostatic head which should be omitted. So the total energy can be written as:

$$H = Z + \frac{p}{\rho g}$$

The total energy can also be expressed in the form of pressure:

$$p_r = \rho g H = p + \rho g Z$$

In the formula above, $p_r$ is the converted pressure at point $M$, it stands for the total energy of fluid point $M$, and $p$ stands for the magnitude of pressure energy only.

Before oil and gas reservoir development, the pressure that is converted to the initial oil-water interface of every well should be equal. Then the fluid flows from the place with high converted pressure to the place with lower converted pressure after oil and gas reservoir development. Because the uplift amplitude of the reservoir is always much lower than its lateral extension, for the sake of convenience for analysis, three-dimensional seepage is usually simplified to the planar seepage. Hence, the concept of converted pressure must be used. When studying the seepage of the whole reservoir in future, all the pressures are converted pressures if there are no added illustrations.

**Example 1-1** The mid-depth of formation at a given well is -940m above sea-level, the formation crude oil density is 800kg/m³, its oil-water interface altitude is -1200m above sea-level. The reservoir middle pressure actually measured is 12.5 MPa, and then what is the converted pressure on the oil-water interface?

**Solution**: the elevation of middle reservoir of this well is:

$$Z = 1200 - 940 = 260(\text{m})$$
$$p_r = p + \rho g Z = 12.5 + 800 \times 9.81 \times 260 \times 10^{-6} = 14.54(\text{MPa})$$

In the course of well performance analysis, the formation pressure of each well need to be compared, and at this moment only use converted pressure can we get right conclusions.

## 1.2 Driving forces and driving modes in reservoir

When the bottom hole pressure of a well in reservoir decreases, the oil or gas will flow through the reservoir to the bottom of well. Well, what is the energy driving the oil in reservoir? Some kinds of energy are summarized as follows.

### 1.2.1 Hydrostatic pressure

When there is the existence of water supply area which is connected with oil and gas reservoir, the hydrostatic pressure of water supply area is the driving force. The fluid will flow as long as the bottom hole pressure is lower than the pressure of supply area. When the water supply is ample, the pressure on the oil-water interface can be considered unchanged. If the depth of oil-water interface is $H$ (Figure 1-4), then the hydrostatic pressure on the oil-water interface is:

$$p = 10^{-6}\rho_w gH$$

Where: $\rho_w$—The density of water, kg/m$^3$;

$g$—Acceleration of gravity, m/s$^2$;

$H$—The depth of oil-water interface, m;

$p$—The pressure on the oil-water interface, MPa.

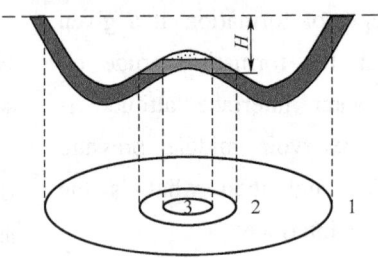

Figure 1-4 Pressure on oil-water interface

1—supply boundary; 2—oil-bearing boundary; 3—gas boundary

This kind of driving force that relies on hydrostatic pressure of supply area to displace oil is called water drive. For most reservoirs, water supply is always less than the oil production. Apart from a few small oil and gas reservoirs, in the whole development stage, it is rare to see the reservoirs with natural water drive, and artificial water flooding is the primary development scheme. The majority of oil and gas reservoirs in China are developed by water flooding.

## 1.2 油气藏中的驱动力和驱动方式

在油气藏中钻了一口井，当降低井底压力时，原油就会经油层流到井底。那么，原油是靠什么能量在油层中流动的呢？归纳起来主要有以下几种。

### 1.2.1 静水压力

当油气层周围存在与含油气区连通的供水区时，供水区的静水压力就是驱动力，只要井底压力低于供水区压力就会产生流动，供水充足时，可以认为油水界面上的压力保持不变。如果油水界面深度为$H$(图1-4)，则油水界面上的静水压力为：

式中：$\rho_w$ = 水的密度，kg/m$^3$；

$g$ = 重力加速度，m/s$^2$；

$H$ = 油水界面深度，m；

$p$ = 油水界面上的压力，MPa。

这种靠供水区的静水压力驱油的驱动方式称为水压驱动。对于多数油气藏，供水量总是小于采油量的。除了少数小油气藏外，在整个开发阶段，都是天然水驱的油气藏很少见，人工注水是主要开发方式，我国多数油气藏都是靠注水开发的。

## 1.2.2 Elastic energy of formation and the fluid in it

The crude oil in the formation is in compression state under the original reservoir pressure long term. After the well is put into production, the reservoir fluid will expand with the drop of reservoir pressure.

The physical quantity that characterizes the magnitude of elastic energy of fluid is compacting factor of liquid $C_L$, it is the reciprocal of modulus of volume elasticity, and it shows the relative change of fluid volume when unit pressure changes. It is expressed as the following formula:

$$C_L = -\frac{1}{V}\frac{\partial V}{\partial p} \quad (1-1)$$

In the formula above, $V$ is volume of fluid, because the fluid volume $V$ is a decreasing function of pressure $p$, so its derivative is negative. In order to make $C_L$ positive, a negative sign need to be added in front of the formula. Also, because fluid volume is a function of temperature $T$, marked as $V(p, T)$, so there appears a partial derivative in the formula, and $C_L$ is isothermal compressibility.

Usually, the temperature of the reservoir changes little during development, so the impact of temperature on the liquid volume is not considered generally. The coefficient of compressibility of water is considered as a constant within the reservoir pressure variation range. And its magnitude is about $(3.7 \sim 5) \times 10^{-3} \mathrm{MPa}^{-1}$. The coefficient of compressibility of oil is relevant to the natural gas content dissolved in it, and it has a big range, its magnitude usually is $(7 \sim 140) \times 10^{-3} \mathrm{MPa}^{-1}$. While the coefficient of compressibility of gas varies more widely than that of oil and water, and it can't be considered as a constant with no relations to pressure. When the temperature of reservoir is a constant, formula (1-1) can be simplified as:

$$C_L = -\frac{1}{V}\frac{dV}{dp} \quad (1-2)$$

When the coefficient of compressibility of fluid changes little, and it can be approximately considered as a constant,

## 1.2.2 地层及其中所含流体的弹性能

地层中的原油长期处在原始地层压力下,当油井投产以后,由于油层压力下降,油层中原来受压缩的液体就会膨胀,从而将部分石油驱向井底。

表征流体弹性能大小的物理量是流体的压缩系数 $C_L$,它是体积弹性模量的倒数,表示改变单位压力时的流体体积的相对变化量,用公式表示为:

$$(1-1)$$

式中,$V$ 为流体的体积,由于流体体积 $V$ 是随压力 $p$ 增加而减小的递减函数,所以其导数为负值。为了使 $C_L$ 值为正,前面需加负号。又因为流体体积也是温度 $T$ 的函数,记作 $V(p,T)$,所以上述公式中会出现偏导数。$C_L$ 为等温压缩系数。

通常在开发过程中,油气藏的温度变化不大,所以一般不考虑温度对液体体积的影响。在油气藏压力变化范围内可以认为水的压缩系数为常数,其值为 $(3.7 \sim 5) \times 10^{-3} \mathrm{MPa}^{-1}$。油的压缩系数与原油中所溶解的天然气量有关,且变化范围较大,一般为 $(7 \sim 140) \times 10^{-3} \mathrm{MPa}^{-1}$。而气体的压缩系数变化范围更大,不能视作与压力无关的常数。当油层温度为常数时,式(1-1)简化为:

$$(1-2)$$

当流体的压缩系数变化较小,可以近似当作常数时,该公式

the formula above can also be expressed as:

$$C_L = -\frac{1}{V}\frac{\Delta V}{\Delta p} \qquad (1-3)$$

还可以用下列差分形式表示：

If volume factor of oil $B_o$ is introduced, and another formula of compressibility of oil $C_o$ can be educed from formula (1-2) as:

如果引入原油的体积系数 $B_o$，则由式(1-2)可导出原油的另一个压缩系数 $C_o$ 的表达式：

$$C_o = -\frac{1}{B_o}\frac{dB_o}{dp} \quad (p > p_b) \qquad (1-4a)$$

$$C_o = \frac{1}{B_o}\frac{dB_o}{dp} \quad (p < p_b) \qquad (1-4b)$$

This is why the formation pressure is lower than the saturation pressure, volume factor is the increasing function of pressure, so there is no negative sign in front of formula. The negative sign is needed when the formation pressure is higher than the saturation pressure.

这是因为当地层压力低于饱和压力时，体积系数是压力的增函数，前面不要负号；地层压力高于饱和压力时则仍需加负号。

Similarly, the rock particles are under uniform compression of formation pressure, when the formation pressure drops, the rock particles will expand. But coefficient of compressibility of solid is minimal compared with that of liquid and gas so that it can be omitted. The formation lies deep underground. It bears overburden force $F$, and this force is balanced by the sum that rock frame stress $\sigma_m$ multiplied by the contact area of cap formation and reservoir bed $A_m$, and add the product of formation pressure (hydrostatic pressure of fluid in reservoir pores) $p$ multiplied by the cross-section area of pores $A_p$, the formula is shown below:

同样岩石颗粒处于地层压力均匀压缩之下，地层压力下降时，岩石颗粒体积也会有所膨胀，但固体压缩系数很小，与液体、气体的压缩性比可以忽略不计。储层深埋地下，承受着上覆岩层的作用力 $F$，这一作用力为岩石骨架的应力 $\sigma_m$ 乘以储层与盖层的接触面积 $A_m$，加上地层压力(储层孔隙中流体的静水压力)$p$ 乘以孔隙截面积 $A_p$ 所平衡，即：

$$F = \sigma_m A_m + pA_p$$

In the process of oil recovery, the formation pressure will drop; especially around the bottom hole, the pressure drop even more. Because the overburden acting force $F$ is a constant, the stress that the rock frame bears will increase which causes its deformation, the decreasing pore volume and the reduction of porosity and permeability. Figure 1-5 (Dynamic graph 1-4) is the deformation figure of rock frame.

采油过程中，地层压力 $p$ 必然会下降，特别是井底周围，下降得更多。由于上覆岩层的作用力 $F$ 是一个常数，岩石骨架所受的应力必然会增加，于是引起它的变形，结果造成孔隙体积的减小，孔隙度和渗透率降低。图1-5(动态图1-4)就是岩石骨架变形图。

Figure 1-5  The deformation figure of rock frame
图 1-5  岩石骨架变形图

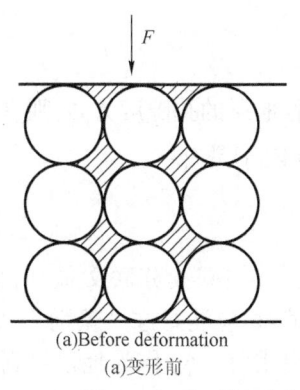

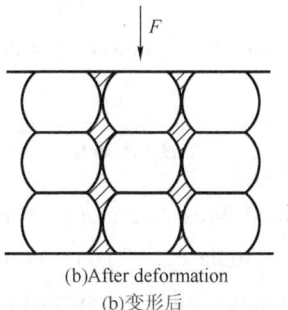

(a) Before deformation / (a)变形前
(b) After deformation / (b)变形后

Dynamic graph 1-4 Deformation of rock frame
动态图 1-4 岩石骨架的变形

If the water injection pressure increases, the stress that the rock frame bears will decrease and rock frame will recover which makes the pore volume increase. In fact, this process is irreversible, the increase of rock frame stress will cause plastic deformation, such as the destruction of cementation texture, and it will not recover after pressure recovery. To simplify, if the reservoir is not deep, the process can be considered as reversible, i.e. elastic, and the error is not significant. If the depth of reservoir is more than 4000m or the porosity and permeability of the abnormal high pressure layer change with pressure cannot be ignored. The experience of low permeability oil and gas field development has shown that the change of permeability with the pressure of low permeability oil and gas field is noticeable, and this problem can not be ignored.

It can be seen from the above analysis when the formation pressure decreases, the pore volume $V_p$ decreases; while the formation pressure increases, the pore volume $V_p$ increases. So the compression coefficient of formation pores (formation compressibility for short) $C_f$ is defined as:

$$C_f = -\frac{1}{V_p}\frac{\partial V_p}{\partial p} \qquad (1-5)$$

Assuming that the process of seepage is isothermal, and then the partial derivative form can be written as the total derivative form. In former Soviet Union literatures, the formation coefficient of compressibility $\beta_n$ is defined by rock bulk volume $V$ as denominator as shown:

如果注水压力升高,岩石骨架所受应力减小,岩石骨架又会复原,使孔隙体积增加。实际上这一过程是不可逆的,岩石骨架应力增加将引起塑性变形,如胶结结构的破坏,在压力恢复后是不能复原的。为了简化,在油层深度不大时,可以认为这个过程是可逆的,即弹性的,这样做实际上误差不大。油层深度超过4000m或异常高压层的孔隙度和渗透率随压力的变化就不容忽视。低渗透油气田的开发经验证明,低渗透油气田的渗透率随压力变化比较显著,而这一问题是不容忽视的。

由以上分析得知,地层压力减小,孔隙体积 $V_p$ 减小;地层压力增加,孔隙体积 $V_p$ 增加。所以地层孔隙压缩系数(简称地层压缩系数)$C_f$ 定义为:

假设渗流过程是等温的,则可以将偏导数写成全导数形式。在苏联文献中,用岩石的表观体积 $V$ 作分母来定义地层的压缩系数 $\beta_n$,即:

$$\beta_n = \frac{1}{V}\frac{dV_p}{dp}$$

If the porosity of formation is $\phi$, and the equation $V_p = \phi V$ can be obtained, then:

$$\beta_n = \phi C_f$$

From former Soviet Union and West literatures, this difference should be noticed. The formula (1 – 5) is used as the definition of coefficient of compressibility of formation pores.

In actual oil fields, the situation that oil-water two-phase exists or oil-gas-water three-phase co-exists, and then a total coefficient of compressibility of rock and fluid $C_t$ can be introduced now. It shows the total fluid driven from unit pore volume depends on the expansion of oil, gas and water and the reduction of pore volume when the formation pressure drops one unit. When the saturation of oil, gas and water is respectively $S_o$, $S_g$ and $S_w$, then:

$$C_t = C_f + C_o S_o + C_w S_w + C_g S_g \qquad (1-6)$$

In the formula above, $C_o$, $C_g$ and $C_w$ are respectively the coefficient of compressibility of oil, gas and water.

The oil and gas flow from reservoir to bottom hole depending on elasticity of formation only is called elastic driving. Generally speaking, as a driving force, elastic drive accounts for less than 10% of whole drive, but it exists in the entire developing process, especially when the work system is changed.

### 1.2.3 Elastic energy of the dissolved gas

There is a large number of natural gas dissolved in crude oil, and when formation pressure drops below the oil saturation pressure, the previously dissolved gas escape will from oil and become free gas, the elastic expansion of free gas will drive the oil from the formation to the hole bottom. This kind of driving mode relying on only dissolved gas without any other energy is called dissolved gas drive. The driving energy of dissolved gas

若地层的孔隙度为 $\phi$，则有 $V_p = \phi V$，于是有：

在阅读苏联文献和西方文献时，应注意这一差别。我国采用式(1-5)作为地层孔隙压缩系数的定义。

实际油田往往油水两相或油气水三相并存，这时还可以引入一个流体与岩石的总压缩系数 $C_t$。它表示地层压力下降一个单位从单位孔隙体积中依靠油气水的膨胀以及孔隙体积的减小所驱出的流体总量。当油气藏中油气水的饱和度分别为 $S_o$、$S_g$ 和 $S_w$ 时，则有：

式中，$C_o$、$C_g$ 和 $C_w$ 分别为油、气、水的压缩系数。

油气靠地层及本身的弹性由储层流向井底称为弹性驱动，一般弹性作为驱动力在整个开发过程中所占比例不超过10%，但在整个采油过程中一直存在，特别是油井的工作制度变化的时候。

### 1.2.3 溶解气的弹性能

一般地层原油中都溶有大量的天然气，当地层压力降到饱和压力以下时，原先溶解的天然气从原油中逸出成为自由气，自由气的弹性膨胀会把油从地层驱向井底。没有其他驱动能量，仅依靠分离出的溶解气驱油的驱动方式

drive and elastic drive is uniformly distributed in the oil reservoirs, and it is different from the energy of water drive from outer boundary.

### 1.2.4 Elastic energy of gas cap

As for the saturated oil and gas reservoir with the gas cap, when wells are put into production, the decrease of formation pressure will inevitably cause the pressure decline of gas cap, which leads the natural gas in gas cap to expend and drive the oil to the bottom hole. Apparently, the gas cap drive must be along with dissolved gas drive.

### 1.2.5 Action of gravity

Oil will naturally flow from the higher place of the structure to the lower place, from the top of oil layer to the bottom; the gravity of oil is also a driving force when there is no other driving energy.

The driving forces listed above are the driving energy which is met often. As for a particular reservoir, several driving energies can exist simultaneously, but in different stages of production, one of them must play a dominant role and others are in secondary status. Which energy will be depended on to drive oil in the production process is called the driving mode of reservoir. Therefore, according to the different driving energies above, the driving modes can be divided into water drive, elastic drive, dissolved gas drive, gas cap drive and gravity drive.

However, the driving mode is not unchangeable; it can be converted from one driving mode to another under a certain condition. For example, if the developing method used is unreasonable, and there is big recharge area around the reservoir, however, because the production of reservoir is too much making. The energy supplement of edge water can't catch up with the consumption of reservoir, then the pressure will decrease and below saturation pressure in some areas of reservoir, and now in this part of area the water drive with high efficiency change to the dissolved gas drive with low efficiency. Conversely, if there is no supply of natural recharge area, and we adopt the method of artificial water

称为溶解气驱。溶解气驱和弹性驱的驱油能量是均匀分布在油气藏中的，它们与来自外边界的靠静水压力驱动的能量不同。

### 1.2.4 气顶的弹性能

对于存在气顶的饱和油气藏，油井投产后，地层压力的下降必然会引起气顶压力的下降，从而导致气顶中的天然气膨胀驱动原油流向井底，显然气顶驱必然伴随着溶解气驱。

### 1.2.5 重力作用

油会从构造高部位流向构造低部位，自油层顶部流向底部，当没有其他驱油能量时，油本身的重力也是一种驱动力。

以上所列举的是几种常见的驱油能量。对于某一具体的油气藏，可以同时存在几种驱油能量，但是在不同的生产阶段，其中必有一种能量起着主导作用，而其他能量处于次要地位。在生产过程中主要依靠哪一种能量来驱油，称为油气藏的驱动方式。因此，根据以上几种不同的驱动能量，可以将驱动方式划分为水压驱动、弹性驱动、溶解气驱动、气顶驱动和重力驱动。

然而驱动方式并非一成不变，在一定条件下可以由一种驱动方式转化成另一种驱动方式。例如，如果所采用的开采方式不合理，油气藏周围虽然存在较大的供水区，但由于油气藏内部产油量过大，边水能量的补充远跟不上油气藏内能量的消耗，在油气藏的局部地区压力迅速下降，低于饱和压力，则此局部地区就由高效率的水压驱动转化成低效率的溶解气驱动。

flooding, then the dissolved gas drive and elastic drive can be changed to water drive with high efficiency.

If the driving mode is different, then the seepage mechanism is different and also the seepage process, even the seepage fluid. In order to develop reservoir rapidly and efficiently, the seepage law under different driving modes must be studied to direct reservoir development practice.

There is no doubt that oil must overcome the resistance when flowing in formation. The flow resistance is mainly viscosity resistance because of the complex pore structure, uneven distributed and narrow porous channel and extremely rough channel surface. However, because of the existence of local loss caused by inertia, the Jamin effect will cause extra resistance when there is multiphase flow.

## 1.3 Basic laws of Darcy flow

### 1.3.1 Continuous media

All objects are composed of molecule. Even though its appearance is static, the molecule is constantly in motion. Strictly speaking, all the objects are not continuous, although molecular motion laws can be predicted when we know the initial state and moment theoretically. In fact, it is full of trouble when studying the movement of more than three molecules, so it is impossible to study the motion law of fluid at molecular level.

What we care about is the assembly of many molecules and the study of the movement of this assembly. All the properties that characterize the movement of fluid (velocity for example) are the average value of every molecule in the assembly. If the assembly is taken as the basic unit, then fluid can be considered to be continuous, and the basic unit is called mass point, while it is not a mathematical point. The mass point must be as small as possible to fully reflect the flow property. Also, the mass point must include sufficient molecules to guarantee the steady state of the average value. The so-called density of one point $\rho$ expressed with formula is:

$$\rho = \lim_{\Delta V \to 0} \frac{\Delta m}{\Delta V}$$

Where: $\Delta V$—The volume around given point, m$^3$;
$\Delta m$—Quality of fluid in $\Delta V$, kg.

According to the continuum theory, $\Delta V \to 0$ is to approach one mass point, while there are plenty of molecules in the mass point and it is not the infinitesimal in mathematics. To study the fluid movement is not starting from the molecular level but from the mass point of continuous media, it is called microscopic level.

Seepage is the movement of fluid in porous media. Porous media includes not only solid but also pores. So one mass point of porous media must include enough solid and pores. It can be imaged that the mass point of porous media is bigger than that of fluid or solid mass point. The so-called property of one point in porous media (pressure and velocity for example) is the average value at this point. Because of the extreme complexity of the porous media structure, in most cases, macroscopic phenomena are studied through the seepage mechanics.

### 1.3.2 Darcy's law

Assuming that the porous media is composed of isodiametric ball type particles, whose radius is the average value derived from grain size analysis. According to the result of ball type particles packing study, the porosity of porous media has nothing to do with the size of particles but the mode of arrangement of particles. The porosity is the smallest when the four center of sphere arrange as rhombus and the value is 0.259, and the stable filling mode with the biggest porosity is not found until today, while the biggest porosity found already with stable arrangement is 0.875.

The actual shape of rock particles is not globular, there is cementing material except particles. It's hard to explain the real issues with globular particle model. The smallest reservoir porosity around the world is less than 0.259. So far the effective way to simplify the porous media has not been found yet.

All the complicated mechanical phenomena and movement in macro-world comply with Newton's law without any exception. Although seepage form and fluid of seepage are vastly different, they have general characters. And these general characters are the basic laws of seepage. Studying seepage mechanics in a porous media is to study the properties of every type of seepage under common conditions—the special seepage law adapts to a given specific case.

The nature of seepage is the actual fluid flow through the porous media. The diameter of pores is minimal and the contact surface is very big when fluid flow through porous media deduced from the property, the conclusion can be obtained that the viscous loss is the primary loss. Therefore, if a kind of liquid with viscosity $\mu$ passes through a formation with the cross-section area $A$ and length $L$ with flow rate $Q$ and causes the differential pressure of formation is $\Delta p$, it is easy to know:

实际岩石颗粒形状不是球形,除了颗粒之外还有胶结物。很难用球形颗粒模型来说明实际问题。世界各国的油气藏的最小孔隙度都小于0.259。到目前为止,还没有发现有效的简化多孔介质的方法。

宏观世界纷繁的力学现象、复杂的运动过程无不遵守牛顿定律。渗流的形式、渗流的流体尽管千差万别,也应具有共性。这一共性就是渗流的基本定律,研究渗流力学就是在这一共性指导下去研究每一种渗流方式的特性——适应某一具体情况的特殊渗流规律。

渗流的本质是实际流体流过多孔介质的流动,根据孔隙直径很小、渗流时流体与孔隙表面接触面很大这一特性推断,黏滞损失是主要的。因此,若黏度为$\mu$的液体以流量$Q$通过截面积为$A$、长度为$L$的地层产生的压力降为$\Delta p$,易知:

$$Q \propto \frac{A\Delta p}{\mu L}$$

If the formula above is written in the form of an equation:

如果写成等式则为:

$$Q = \frac{KA\Delta p}{\mu L} \qquad (1-7)$$

This is Darcy's law. In 1856 a French engineer Darcy proved this law by laboratory experiments. Darcy made a straight metal cylinder about 1m tall with loose sand in it, and sealed the top and bottom of the tube with screen, then fixed piezometer tube on both top and bottom, the water flows from top to bottom (Figure 1-6, Dynamic graph 1-5). He found that no matter how the rate of flow changed, the seepage velocity $v$ is directly proportional to the hydraulic slope:

这就是达西定律。1856年法国工程师达西通过实验证明了这一规律。达西做了一个约1m高的直立铁桶,内装松散的砂子,两端用滤网封住,在砂桶的上下两端装上测压管,水自上向下流(图1-6、动态图1-5),他发现不管流量如何变化,渗流速度$v$与水力坡度成正比:

$$v = K_s \frac{h_1 - h_2}{L} = K_s \frac{\Delta h}{L}$$

Where: $v$ —The seepage velocity, it is the quotient of flow rate dividing by cross-section area of the sand column, m/s;

$\Delta h$—The height difference of water column in the piezometer, m;

$L$—The length of the sand column, m;

$K_s$—The seepage coefficient, m/s.

式中：$v$ = 渗流速度,它等于流量除以砂柱的横截面积, m/s;

$\Delta h$ = 测压管内水柱高差,m;

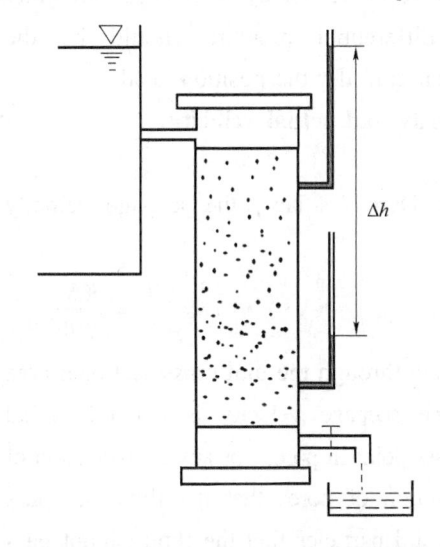

$L$ = 砂柱长度,m;

$K_s$ = 渗流系数,m/s。

Dynamic graph 1-5 Darcy flow
动态图1-5 达西渗流

Figure 1-6 The flow diagram of Darcy's experiment
图1-6 达西实验流程图

If only the seepage coefficient $K_s$ is changed and the sand type remains unchanged, the relationship between hydraulic slope and velocity is unchangeable. Darcy only cared about water at that time, so he used hydraulic slope instead of the pressure gradient. As for every kind of sand, the seepage coefficient of water is a constant, Darcy's law of this form is very convenient on studying the seepage of water; it is still used in groundwater seepage mechanics until today. It is not convenient to use hydraulic slope and seepage coefficient to study the movement of oil, gas, water and their mixture in formation, and using pressure gradient and the seepage coefficient which get rid of the impact of fluid property is better. It is apparently to see that：

如果不改变装砂类型,只改变渗流系数$K_s$,渗流速度与水力坡度关系不变。当时达西关心的只是水,所以使用水力坡度而不用压力梯度。而对于每一种砂子,水的渗流系数是一常数。这个形式的达西定律对于研究水的渗流非常方便,至今仍在地下水渗流力学中广泛使用。要研究油气水及其混合物在地层中的运动,使用水力坡度和渗流系数就不方便了,不如使用压力梯度和剔除液体性质影响的渗流系数,明显可以看出：

$$K_s = K\frac{\rho g}{\mu}$$

In the formula above: scale coefficient is $K$ called permeability, it entirely depends on the properties of porous media and has nothing to do with the properties of fluid. Then the Darcy's law becomes:

$$v = \frac{K\Delta p}{\mu L} \tag{1-8}$$

As the result imagined, many people repeated Darcy's experiment after him. No matter the cylinder is put straightly or inclinedly, the differential pressure should be the converted pressures that consider the position head.

### 1.3.3 Seepage velocity and actual velocity

According to the Darcy's law, the seepage velocity $v$ is:

$$v = \frac{Q}{A} = \frac{K\Delta p}{\mu L}$$

It shows the rate of flow through the unit cross-sectional area of rock. However, the seepage velocity is not the actual velocity of the fluid mass point in pores, for any cross-section of rock, there are the section of pores that the fluid can pass through and also the solid particles that the fluid cannot pass through. Taking any cross section of core and assuming its pore area is $A_p$, then:

$$n = \frac{A_p}{A}$$

In the formula above, $n$ is called the transparence of the cross section, also called surface porosity. Apparently, the actual velocity on any cross section is:

$$v_t = \frac{Q}{A_p} = \frac{Q}{nA}$$

Because the heterogeneity of rock, even though the area $A$ of each section remains unchanged, the pore area of each section is different, and the transparency is not the same, therefore, the actual velocity is also different. In order to study efficiently, the average value $\bar{n}$ of transparence along the flow route is taken to determine the actual velocity of fluid as shown below:

$$v_t = \frac{Q}{nA}$$

$$\bar{n} = \frac{1}{L}\int_0^L n\,\mathrm{d}x$$

And the pore volume of rock can be written as:

$$V_p = \int_0^L An\,\mathrm{d}x = AL\left(\frac{1}{L}\int_0^L n\,\mathrm{d}x\right) = A\bar{n}L$$

According to the definition of porosity $\bar{n} = \phi$, so the actual velocity of fluid along the flow route $L$ is:

$$v_t = \frac{Q}{\phi A} = \frac{v}{\phi} \qquad (1-9)$$

Apparently, the value of actual velocity of fluid is bigger than seepage velocity.

If the bound water saturation $S_{wc}$ is known, then:

$$v_t = \frac{v}{\phi(1 - S_{wc})}$$

Darcy's law is obtained through the experiment of steady seepage using homogeneous liquid in uniform cross section of sand layer. Under normal circumstances, the cross-section of seepage is changing, the formation is not only heterogenous but also anisotropic, and the fluid property sometimes changes with location. It is considered in this book that the Darcy's law is right on the representative elemental volume, it is generally written as:

$$v = -\frac{K}{\mu}\mathrm{grad}(p + \rho g z) = -\frac{K}{\mu}\nabla(p + \rho g z)$$

Dividing the velocity into three components along three directions of $x$, $y$ and $z$, then:

$$v_x = -\frac{K}{\mu}\frac{\partial}{\partial x}(p + \rho g z);\quad v_y = -\frac{K}{\mu}\frac{\partial}{\partial y}(p + \rho g z);\quad v_z = -\frac{K}{\mu}\frac{\partial}{\partial z}(p + \rho g z)$$

The negative sign in the above formula means the direction of seepage velocity always opposite with that of pressure gradient, which means the direction of seepage velocity is always along the direction of pressure decline. The thing that needs to be noticed: the representative elemental volume must include enough sand particles and enough fluid, if the representative elemental volume is really considered as infinitely small, maybe there are only solid particles or fluid mass point in it, then it is not seepage, and this is the statistical property of the Darcy's law.

### 1.3.4 Dimension of permeability

Permeability is a valuable physical property of reservoir; it is directly related to oil well productivity and is an essential parameter in the reservoir evaluation and development. According to the principle of dimension compatibility and use formula (1-7), the dimension of permeability can be obtained as:

$$[K] = \frac{[\mu][L][Q]}{[A][\Delta p]} \quad (1-10)$$

Mass $M$, time $T$ and length $L$ are taken as three basic dimensions, then the right end of the formula above changes to be:

$$[Q] = L^3/T; \quad [\mu] = \frac{M}{TL}; \quad [\Delta p] = \frac{M}{T^2 L}$$

Substitute the equations above into formula(1-10), then:

$$[K] = \frac{M}{TL}L \cdot \frac{L^3/T}{L^2 \cdot M/T^2 L} = L^2$$

It can be known from above that permeability $K$ has the dimension of area, because it shows the ability of the fluid passing through the porous media, so it is reasonable for permeability to have the dimension of area. The unit of area is $m^2$ in SI; it is too big for use. Therefore, the second power of micrometer ($\mu m^2$) is taken as the unit of permeability $K$ in reservoir engineering SI system.

For example, a kind of liquid whose viscosity is $1mPa \cdot s$ passes through a rock sample whose length is $1cm$, cross-section area is $1cm^2$, and its differential pressure of both ends is $10^5 Pa$, if the rate of flow of the liquid is $1cm^3/s$, then the permeability of rock sample is:

$$K = \frac{Q\mu L}{A\Delta p} = \frac{10^{-6} m^3/s \times 10^{-3} Pa \cdot s \times 10^{-2} m}{10^{-4} m^2 \times 10^5 Pa} = 10^{-12} m^2 = 1\mu m^2$$

## 1.4 The limitations of Darcy flow and non-Darcy flow

The Darcy's law shows that the pressure loss is entirely determined by the viscous resistance which is consistent with the characteristic that the porous media has big specific area. The bigger contact area of liquid and solid is, the more the viscous loss is. Therefore, in most cases, Darcy's law is obeyed. However, the second characteristic of the porous media is that its channel diameter has a big varying range and it has lots of branches, mergence and turnings, so the inertia loss (local loss) is big. The relationship between viscous loss and velocity is proportional, while the relationship between inertia loss and the second power of velocity is proportional. When the seepage velocity is minor, $v \gg v^2$, the inertia loss can be ignored. When the seepage velocity is more significant, the inertia loss cannot be omitted. Therefore, the Darcy's law will lose effectiveness with the increase of seepage velocity.

The failure of Darcy's law does not mean the occurrence of turbulent flow in porous media, because the possibility of turbulent flow is rather small for the very small channel diameter of porous media. In fact, the failure of Darcy's law is mainly because of the large inertia loss under laminar flow condition, and experiment also confirmed this.

The Darcy's law is valid when the flow pattern is laminar flow. The laminar flow is a kind of flow patterns. The three flow patterns can be defined as laminar flow with low speed, inertial flow with medium speed and turbulent flow with high speed.

In fluid mechanics, the criterion judging the flow pattern is the Reynolds number $Re$ (ratio of viscosity force and inertia force). When studying whether the fluids flow in porous media obeys Darcy's law, the Reynolds number $Re$ is also adopted. When the Reynolds number reaches a certain value, the Darcy's law loses effectiveness. It can be known from the knowledge of fluid mechanics:

$$Re = \frac{\rho d v_t}{\mu} \qquad (1-11)$$

Where: $\rho$—the density of the liquid, kg/m³;

$\mu$—viscosity of liquid, Pa·s;

$d$—hydraulic diameter of porous media channel, m;

$v_t$—actual velocity of liquid, m/s.

Although the above formula is got, however, because

the seepage channel is very irregular, even it is very difficult to give a certain definition of the pore diameter, and also to measure the pore diameter directly, so an ideal model is adopted usually to calculate the average channel diameter. The ideal model is the simplest simulation of actual core, and it is far from reality. As shown in Figure 1-7, (a) is the actual core, (b) is the ideal model. Assuming: (1) The geometry of the actual core and the ideal model is the same. (2) The ideal model consists of a cluster of isodiametric capillary and pore volume of actual core and the ideal model is equal; (3) Under the same pressure difference, the flow rate of liquid passing through actual core and the ideal model is the same. The diameter of capillary $d$ meeting the above conditions of the ideal model is the equivalent average diameter of formation pore.

由于渗流的孔道极不规则,甚至连孔道直径都难以准确定义。直接测量孔道直径是非常困难的,通常采用一个理想模型来求孔道的平均直径。理想模型是对实际岩心的最简单的模拟,与实际相差甚远。如图1-7所示,(a)为实际岩心,(b)为理想模型。假设:(1)实际岩心与理想模型的几何尺寸相同;(2)理想模型由一簇等直径的毛细管所组成,且理想模型与实际岩心的孔隙体积相等;(3)在相同的压差作用下,通过实际岩心与理想模型的液体流量相等。满足以上条件的理想模型的毛细管直径$d$就是地层孔道的等效平均直径。

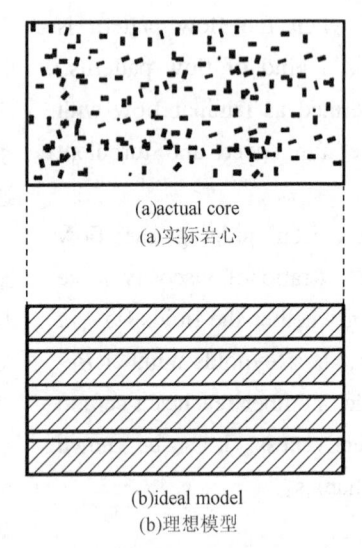

Figure 1-7 The actual core and ideal model
图1-7 岩心和理想模型

Under the condition of laminar flow, the rate of flow passing throgh a single capillary tube is:

$$q' = \frac{\pi r^4 \Delta p}{8\mu L}$$

In the formula above, $r$ is the radius of the capillary. Assuming that the number of capillary tubes is $N$ in the unit cross section of porous media, and the total amount of the

在层流条件下通过单根毛细管的流量为:

式中,$r$为毛细管半径,设多孔介质单位截面上有$N$根毛细管,则毛细管总数为$NA$,通过理想模型

capillary is NA, then the total rate of flow passing the ideal model is:

$$Q = NAq' = \frac{NA\pi r^4 \Delta p}{8\mu L} \quad (1-12)$$

Also from the equal of the pore volume of the actual core and the ideal model, the following equation can be obtained:

$$NA\pi r^2 L = \phi AL$$

After simplifying the equation above, the following equation can be obtained:

$$N = \frac{\phi}{\pi r^2}$$

Bring the equation above to the formula (1-12), then:

$$Q = \frac{\phi A r^2 \Delta p}{8\mu L} \quad (1-13)$$

While according to Darcy's law, the rate of flow passing through the actual core is:

$$Q = \frac{KA\Delta p}{\mu L} \quad (1-14)$$

According to the assuming condition (3), make the formula (1-13) equal to formula (1-14), then:

$$K = \frac{\phi r^2}{8} = \frac{\phi d^2}{32}$$

$$d = 4\sqrt{\frac{2K}{\phi}} \quad (1-15)$$

Bring the formula (1-15) and $v_t = \frac{v}{\phi}$ into formula (1-11), then:

$$Re = \frac{4\sqrt{2}\rho v \sqrt{K}}{\mu \phi^{3/2}} \quad (1-16)$$

If kg/cm³ is used as the unit of fluid density, m/s is used as the unit of seepage velocity, μm² is used as the unit of permeability, mPa·s is used as the unit of viscosity, then the formula (1-16) changes to:

$$Re = \frac{4\sqrt{2} \times 10^{-3} \rho v \sqrt{K}}{\mu \phi^{3/2}} \quad (1-17)$$

The Reynolds number is called critical Reynolds number ($Re_{kp}$) when seepage law changes from linearity to nonlinearity.

For intergranular porous media, the $Re_{kp}$ is 0.2 ~ 0.3.

For pure fracture media, the $Re_{kp}$ is 300 ~ 500.

For big cave (close to pipe flow), the $Re_{kp}$ is 2000 ~ 3000.

At the same time, confirming the flow pattern through calculating the Reynolds number. A smaller value of Reynolds number is for laminar flow. Larger value of Reynolds number is for turbulent flow. T. Govier (1978) raised that using Reynolds number to explain the classification of flow patterns, as shown in Table 1 - 1.

**Table 1 - 1 Classifications of flow patterns**

| Flow patterns | Laminar flow | Inertial flow | Turbulent flow |
| --- | --- | --- | --- |
| Illustration | Low speed flow $Re < 1$ | Medium speed flow $1 \leqslant Re \leqslant 600$ | High-speed flow $Re > 600$ |

Many scholars are trying to find a law that takes both viscous loss and inertia loss into consideration, and Darcy's law is treated as its special case. The binomial seepage law is considered not only to think about the viscous loss but also inertia loss, and it is generally accepted as the equation that can reflect the reality:

$$\text{grad } p = -\frac{\mu}{K}v + b\rho v^2 \tag{1-18}$$

In the formula above: grad $p$ is the pressure gradient. $b$ is a constant depending on the properties of porous media and is confirmed from experiment. $\rho$ is the density of the fluid. The first item of the right part of the formula (1 - 18) means viscous loss, and the second item means inertia loss. When the seepage velocity is very small and $v \gg v^2$, the binomial law will change to be the Darcy's law. If a third power column is added after the binomial law, the result will be accordant with the experimental outcome, but the physical meaning becomes vague. Sometimes, exponent seepage law is also used for the convenience of mathematical handling:

$$v = C(\text{grad } p)^n \quad (0.5 \leqslant n \leqslant 1)$$

With the increase of seepage velocity, the value of $n$ changes from 1 to 0.5 continuously, although the formula is simple, $n$ and $C$ are not constants in the process of seepage, they will also change with the seepage velocity. If $n$ is assumed as a constant, this seepage law is very convenient to use even though its physical meaning is vague, it is still in use.

What should be pointed out, both the binomial law and exponent seepage law are nonlinear seepage law.

The pores of low permeability reservoir are very small, the thickness of the adsorption layer on the pore wall is as big as the size of pore, and the flow resistance of adsorbed layer is much bigger, so fluid is hard to flow when pressure gradient is not big enough, which causes low permeability reservoir does not obey the Darcy's law when pressure gradient is very small. Only if the pressure gradient reaches a specific limit value (trigger pressure gradient) $\lambda$ in low permeability reservoir the Darcy's law is obeyed (Figure 1-8), that is:

$$v = -\frac{K}{\mu}(\nabla p - \lambda); \quad \nabla p < \lambda, v = 0$$

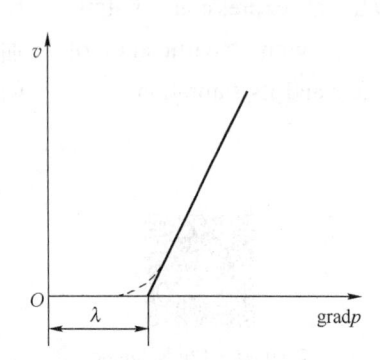

Figure 1-8  The relational diagram of the seepage velocity and pressure gradient
图 1-8 渗流速度与压力梯度关系图

### Exercises

1-1  A cylindrical rock sample, its diameter is $D = 6\text{cm}$, length $L = 10\text{cm}$, permeability $K = 2\mu\text{m}^2$, porosity is $\phi = 0.2$ oil flow through the rock sample along its axial direction, and the viscosity of oil is $4\text{mPa} \cdot \text{s}$, its density is $800\text{kg/m}^3$, the pressure of the inlet end is $p_e = 0.3\text{MPa}$, the pressure of the exit end is $p_w = 0.2$MPa. Please solve the following problems: (1) The flow quantity of every minute. (2) The Reynolds number. (3) The Reynolds number of fluid with the viscosity $\mu_w = 162$mPa·s, density $\rho = 1000$kg/m³ when passing though rock sample (the rest conditions are unchanged).

1-2  Assuming a kind of liquid passes through a sand pipe with the $D = 10$cm, $L = 30$cm, and $\phi = 0.2$, $\mu_0 = 0.65$mPa·s, $\Delta p = 0.7$MPa, $S_{wc} = 0.3$, $K_0 = 0.2\mu m^2$. Please solve the problems: the production $Q$, the seepage velocity $v$ and the average true seepage velocity $v_t$.

1-3  Assuming that the length of sand layer is $L = 500$m, width $B = 100$m, thickness $h = 4$m, $K = 0.3\mu m^2$, porosity $\phi = 0.32$, $\mu_0 = 3.2$mPa·s, $Q = 15$m³/d, $S_{wc} = 0.17$. Please solve the following problems: (1) The pressure difference $\Delta p$, the seepage velocity $v$ and the average true seepage velocity $v_t$. (2) If $Q = 30$m³/d, what is $\Delta p$, $v$ and $v_t$? (3) The needed time $T_1$ and $T_2$ when oil passing through the oil sand of both conditions.

1-4  Please derive the relational expression of total coefficient of compressibility $C_t$ with coefficient of compressibility of oil, gas and water and its saturation.

端压力 $p_e = 0.3$MPa,出口端压力为 $p_w = 0.2$MPa。(1)求每分钟渗过的液量。(2)求雷诺数 $Re$。(3)求黏度 $\mu_w = 162$mPa·s、密度 $\rho = 1000$kg/m³ 的水通过岩样时的雷诺数(其余条件不变)。

1-2  设液体通过直径 $D = 10$cm、长 $L = 30$cm 的砂管,已知 $\phi = 0.2$,$\mu_0 = 0.65$mPa·s,$\Delta p = 0.7$MPa,$S_{wc} = 0.3$,$K_0 = 0.2\mu m^2$,求产量 $Q$、渗流速度 $v$ 和平均真实渗流速度 $v_t$。

1-3  设砂层 $L = 500$m,宽 $B = 100$m,厚 $h = 4$m,$K = 0.3\mu m^2$,孔隙度 $\phi = 0.32$,$\mu_0 = 3.2$mPa·s,$Q = 15$m³/d,$S_{wc} = 0.17$。(1)求压差 $\Delta p$、渗流速度 $v$ 和平均真实渗流速度 $v_t$。(2)若 $Q = 30$m³/d,则 $\Delta p$、$v$ 和 $v_t$ 又为多少?(3)求两种情况原油经过砂层所需的时间 $T_1$、$T_2$。

1-4  试推导总压缩系数 $C_t$ 与油气水的压缩系数及其饱和度的关系式。

Chapter 1 Study Guide
第一章学习指南

# Steady-state seepage in porous media of single-phase incompressible fluid

## 单相不可压缩液体的稳定渗流

# 第二章知识图谱

## 基本概念
- 渗流场图
- 表皮效应与表皮系数
- 井的不完善程度及不完善系数
- 折算半径
- 附加阻力
- 试井
  - 稳定试井
  - 不稳定试井
- 指示曲线
- 采油、采液、吸水指数
- 采油、采液强度
- 导压系数
- 压降漏斗

## 单相不可压缩液体的稳定渗流

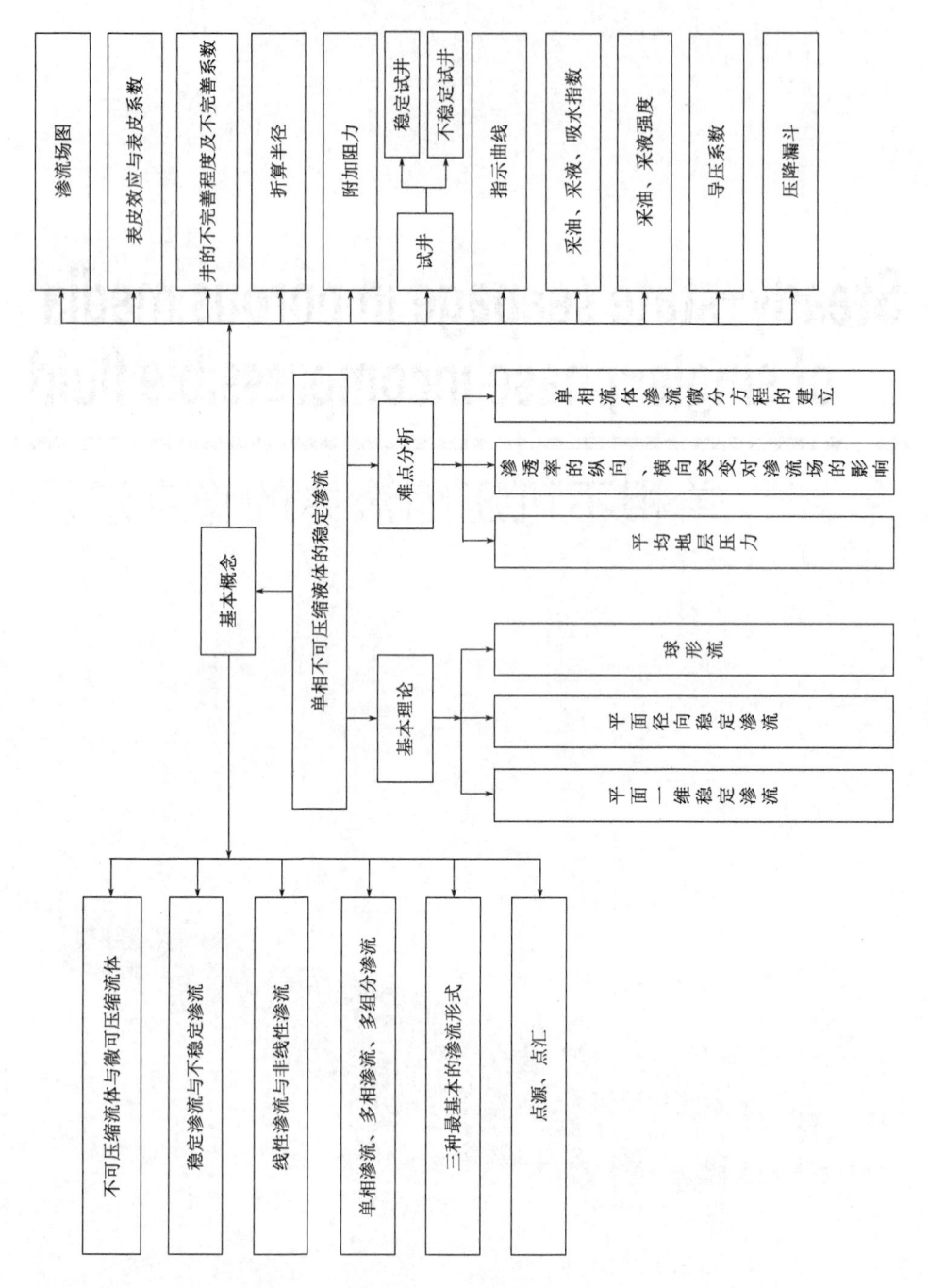

### 难点分析
- 单相流体渗流微分方程的建立
- 渗透率的纵向、横向突变对渗流场的影响
- 平均地层压力

### 基本理论
- 球形流
- 平面径向稳定渗流
- 平面一维稳定渗流

## 基本概念
- 不可压缩流体与微可压缩流体
- 稳定渗流与不稳定渗流
- 线性渗流与非线性渗流
- 单相渗流、多相渗流、多组分渗流
- 三种最基本的渗流形式
- 点源、点汇

— 40 —

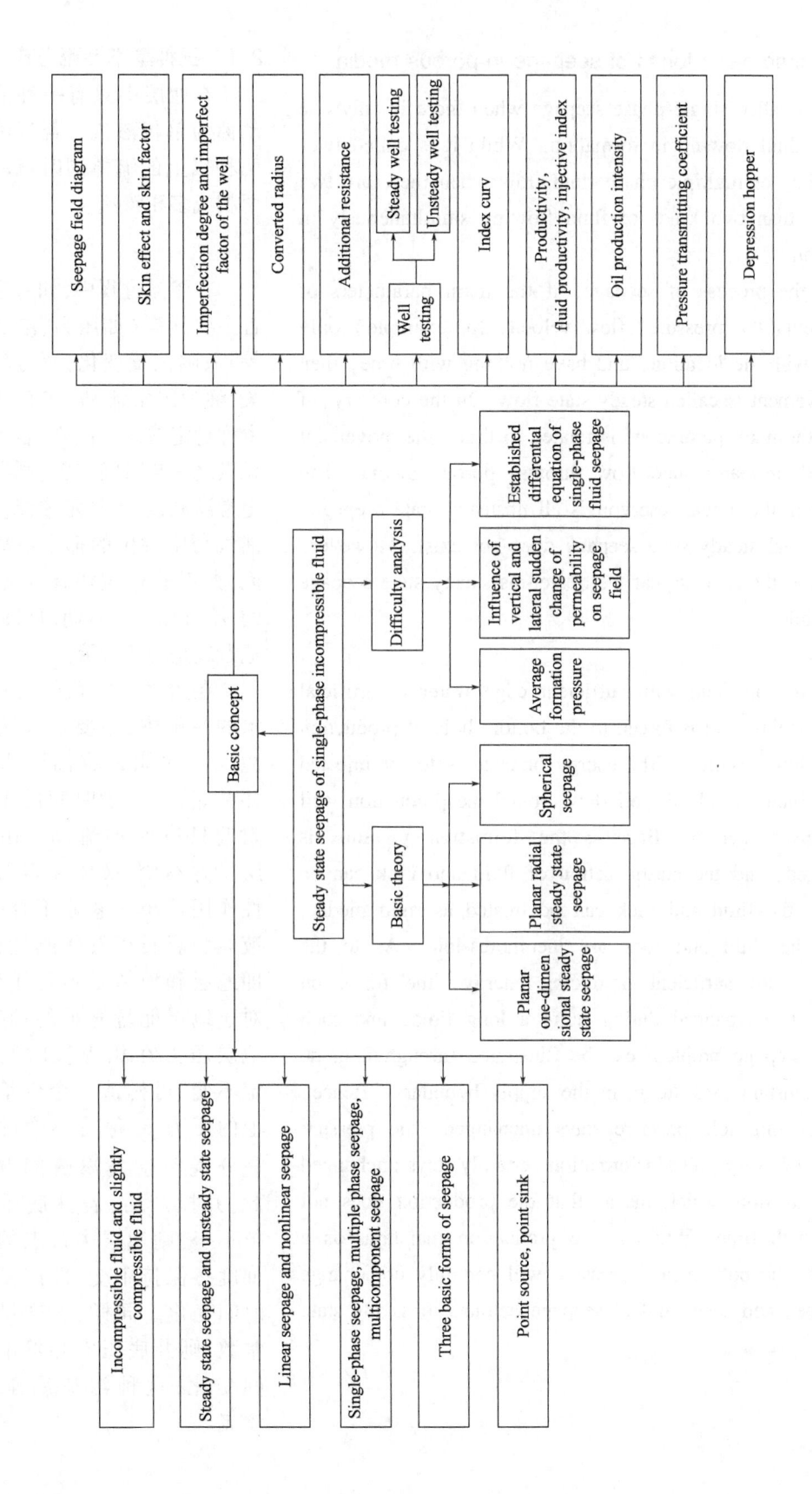

## Chapter 2 Knowledge Graph

**Basic concept**

Steady state seepage of single-phase incompressible fluid

- Incompressible fluid and slightly compressible fluid
- Steady state seepage and unsteady state seepage
- Linear seepage and nonlinear seepage
- Single-phase seepage, multiple phase seepage, multicomponent seepage
- Three basic forms of seepage
- Point source, point sink

**Basic theory**
- Planar one-dimensional steady state seepage
- Planar radial steady state seepage
- Spherical seepage

**Difficulty analysis**
- Average formation pressure
- Influence of vertical and lateral sudden change of permeability on seepage field
- Established differential equation of single-phase fluid seepage

**Applications:**
- Seepage field diagram
- Skin effect and skin factor
- Imperfection degree and imperfect factor of the well
- Converted radius
- Additional resistance
- Well testing → Steady well testing, Unsteady well testing
- Index curv
- Productivity, fluid productivity, injective index
- Oil production intensity
- Pressure transmitting coefficient
- Depression hopper

— 41 —

## 2.1 Three basic forms of seepage in porous media

It is called single-phase seepage when there is only one kind of fluid flowing in formation. While it is called two-phase flow or multiple phase fluid flow when there are two or more than two kinds of fluid flowing simultaneously in formation.

In the process of seepage, if the main parameters of movement (the pressure, flow velocity for example) only change with the location, and have nothing with time, then the movement is called steady state flow. On the contrary, if one of the main parameters is related to time, the movement is called unsteady state flow through porous media. The seepage in the actual reservoir is all unsteady state seepage, and the real steady state seepage does not exist. However, sometimes the seepage can be treated as steady state seepage for a while.

In the oil field with sufficient edge water or artificial water flooding, oil is forced to the bottom hole of production well mainly relying on the energy of edge water or injected water. There is only the oil flow around the production well in a given period. Because the formation pressure is unchanged, and the elastic action of fluid and rock can be omitted, the fluid and rock can be treated as rigid media, that is the fluid and rock are incompressible. As to the reservoir with sufficient producing energy, the formation pressure is unchanged during quite a long time, and such kind of seepage problem can be illustrated through a mode with a constant pressure $p_e$ in the supply boundary. Hence, if the bottom hole pressure stays unchanged, the pressure gradient of every point in formation certainly stays unchanged after production, which means that the production does not change with time. While if the production maintains as a constant, the bottom hole pressure will certainly not change with time, and such kind of seepage belongs to steady state flow.

## 2.1 三种基本渗流方式

在地层中只有一种流体流动称为单相渗流。若有两种或两种以上的流体同时流动称为两相或多相渗流。

在渗流过程中,如果运动的各主要元素(如压力、渗流速度等)只随位置变化,而与时间无关,称为定常流动,在石油工程中称稳定渗流。反之,若各主要元素之一与时间有关,则称为非定常渗流,或不稳定渗流。实际油气藏中发生的都是不稳定渗流,真正的稳定渗流是不存在的,不过在一个时期内可以近似地看成是稳定渗流。

在边水供应充足或人工注水的油田中,主要依靠边水或注入水的能量将油驱入生产井井底,在一段时间内生产井附近只有原油流动。由于地层压力不变,液体和岩石的弹性作用很小而忽略不计,此时液体和岩石可看作刚性介质,即地层和液体是不可压缩的。对于地层能量充足的油气藏,地层压力在相当长时间内稳定不变,可构造一个供给边缘上的压力 $p_e$ 保持不变的模型充分说明这类渗流问题。因此,井投产后,若井底压力保持不变,则地层中每点的压力梯度必保持不变,即产量不随时间变化。同样,若产量维持常数,则井底压力必然不随时间变化,这种渗流属于稳定渗流。

The shape of the actual reservoir is irregular, and there are many ways to arrange wells. As for the horizontal reservoir with sealed fault as the boundary from three sides as shown in Figure 2−1(Dynamic graph 2−1), beyond the area that is half of the well spacing, the streamlines are basically parallel. And this area is like a big core, the unique features of the rectangular reservoir are reflected intensively. The flow velocity and the flow rate are the same on the every section, and the pressure is only in connection with one coordinate. Such kind of flow is called planar one-dimensional seepage. Most reservoirs are rarely real rectangular, but just the shape is analogous. The conclusion here is valid.

As for the similar round horizontal reservoir shown in Figure 2−2(Dynamic graph 2−2), beyond the area that is half of the well spacing, the streamlines run to the well array along the direction of the radius of well array circumference. Such kind of seepage along the direction of radius is called planar radial fluid flow. Of cause, it is simplification of similar round reservoir. Many problems can be illustrated using this simplified mode, and it is also of high accuracy. No matter how complicated the well network is, it can always be treated as radial fluid flow around a well.

When a part of oil layer of a given well is drilled or the perforation completion, the similar globe radial fluid flow will happen around the bottom hole within a small area.

In all, the planar one-dimensional seepage, planar radial seepage and the globe radial fluid flow are the basic seepage modes, and the planar radial fluid flow is the main research point.

实际油气藏的形状都是不规则的,布井方式也多种多样。对于如图2−1(动态图2−1)所示的三面以封闭性断层为界的水平油气藏,在离井排为井距一半的范围以外的区域,流线基本上彼此平行。该区域就像一块大岩心,集中反映了长方形油气藏的渗流特点,各过水断面上的渗流速度、流量相等,压力只和一个坐标有关,这种流动称为平面一维渗流。实际油气藏虽然很少是真正的长方形,只要大体近似,这里的结论仍然有效。

对于如图2−2(动态图2−2)所示的近似圆形的水平油气藏,在离井排为井距一半的距离以外的区域,流线沿井排圆周的半径方向趋向井排,这种沿半径方向的渗流称为平面径向渗流。当然这也是对实际近似圆形油气藏的一种简化,以后会看到这种简化模型说明的实际问题是很多的,精度也比较高。而且不管井网多么复杂,在一口井的附近总可以看作是平面径向渗流。

当某一油井只钻开一小部分油层或射孔完成时,在井底附近的小范围内会发生近似球面径向渗流的空间流动。

总之,平面一维渗流、平面径向渗流和球面径向渗流是渗流的基本方式,而平面径向渗流将是重点研究对象。

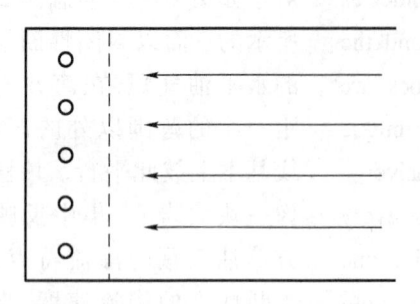

Figure 2-1　The form of planar
one-dimensional seepage

图 2-1　平面一维渗流模型

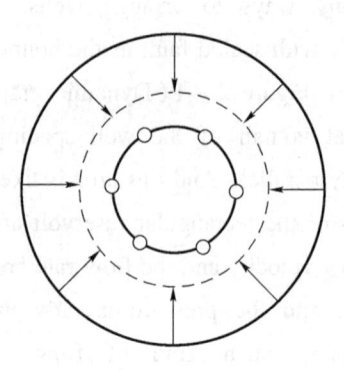

Figure 2-2　The form of planar
radial fluid flow

图 2-2　平面径向渗流模型

Dynamic graph 2-1 Planar one-dimensional seepage
动态图2-1　平面一维渗流

Dynamic graph 2-2 Planar radial fluid flow
动态图2-2　平面径向渗流

## 2.2　Planar one-dimensional steady seepage of single-phase incompressible fluid

Assuming that there is a horizontal strip oil layer, and its length is $L$, width is $B$, height is $h$, the reservoir is rigid water drive, and the fluid and rock are incompressible. And the formation is homogeneous, and isotropic. The pressure of the supply boundary is $p_e$, and the delivery end is a drainage pit whose pressure is $p_w$. And the flow of the fluid in the oil layer is planar one-dimensional steady state seepage.

A coordinate is set up as shown in Figure 2-3, the flow velocity of an arbitrary point $x$ in formation can be shown with differential form of Darcy's law:

## 2.2　单相不可压缩液体的平面一维稳定渗流

假设有如图 2-3 所示长为 $L$、宽为 $B$、厚度为 $h$ 的水平条带状的油层,刚性水压驱动,即岩石和液体是不可压缩的,地层是均质、等厚、各向同性的。供给边界上的压力为 $p_e$,出口端为一排液坑道,其压力为 $p_w$。显然流体在油层中作平面一维渗流。

建立如图 2-3 所示的坐标,该地层中任一点 $x$ 处的渗流速度可用达西定律的微分形式表示:

$$v = -\frac{K}{\mu}\frac{\mathrm{d}p}{\mathrm{d}x} \qquad (2-1)$$

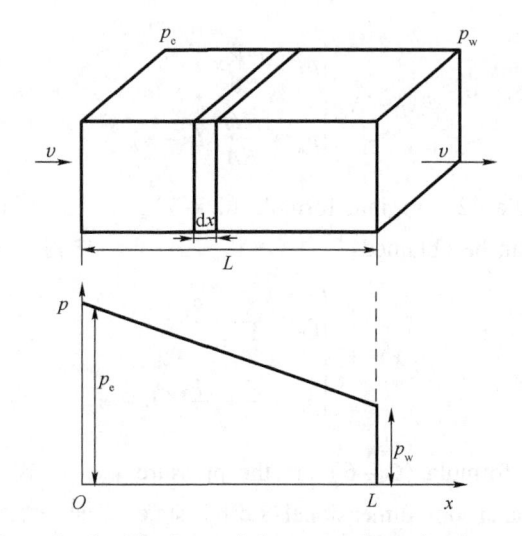

Figure 2-3  The form of planar one-dimensional seepage

图2-3  平面一维渗流模型

Choosing two random sections of the oil layer, and assume the volume flow rate of the two sections is selectively $Q_1$ and $Q_2$. Because the formation and the fluid is incompressible, the volume of fluid in the layer between the two sections can not increase or decrease. Then the equation $Q_1 = Q_2$ can be obtained. That is the volume flow rate passing through a random section in formation is a constant, marked as $Q$, from the formula (2-1), the following can be obtained:

任取该油层的两个过水断面,并设通过这两个断面的体积流量分别为 $Q_1$ 和 $Q_2$,由于地层和流体是不可压缩的,因此两断面间孔隙中的液体体积不可能增加或减少,所以必有 $Q_1 = Q_2$,即通过地层中任一过水断面的体积流量为一常数,记作 $Q$,利用式(2-1)有:

$$Q = -\frac{K}{\mu}A\frac{dp}{dx} \quad (2-2)$$

$$A = Bh$$

In the formula above, $A$ is cross-section area. Separating the variables of the formula (2-2) and integrating, the integrating interval of $x$ is $0 \to L$, and the integrating interval of pressure $p$ is $p_e \to p_w$, then the following equation can be got:

式中,$A$ 为截面积。将式(2-2)分离变量并积分,$x$ 的积分区间为 $0 \to L$,压力 $p$ 的积分区间为 $p_e \to p_w$,则:

$$\frac{\mu Q}{KA}\int_0^L dx = -\int_{p_e}^{p_w} dp \quad (2-3)$$

Then we can obtain the formula (2-4) as shown below:

于是有:

$$Q = \frac{KA(p_e - p_w)}{\mu L} \quad (2-4)$$

The formula (2-4) is the production formula of planar one-dimensional seepage, if the integrating interval $(0, L)$ of formula (2-3) is changed to $(0, x)$ or $(x, L)$, the following expression can be obtained:

式(2-4)为平面一维渗流的产量公式,如果将式(2-3)的积分区间 $(0, L)$ 变为 $(0, x)$ 或 $(x, L)$,则有:

$$p(x) = \begin{cases} p_e - \dfrac{\mu Q}{KA}x \\ p_w + \dfrac{\mu Q}{KA}(L-x) \end{cases} \quad (2-5)$$

Substituting the formula (2-4) into formula (2-5), the following expression can be obtained:

将式(2-4)代入式(2-5)中得:

$$p(x) = \begin{cases} p_e - \dfrac{p_e - p_w}{L}x \\ p_w + \dfrac{p_e - p_w}{L}(L-x) \end{cases} \quad (2-6)$$

Formula (2-5) and formula (2-6) are the pressure distribution formula of planar one-dimensional steady state seepage. When the bottom hole pressure $p_w$ and the production $Q$ stay unchanged, the pressure $p$ and $x$ has a linear relationship. On the other hand, the pressure $p$ of an arbitrary point in formation is only in connection with coordinate $x$; when the value of $x$ is equal, the value of pressure is equal. The line (surface) connected by the points with equal pressure is called isobar (surface), and the lines that are vertical with the isobar (surface) is called streamline. The vertical grids constituted by constant pressure lines and streamlines is called stream field which is also called stream net. For the planar one-dimensional seepage, numerous constant pressure lines can be obtained. In order to see the pressure distribution in formation clearly, like drawing the electric field diagram, the rule is regulated when drawing seepage field diagram: the differential pressure between two arbitrary adjacent isobars must be equal. At the same time, the flow rate between two arbitrary streamlines also must be equal. The seepage field diagram can illustratively reflect the behavior of seepage, the pressure changes steeply in the place where the isobars are condensed, while the pressure changes slowly in the place where the isobars are rare. According to the spacing of the streamlines, the flow velocity can also be decided.

式(2-5)、式(2-6)为平面一维稳定渗流的压力分布公式。当井底压力 $p_w$ 或产量 $Q$ 保持不变时,压力 $p$ 与 $x$ 呈线性关系。地层中任一点的压力 $p$ 仅与坐标 $x$ 有关,凡是 $x$ 值相等的各点其压力值也相等。由这些压力相等的点连成的线(面)称为等压线(面),与等压线(面)相垂直的线称为流线。由等压线和流线构成的正交网格称为渗流场,也称渗流网。对于平面一维渗流,可以得到无数条等压线,为了清晰地看出地层中的压力分布,像绘制电场图一样,在绘制渗流场图时规定了这样的原则:任何相邻两条等压线之间的压差必须相等。同时,任何相邻两条流线之间的流量必须相等。渗流场图可以直观地反映渗流情况,等压线密集处,压力变化急剧;等压线稀疏处,压力变化缓慢。根据流线的疏密,也可判断渗流速度的大小。

Differentiating both sides of the formula (2-6) concerning $x$, the following expression can be got:

将式(2-6)两边对 $x$ 求导得:

$$\frac{dp(x)}{dx} = -\frac{p_e - p_w}{L} \quad (2-7)$$

Under the condition that $p_e$, $p_w$ or the production is a constant, the pressure gradient of an arbitrary point is a constant, which means that the pressure changes on unit length is equal. So the isobars of planar one-dimensional steady state seepage of an incompressible fluid is a tuft of equal-spaced parallel lines.

Substituting the formula (2-7) into formula (2-1), then the flow velocity of an arbitrary point in formation is:

$$v = \frac{K(p_e - p_w)}{\mu L} \qquad (2-8)$$

We can see that the seepage velocity of an arbitrary point is also a constant, so the streamlines are also a tuft of equal-spaced parallel lines. The Figure 2-4 is the seepage field diagram of steady seepage of one-dimensional planar of an incompressible fluid.

在 $p_e$ 和 $p_w$ 或产量为定值的条件下,任一点的压力梯度为常数,即单位长度上的压力变化量相等,所以不可压缩液体平面一维稳定渗流的等压线是一簇等距的相互平行的直线。

将式(2-7)代入式(2-1),则地层中任一点的渗流速度为:

可见,任一点的渗流速度也是常数,因而流线也是一簇等距的互相平行的直线。图2-4是不可压缩液体作平面一维稳定渗流的渗流场图。

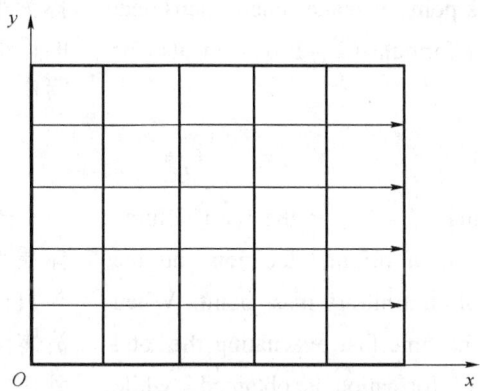

Figure 2-4　The seepage field diagram of planar one-dimensional steady state seepage

图2-4　平面一维稳定渗流场

The average true seepage velocity of fluid mass point moving in the porous media $v_t$ is:

$$v_t = v/\phi$$

And from formula (2-8), the following equation can be obtained:

$$v_t = \frac{K(p_e - p_w)}{\phi \mu L}$$

Obviously, $v_t = \dfrac{\mathrm{d}x}{\mathrm{d}t}$, so:

$$\mathrm{d}x = \frac{K(p_e - p_w)}{\phi \mu L}\mathrm{d}t \qquad (2-9)$$

对于液体质点在多孔介质中运动的平均真实渗流速度 $v_t$ 有:

并利用式(2-8),则:

显然,$v_t = \dfrac{\mathrm{d}x}{\mathrm{d}t}$,于是有:

Assuming that the location of an arbitrary fluid mass point is $x = x_0$ when the time is $t = 0$, after time $t$, the location of the mass point is $x$, then integrate the formula (2-9), the following expression can be obtained:

$$x - x_0 = \frac{K(p_e - p_w)}{\phi \mu L} t \qquad (2-10a)$$

According to the production formula (2-4), the following expression can be obtained:

$$\frac{K(p_e - p_w)}{\mu L} = \frac{Q}{A}$$

Substituting the expression above into formula (2-10a), the following expression can be obtained:

$$x - x_0 = \frac{Qt}{\phi A} \qquad (2-10b)$$

The formula (2-10a) or the formula (2-10b) is called the moving law of fluid mass point of planar one-dimensional steady state seepage, and the formula (2-10b) can also be written as:

$$t = \frac{\phi A(x - x_0)}{Q} \qquad (2-11)$$

The numerator of formula (2-11) is the fluid volume between the flow cross-section of original location and the present flow cross-section of an arbitrary mass point. When dividing it by flow rate $Q$, the time that evacuating the total fluid between two sections of formation is obtained, while the time is just equal to the time that the fluid mass point moving from $x_0$ to $x$.

Especially, when the fluid mass point studied moves from the original location $x_0 = 0$ to $x = L$ after the time $T$, after substituting the value into equation (2-11), the following expression can be obtained:

$$T = \frac{\phi A L}{Q} \qquad (2-12)$$

**Example 2-1** Assuming that there is a rigid water drive oil layer as shown in Figure 2-5. The pressure of formation supply boundary is $p_e = 20\text{MPa}$, $p_w = 16\text{MPa}$, the permeability suddenly changes at the length $L_1$; when $x \leq L_1$,

$K = K_1 = 0.5\mu m^2$; when $x > L_1$, $K = K_2 = 2\mu m^2$. Assuming that $L_1 = 500m$, $L = 1000m$, the height of formation is $h = 5m$, width is $B = 100m$, the viscosity of crude oil is $\mu = 4mPa \cdot s$. Please solve the following problems: (1) the production formula of fluid seepage and calculate its daily production. (2) the pressure distribution formula and draw the pressure distribution diagram.

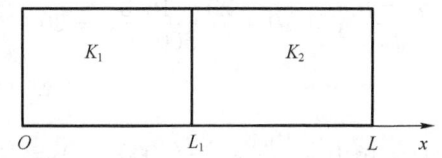

Figure 2-5  The distribution of permeability

**Solution**: (1) Because the permeability $K$ changes, so this formation can not meet the assumption of homogeneous formation, but it can meet the assumption of homogeneous seepage in the area of $0 \leq x \leq L_1$ and $L_1 < x \leq L$, and according to the continuity principle, the flow rate is equal in the two areas. Assuming that the pressure is $p_1$ when $x = L_1$, then the following expression can be obtained:

$$Q = \frac{K_1 A(p_e - p_1)}{\mu L_1} = \frac{K_2 A(p_1 - p_w)}{\mu(L - L_1)} \quad (2-13)$$

Deform the formula (2-13):

$$Q = \frac{p_e - p_1}{\frac{\mu L_1}{K_1 A}} = \frac{p_1 - p_w}{\frac{\mu(L - L_1)}{K_2 A}} \quad (2-14)$$

Then the following production formula can be obtained using proportion by addition and subtraction law to formula (2-14):

$$Q = \frac{A(p_e - p_w)}{\mu\left(\dfrac{L_1}{K_1} + \dfrac{L - L_1}{K_2}\right)} \quad (2-15)$$

Substituting the given datas into formula (2-15):

$$Q = \frac{(20-16) \times 5 \times 100}{4 \times 10^{-9}} \div \left(\frac{500}{0.5 \times 10^{-12}} + \frac{1000-500}{2 \times 10^{-12}}\right)$$
$$= 4.0 \times 10^{-4}(m^3/s) = 34.56(m^3/d)$$

(2) the value of $p_1$ can be obtained from the formula (2-13):

$$p_1 = \frac{K_1(L-L_1)p_e + K_2L_1p_w}{K_1(L-L_1) + K_2L_1}$$

$$= \frac{0.5 \times 10^{-12}(1000-500) \times 20 + 2 \times 10^{-12} \times 500 \times 16}{0.5 \times 10^{-12}(1000-500) + 2 \times 10^{-12} \times 500}$$

$$= 16.8(\text{MPa})$$

When $0 \leq x \leq L_1$, the pressure distribution is:

$$p(x) = p_e - \frac{p_e - p_1}{L_1}x = 20 - \frac{20-16.8}{500}x = 20 - 6.4 \times 10^{-3}x$$

When $L_1 < x \leq L$:

$$p(x) = p_w + \frac{p_1 - p_w}{L - L_1}(L-x) = 16.0 + \frac{16.8-16}{500}(1000-x) = 16.0 + 1.6 \times 10^{-3}(1000-x)$$

The Figure 2-6 is the pressure distribution in the formation.

**Example 2-2** Assuming that there is a homogeneous, and isotropic core angled α with horizontal direction whose length is $L$ (Figure 2-7). A piezometer is respectively connected to the inlet end and delivery end. Fluid in the core moves as incompressible one-dimensional steady state seepage. Assuming that the permeability of the formation is $K$, the fluid viscosity is $\mu$, the density is $\rho$, and the height difference of the two piezometers is $\Delta H$, please derive the flow rate formula of the fluid passing through the core.

（2）由式（2-13）可解出压力 $p_1$ 为：

当 $0 \leq x \leq L_1$ 时，其压力分布为：

当 $L_1 < x \leq L$ 时：

图 2-6 是该地层的压力分布图。

**例 2-2** 设有如图 2-7 所示的一个均质、等厚、各向同性、与水平方向成 α 角的岩心，长度为 $L$，在入口端和出口端分别接一根测压管，液体在岩心内作不可压缩一维稳定渗流，设地层渗透率为 $K$，液体黏度为 $\mu$，密度为 $\rho$，两测压管之间的高度差为 $\Delta H$，试推导液体通过岩心的流量公式。

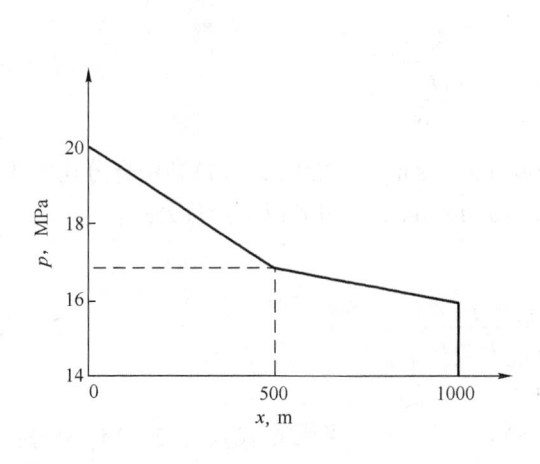

Figure 2-6  The pressure distribution with unequal permeability
图 2-6  渗透率不等的压力分布图

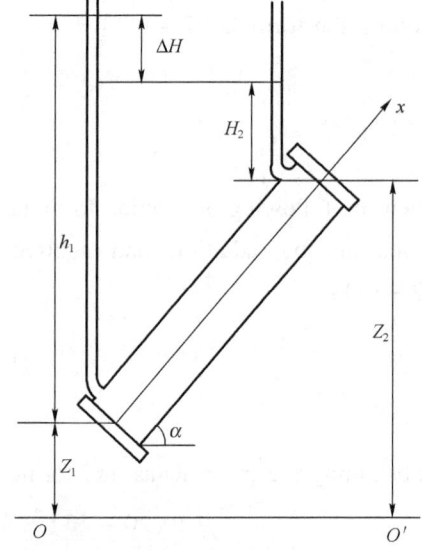

Figure 2-7  The figure of Example 2-2
图 2-7  例 2-2 图

**Solution**: For convenience and without losing the generality, consider the axial direction of the core as the $x$-axes, and $y$-axes is vertical with $x$ and get the origin at the inlet, $OO'$ is the base level. Because the fluid only flow along the $x$-direction, and does not flow along the $y$-direction, so the converted pressure of an arbitrary point on an arbitrary cross-section that is vertical with $x$-axes is equal. According to Darcy's law, the flow velocity of a arbitrary point in the core is:

$$v = -\frac{K \mathrm{d} p_r}{\mu \mathrm{d} x} \qquad (2-16)$$

Assuming that the converted pressures of the inlet end and the delivery end are respectively $p_{r1}$ and $p_{r2}$, substituting $v = Q/A$ into the formula (2-16) and separating the variables and integrating, then the equation below can be obtained:

$$Q = \frac{KA(p_{r1} - p_{r2})}{\mu L} \qquad (2-17)$$

According to the definition of reduced pressure:

$$p_r = p + \rho g Z$$

So:

$$p_{r1} - p_{r2} = \rho g (Z_1 + h_1 - Z_2 - h_2) \qquad (2-18)$$

In the formula above, $h_1$ and $h_2$ are respectively the height from the fluid level of the piezometer at inlet end and delivery end to the axial line of the core; $Z_1$ and $Z_2$ are respectively the height from the base level of the inlet end and delivery end to $OO'$; $(Z_1 + h_1)$ and $(Z_2 + h_2)$ are respectively the height from the fluid level of the piezometer of inlet end and delivery end to the base level $OO'$. The distance between the two fluid levels, marked as $\Delta H$. So from the formula (2-17) and the formula (2-18), the following expression can be obtained:

$$Q = \frac{KA\rho \Delta H g}{\mu L}$$

From the expression, it can be seen that flow rate has nothing to do with the angle of inclination and is only in connection with the height difference of the fluid level of the two piezometers.

## 2.3 Planar radial steady seepage of single-phase incompressible fluid

As previously mentioned, the area around every well in actual reservoir can be approximately considered as radial fluid flow, so the study of radial fluid flow has more significance. In order to set up the model of radial fluid flow, the following assumptions are needed: (1) The formation is homogeneous, isopachous and isotropic. (2) There is only one kind of homogeneous and incompressible fluid flowing in formation, and the compressibility of formation is not considered. (3) There is no physical and chemical action between fluid and rock. (4) The production well is barefoot well and the shape of formation is shown as Figure 2-8. The radius of the supply boundary is $r_e$, the well lies in the center of formation and its radius is $r_w$, the height of formation is $h$, permeability is $K$, porosity is $\phi$, the pressure of supply boundary is $p_e$ and the bottom hole pressure is $p_w$.

## 2.3 单相不可压缩液体的平面径向稳定渗流

如前所述,实际油气藏每一口井的附近都近似为平面径向渗流,因此,研究平面径向渗流将更有实际意义。为了建立平面径向渗流模型,需作如下假设:(1)地层是均质、等厚、各向同性的。(2)地层内只有一种均质不可压缩液体在流动,且不考虑地层的压缩性。(3)流体与岩石无物理化学反应。(4)油井为裸眼井,地层的形状如图2-8所示。供给边缘的半径为$r_e$,井位于地层中心且半径为$r_w$,地层厚度为$h$,渗透率为$K$,孔隙度为$\phi$,供给边缘上的压力为$p_e$,井底压力为$p_w$。

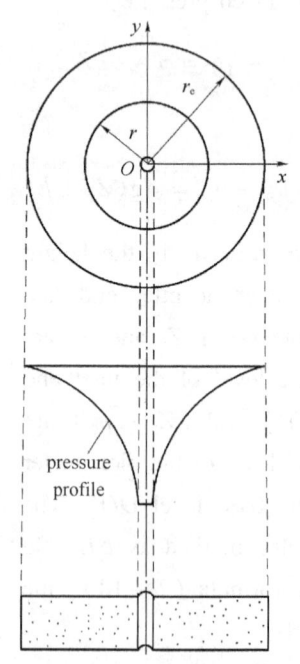

Figure 2-8　Radial fluid flow
图2-8　平面径向渗流

The way to study the radial fluid flow is exactly the same as the way to study the seepage of one-dimensional planar. Because the formation is homogeneous, isopachous and isotropic, so the constant isobaric surface should be the

研究平面径向渗流问题的方法同研究平面一维渗流时完全相同。由于地层是均质、等厚、各向同性的,所以等压面应当是

cylinder surface which is concentric with well. As to the arbitrary point on the cylinder surface which the distance is $r$ from the center of the formation, the seepage velocity can be expressed by the Darcy's law:

$$v = -\frac{K \mathrm{d}p}{\mu \mathrm{d}x} = -\frac{K \mathrm{d}p}{\mu \mathrm{d}(r_e - r)} = \frac{K}{\mu}\frac{\mathrm{d}p}{\mathrm{d}r} \qquad (2-19)$$

Different with the one-dimensional seepage, the cross section is not a constant now. But according to the continuity principle, the flow rate passing through arbitrary cross section should not change, so the seepage velocity of an arbitrary cross section whose radius is $r$ should be:

$$v = \frac{Q}{2\pi h r} \qquad (2-20)$$

Combine the formula (2 – 19) and formula (2 – 20) and separate the variables:

$$\frac{\mathrm{d}r}{r} = \frac{2\pi K h}{\mu Q}\mathrm{d}p \qquad (2-21)$$

Integrate the both sides of the formula (2 – 21), the integrating interval of $r$ is $r_e \to r_w$, the integrating interval of $p$ is $p_e \to p_w$, so the production formula of radial fluid flow is:

$$Q = \frac{2\pi K h(p_e - p_w)}{\mu \ln \frac{r_e}{r_w}} \qquad (2-22)$$

The formula (2 – 22) is also called the Dupuit formula. If the integrating interval of formula (2 – 21) is changed from $[r_e, r_w]$ to $[r_e, r]$ or $[r, r_w]$, the following expression can be obtained:

$$p(r) = \begin{cases} p_e - \dfrac{Q\mu}{2\pi K h}\ln\dfrac{r_e}{r} \\ p_w + \dfrac{Q\mu}{2\pi K h}\ln\dfrac{r}{r_w} \end{cases} \qquad (2-23)$$

Substitute the formula (2 – 22) into formula (2 – 23), the following expression can be obtained:

$$p(r) = \begin{cases} p_e - \dfrac{p_e - p_w}{\ln\dfrac{r_e}{r_w}} \ln\dfrac{r_e}{r} \\[2ex] p_w + \dfrac{p_e - p_w}{\ln\dfrac{r_e}{r_w}} \ln\dfrac{r}{r_w} \end{cases} \quad (2-24)$$

$p(r)$ is the pressure where the radius is $r$, the formula (2-24) is the pressure distribution formula and the corresponding pressure distribution curve is shown in Figure 2-8. Actually, the pressure distribution is a surface of revolution of log curve surrounding borehole axis, and it is also called pressure cone of depression. According to the formula (2-24), the pressure $p$ and the distance logarithm $r$ have a linear relationship, which the distance changes by geometric series and pressure difference changes by arithmetic series. Differentiate both sides of formula (2-24) concerning distance $r$:

$p(r)$ 为 $r$ 处的压力, 式(2-24)就是平面径向渗流的压力分布公式, 相应的压力分布曲线如图 2-8 所示。实际上压力分布是对数曲线绕井轴的旋转面, 通常称为压降漏斗。根据式(2-24)可知, 压力 $p$ 与距离 $r$ 的对数呈线性关系, 即距离成几何级数变化, 压差成算术级数变化。将式(2-24)两边对距离 $r$ 求导得:

$$\frac{dp}{dr} = \frac{p_e - p_w}{\ln\dfrac{r_e}{r_w}} \frac{1}{r} \quad (2-25)$$

In the formula (2-25): the pressure gradient is in inverse proportion to radius. The closer it gets to the bottom hole, the bigger the pressure gradient is. Because the closer it gets to the bottom hole, the smaller the cross section is, the bigger the flow rate is and the denser the isobars are in the seepage field. From the Figure 2-9 we can know, most of the pressure loses around the area of bottom hole. It also illustrates theoretically the importance of protecting the formation from damage in the process of drilling and producing.

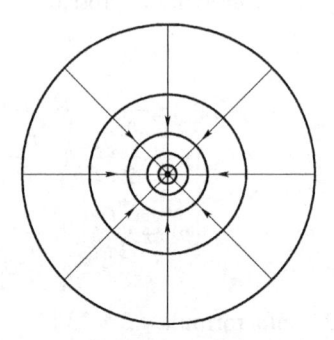

式(2-25)表示: 压力梯度与距离 $r$ 成反比, 即越接近井底, 压力梯度越大。因为越接近井底, 过水断面越小, 流速越大, 在渗流场图中等压线越密集。图 2-9 说明从供给边缘到井底的压差大部分损失在井底附近的区域, 这就从理论上说明了在钻井采油过程中, 保护油层不受伤害的重要性。

Figure 2-9　The seepage field of radial fluid flow

图 2-9　平面径向渗流场图

Substitute the formula (2-25) into formula (2-19), the following expression can be obtained:

$$v = \frac{K(p_e - p_w)}{\mu \ln \frac{r_e}{r_w}} \frac{1}{r} \qquad (2-26)$$

Because the isobars are concentric circles and the streamlines are naturally the rays that converge on original point (Figure 2-9), so the streamlines become denser toward the bottom hole.

**Example 2-3** Assuming that there is a circular reservoir, and the radius of supply boundary is $r_e = 200$m, the radius of bottom hole is $r_w = 0.1$m, the pressure of supply boundary is $p_e$, and the pressure of bottom hole is $p_w$. Please solve the following problems: (1) when the distance from the center of well $r$ = 100m, 50m, 25m, 12.5m, 6.25m, 3m and 1m, what is the ratio of the pressure loss from the distance to bottom hole to total pressure loss? (2) If the well diameter is extended one time, what will the production change correspondingly?

**Solution**: (1) According to the formula (2-24), the pressure loss of an arbitrary point to the bottom hole is:

$$\Delta p(r) = p(r) - p_w = \frac{p_e - p_w}{\ln \frac{r_e}{r_w}} \ln \frac{r}{r_w} \qquad (2-27)$$

While the total pressure loss from the supply boundary to the bottom hole is:

$$\Delta p_t = p_e - p_w$$

Assuming that the ratio of the pressure loss from the distance $r$ to bottom hole to total pressure loss is $f$, the following expression can be obtained:

$$f = \frac{\Delta p_r}{\Delta p_t} \times 100\% = \frac{\ln \frac{r}{r_w}}{\ln \frac{r_e}{r_w}} \times 100\%$$

Substitute $r_e$、$r_w$ and $r$ into the expression above, the results are shown in Table 2-1.

Table 2-1 The calculation results
表 2-1 计算结果表

| $r$,m | 100 | 50 | 25 | 12.5 | 6.25 | 3 | 1 |
|---|---|---|---|---|---|---|---|
| $f$,% | 90.0 | 81.8 | 72.6 | 63.5 | 54.4 | 44.7 | 30.3 |

(2) Assuming that the increasing percent of production is Q if the well diameter is extended one time, the following expression can be obtained:

$$Q = \frac{Q_2 - Q_1}{Q_1} \times 100\% = \frac{\ln\frac{r_e}{r_w} - \ln\frac{r_e}{2r_w}}{\ln\frac{r_e}{2r_w}} \times 100\% = 10.03\%$$

From the example above we can see clearly that the pressure loss of plane radial flow is mainly exhausted around the area of bottom hole. If the radius of bottom hole is extended one time, the increase of drilling cost is far more one time, while the production is only enhanced 10.03%. If the stimulation treatments are adopted such as fracturing, the enhancing times will increase more than 10.03%, if we do this, the cost is far more less than that of extending the well diameter. So extending the well diameter is not treated as a stimulating measure at home and abroad.

It can be known from the formula (2-26) that the average true seepage velocity of the fluid from the well where the distance is $r$ in formation is:

$$v_t = \frac{d(r_e - r)}{dt} = -\frac{dr}{dt} = -\frac{K(p_e - p_w)}{\phi\mu\ln\frac{r_e}{r_w}}\frac{1}{r} \quad (2-28)$$

Assuming that the location is $r_0$ of a given fluid mass point when time is $t = 0$, while the location changes to be $r$ after time $t$, substitute the data into the formula (2-28) and integrate, the following expression can be obtained:

$$r_0^2 - r^2 = \frac{2K(p_e - p_w)}{\phi\mu\ln\frac{r_e}{r_w}}t \quad (2-29a)$$

Combine the formula (2-29a) and formula (2-22), the following expression can be obtained:

$$t = \frac{\pi\phi h(r_0^2 - r^2)}{Q} \quad (2-29b)$$

The formula (2-29a) or the formula (2-29b) shows the motion laws of fluid mass point of radial fluid seepage. The numerator of the formula (2-29b) shows the fluid volume between the cross-section of original location whose radius is $r_0$ of fluid mass point and the cross-section of present location whose radius is $r$, when the numerator divides by the flow rate $Q$, the required time that the flow rate $Q$ exhausts the fluid in the given interval is obtained, and the time is exactly equal to the time that the fluid mass point moves form $r_0$ to $r$. There is always irreducible water in actual reservoir; the ratio of the pore volumes that contain oil to pore volume is called oil saturation, so in order to know the moving time of oil in formation, the right numerator of the formula (2 – 29b) need to multiply the oil saturation.

If the original location of the fluid mass point is $r_0 = r_e$, it moves to the bottom hole ($r = r_w$) after time $T$, and the following expression can be obtained:

$$T = \frac{\pi \phi h (r_e^2 - r_w^2)}{Q}$$

It is not easy to ascertain the boundary pressure $p_e$ of actual reservoir, even if the boundary pressure is a constant, many wells are produced simultaneously, it is impossible to measure the boundary pressure when shutting down one of the wells. Provided there is only a well producing, the boundary pressure can be measured when shutting down the well. Therefore, the average formation pressure is used to substitute the boundary pressure (it is also called supply boundary pressure) when there are numbers of wells producing simultaneously. The so-called average formation pressure is volume weighted average of pressure, that is:

$$\bar{p} = \frac{1}{V_p} \iiint_{V_p} p \, dV_p$$

In the expression above, $p$ is the pressure of a given point in formation; $dV$ is the infinitesimal of volume around the point; $V_p$ is the pore volume of the whole formation. As to the circular, homogeneous, isopachous and isotropic formation, because the isobaric surface is a cylindrical surface, so the following expression can be obtained:

$$V_p = \pi h \phi (r_e^2 - r_w^2)$$
$$dV_p = 2\pi \phi r h \, dr$$

And the average formation pressure is:

$$\bar{p} = \frac{1}{V_p}\iiint_{V_p} p\,dV_p = \frac{2}{r_e^2 - r_w^2}\int_{r_w}^{r_e} pr\,dr$$

Substitute the formula (2-24) into the formula above: 将式(2-24)代入上式得:

$$p = \frac{1}{r_e^2 - r_w^2}\left[\left(p_e - \frac{p_e - p_w}{\ln\frac{r_e}{r_w}}\ln r_e\right)(r_e^2 - r_w^2) + \frac{p_e - p_w}{\ln\frac{r_e}{r_w}}\left(r^2\ln r\bigg|_{r_w}^{r_e} - \int_{r_w}^{r_e} r^2\frac{1}{r}dr\right)\right]$$

$$= p_e - \frac{p_e - p_w}{\ln\frac{r_e}{r}}\ln r_e + \frac{p_e - p_w}{\ln\frac{r_e}{r_w}}(r_e^2\ln r_e - r_w^2\ln r_w)/(r_e^2 - r_w^2) - \frac{1}{2}\frac{p_e - p_w}{\ln\frac{r_e}{r_w}}$$

Because $r_e \gg r_w$, so substitute $r_e^2 - r_w^2 \approx r_e^2$ into the expression above and arrange it, the following expression can be obtained: 由于$r_e \gg r_w$,所以将$r_e^2 - r_w^2 \approx r_e^2$代入上式并整理得:

$$\bar{p} \approx p_e - \frac{p_e - p_w}{2\ln\frac{r_e}{r_w}} \quad (2-30a)$$

The formula (2-30a) can also be written as: 式(2-30a)也可写成:

$$\frac{p_e - \bar{p}}{p_e - p_w} = \frac{1}{2\ln\frac{r_e}{r_w}} \quad (2-30b)$$

$$\frac{\bar{p} - p_w}{p_e - p_w} = 1 - \frac{1}{2\ln\frac{r_e}{r_w}} \quad (2-30c)$$

Usually $\frac{r_e}{r_w} > 10^3$, so $2\ln\frac{r_e}{r_w} > 13.8$, hence, the following expression can be obtained: 通常$r_e/r_w > 10^3$,所以$2\ln\frac{r_e}{r_w} > 13.8$,因而:

$$\frac{\bar{p} - p_w}{p_e - p_w} > 0.93$$

It can be known from the expression above that: the average formation pressure is close to the boundary pressure, that is $\bar{p} \approx p_e$. In actual field condition, the average formation pressure is usually used to replace the supply boundary pressure. 这说明平均地层压力与边界压力接近,即$\bar{p} \approx p_e$,在矿场实际工作中常用平均地层压力代替供给边界压力。

Combine the formula (2-22) and the formula (2-30b) and eliminating ($p_e - p_w$), then the following expression can be obtained: 联立式(2-22)和式(2-30b)并消去$(p_e - p_w)$,可得:

$$\bar{p} = p_w + \frac{Q\mu}{2\pi Kh}\left(\ln\frac{r_e}{r_w} - \frac{1}{2}\right) \quad (2-31a)$$

The formula above can also be written as:

$$Q = \frac{2\pi Kh(\bar{p} - p_w)}{\mu\left(\ln\dfrac{r_e}{r_w} - \dfrac{1}{2}\right)} \quad (2-31\text{b})$$

The formula (2-31b) is the production formula of the plane racial flow expressed by average formation pressure. In practice, the following two points should be paid more attention:

(1) All the derivations are set up under the condition that there is only a producing well in the center of circular formation, but the actual reservoir is irregular and many wells are produced simultaneously, at this time every well is considered as the center, and the connection between the center of neighboring well is considered as the drainage region of the well, and alter oil supply area $A$ to a circular, so the radius of the circular drainage region is supply radius:

$$r_e = \sqrt{\frac{A}{\pi}}$$

In actual field condition, sometimes, the average value of half of the well spacing is simply taken as the supply radius, because the log of supply radius usually appears in formula, although there is error between the two, there is no much influence on production.

(2) The production $Q$ in the formula (2-22) is the production in formation. In actual work, it is easy to measure the surface production $Q_{sc}$, if replace the $Q$ in the formula (2-22) with surface production $Q_{sc}$, the following expression can be obtained:

$$Q_{sc} = \frac{2\pi Kh(p_e - p_w)}{B_o \mu \ln\dfrac{r_e}{r_w}} \quad (2-32)$$

In the formula above, $B_o$ is the volume factor of oil under the condition of formation.

## 2.4 Imperfect well

It is assumed that the oil well is open hole completion discussed in the previous section when we discuss the question of planar radial fluid seepage, this type of oil well is called seepage perfect penetrating well. In actual work, due to the reasons of technology, all oil (gas) layers can not be drilled out by well or even all the oil layers are drilled out, but the bottom hole structure is changed by perfection after well cementation. Otherwise, because the mud plug during the process of drilling or the stimulation treatments (acidize and fracturing technology for example) adopted in the process of producing, the reservoir properties are changed in partial area around the bottom hole. That the bottom hole structure and reservoir properties have changed are called seepage imperfect well, and also called imperfect well for abbreviation. Most of oil wells are imperfect well actually.

### 2.4.1 Classification of imperfect well

There are many types of imperfect well, as to the wells (including oil wells, gas wells and water wells) with changed bottom hole structure, according to the degree of imperfection of the wells, the imperfect well can be classified as three types as below:

(1) The partial penetration well. The whole layer is not drilled out, and the drilled part of the oil layer is totally exposed as shown in Figure 2-10(a) [Dynamic graph 2-3 (a)]. The incompletion of well is called imperfect penetrating; such type of wells is mostly seen in reservoir with active bottom water.

工作中,由于技术上的原因,井不能全部钻穿油层,或者虽钻穿了全部油层,但又经过固井后用射孔方法完井,改变了井底结构。此外,由于钻进过程中的钻井液堵塞或者在油井生产过程中采取了增产技术措施(如压裂和酸化技术等),使井底附近的局部范围内的油层性质发生变化。这些井底结构和井底附近地区油层性质发生变化的井称为渗流不完善井,简称不完善井。实际油井绝大多数都是不完善井。

### 2.4.1 不完善井的分类

不完善井的类型很多,对于井底结构发生了变化的油气井,按照井底结构不同,可以将不完善井划分为以下3类:

(1)打开程度不完善井。油井没有钻穿整个油层,而且钻开部分的井壁是全部裸露的,如图2-10(a)[动态图2-3(a)]所示。油井的这种不完善称为打开程度不完善,这种油井多见于存在活跃底水的油气藏。

(a)Imperfect penetrating well
(a)打开程度不完善井

(b)Imperfect characteristic well
(b)打开性质不完善井

(c)Double Imperfect well
(c)双重不完善井

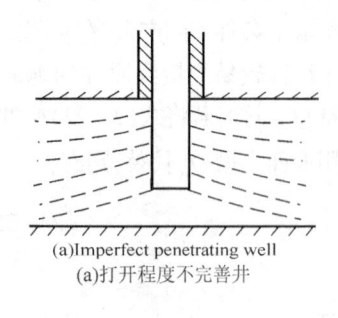

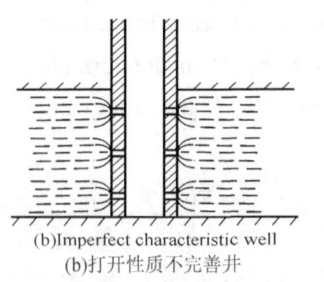

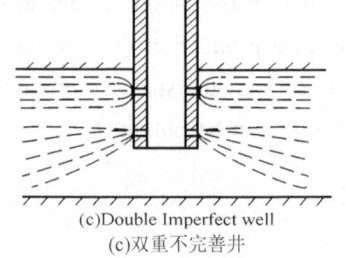

Figure 2-10   The imperfect well
图 2-10   不完善井

Dynamic graph 2-3(a) Partial penetration well
动态图2-3(a)  打开程度不完善井

Dynamic graph 2-3(b) Imperfect characteristic well
动态图2-3(b)  打开性质不完善井

Dynamic graph 2-3(c) Double imperfect well
动态图2-3(c)  双重不完善井

(2) The imperfect characteristic well. The whole layer is drilled out, but the perforation method is used to complete the well, so the fluid in formation can only flow into bore hole through perforation, such imperfect of well is called imperfect characteristic as shown in Figure 2-10(b)[Dynamic graph 2-3 (b)]. Such type of oil wells is mostly seen in sandstone reservoir.

(3) The double imperfect well. The well does not drill out the whole oil layer, and the drilled part is perfect by perforation, such incompletion of well is called double imperfect well as shown in Figure 2-10(c)[Dynamic graph 2-3(c)].

It can be found by theoretical analysis and simulated experiment, a common unique feature of the three types of the imperfect well is that the streamlines bend near the bottom hole. Generally, as to the partial penetration well, the streamlines are basically parallel with each other once ~ twice the thickness of the formation from the well. As to the incomplete characteristic well. The seepage of fluid around the bottom hole is three-dimensional flow, the seepage direction changes constantly, so the inertia loss and flow resistance will increase because of this. Under same conditions that the formation properties, fluid properties, producing pressure difference and the bottom hole radius, the seepage of these imperfect wells must be lower than that of the perfect well.

Generally speaking, the mud contamination in the process of drilling will cause the increase of flow resistance near the area of an oil well. Recent years, because of the wide application of the fracturing technologies and acidizing and the improvement of perforation, the production of imperfect well is more than that of perfect penetrating well. So, the concept of imperfect cannot be interpreted as that its production is lower than that of the imperfect well.

（2）打开性质不完善井。油层全部钻穿，但采用射孔方法完井，这时地层内的流体只能通过孔眼流入井筒，油井的这种不完善称为打开性质不完善，如图2-10(b)[动态图2-3(b)]所示。这种油井在砂岩油田上最常见。

（3）双重不完善井。油井既没钻穿整个油层，而且所钻开的部分又是通过射孔完井的，油井的这种不完善称为双重不完善，如图2-10(c)[动态图2-3(c)]所示。

从理论分析和模拟实验发现，以上3种不完善井有一个共同的特点就是流线在井底附近发生弯曲。一般对于打开程度不完善，离井1~2倍的地层厚度以外流线基本上是彼此平行的。对于打开性质不完善井，流体在井底附近的渗流为空间（三维）流动，渗流速度的方向不断变化，由此而引起的惯性损失将增大，渗流阻力也增大。在地层性质、流体性质、生产压差和井底半径相同的情况下，这些不完善井的产量一定低于完善井的产量。

一般的，在钻井过程中的钻井液伤害将会引起油井附近地区的渗流阻力增大。近年来，由于压裂、酸化技术的广泛应用，以及射孔方法的改善，在相同生产压差作用下油井的产量反而较完善井高。所以"不完善"这个概念不能理解为其产量比完善井低。

## 2.4.2 Research method of imperfect well

As to the research of imperfect wells, lots of scholars have done considerable works. For the imperfect wells, the following methods are mainly used at present:

(1) Skin factor $S$. Because the additional resistance will be caused by the imperfect well near the bottom hole, when talked about the imperfect wells, the formula (2-22) is usually written as:

$$Q = \frac{2\pi Kh(p_e - p_w)}{\mu\left(\ln\dfrac{r_e}{r_w} + S\right)} \qquad (2-33)$$

The $S$ in the formula (2-33) is called skin factor (skin effect) which is an important index when measuring the degree of the imperfect well. As to the partial penetration well, because it has a connection with the drilling depth $b$, bottom-hole radius and the formation height $h$, that is $S(r_w, b, h)$. As to the imperfect characteristic well, the skin factor is the function of perforating density $n$, the perforated depth $l$ in formation, the diameter of perforation $d$ and bottom hole radius $r_w$, that is $S(n, l, d, r_w)$. Although the curve of $S(n, l, d, r_w)$ can be obtained through electrolysis analogy, but because $n$, $l$ (especially $l$) is very difficult to ascertain, so it is not easy to obtain $S$ using this method. The skin factor is usually obtained through well testing. Generally speaking, when the well is contaminated $S > 0$, after fracturing and acidizing $S < 0$.

It also can be seen from the formula (2-33), when the imperfect well is produced with the production $Q$, the flow resistance needed to overcome is different from that of the perfect penetrating well with the same production needed to overcome, and the difference of the two is:

$$\Delta p_S = \frac{Q\mu}{2\pi Kh}S \qquad (2-34)$$

In the formula above, $\Delta p_S$ is called the additional resistance caused by skin factor $S$.

(2) Converted radius $r_{we}$. Because the imperfect well usually happens the spatial flow near the well and planar radial fluid flow a little far from the oil well, so the imperfect wells are always treated as the completely penetrating well. The radius of the completely penetrating well is expressed with $r_{we}$ and is called converted radius (or effective radius). Hence, the production formula of the imperfect well can also be written as:

$$Q = \frac{2\pi Kh(p_e - p_w)}{\mu \ln \dfrac{r_e}{r_{we}}} \qquad (2-35)$$

If make the formula (2-33) and the formula (2-35) equal, then the skin factor $S$ and the converted radius $r_{we}$ have a relationship as:

$$r_{we} = r_w e^{-S} \qquad (2-36)$$

The change of the flow resistance caused by imperfect well equals the well radius changes from $r_w$ to $r_{we}$. And because the change of well diameter has little influence on seepage resistance, in order to make the effect of reducing well radius in line with the resistance change caused by imperfect, the reduced radius is always a very little value.

(3) The flow efficiency (imperfect factor) $\eta$. It is the proportionality of the production of imperfect well and that of completely penetrating well under the same pressure difference, that is:

$$\eta = \frac{\ln \dfrac{r_e}{r_w}}{\ln \dfrac{r_e}{r_w} + S} \qquad (2-37)$$

The flow efficiency connects $\eta$ not only with $S$, but also with $r_e/r_w$. Although it cannot reflect the influence of imperfection accurately, it is very illustrative, so it is still used in field.

(2)折算半径$r_{we}$。由于不完善井主要在油井附近发生空间流动,而在距油井稍远处流体作平面径向渗流,因此,通常把不完善井当作完善井来处理。完善井的井半径用$r_{we}$表示,称为折算半径(或有效半径),因而不完善井的产量公式也可写成:

如果令式(2-33)和式(2-35)相等,则表皮系数$S$与折算半径$r_{we}$存在以下关系:

不完善井引起的渗流阻力的改变量,相当于井半径由$r_w$变化到$r_{we}$所引起的阻力变化,由于井径对渗流阻力的影响不大,所以,要使缩小井径的效果同不完善性引起的阻力变化一致,折算井径往往缩到很小的数字。

(3)流动效率(不完善系数)$\eta$。流动效率指在相同压差作用下,不完善井的产量与完善井的产量之比,即:

流动效率$\eta$不仅与$S$有关,而且与$r_e/r_w$有关,它不能准确反映不完善性的影响,但是它比较直观,所以现场还在用。

## 2.5　Steady well testing

A very important element that determines the commercial value of reservoir is the well production and the injectivity of injection well. With the development of reservoir, the two capacities change all the time. To understand precisely the two capacities is very important in the phases of reservoir evaluation, development and production.

Provided the properties of formation and the fluid in it are known, the oil production and the water injection rate can be obtained using the theoretical formula that we've already talked about (the Dupuit Formula for example). Actually it is very difficult to understand the properties of formation. For example, the permeability measured by core in laboratory is absolute permeability, while there are various degrees of irreducible water in actual oil layer, it is relative permeability that affects deliverability. Secondly, the heterogenetic of oil layer is usually very serious, the difference in permeability is very big between well points. Thirdly, the number of core well is usually quite few and generally less than one-tenth. So far the permeability measured with geophysical means is all indirect whose error is more than 50%. Even the division of net pay thickness also has many human factors. For example, in the early stage of development, the oil layer with high permeability yields oil while the oil layer with low permeability does not. After plugging the high water content ratio layer, the non-effective pay layer without oil changes to be effective pay layers, It is very difficult to ascertain the skin factor $S$ theoretically. It is also not easy to obtain the parameters when using theoretical formula.

The best way to obtain the formation parameters is to directly use the measured production and pressure data and then to predict the production and pressure under new working condition with the obtained formation parameters. This kind of method is called well testing, and it is also called anti problem of seepage mechanics in porous media theoretically. First, well testing is not confined in the development phase. It can be run in the early, middle and late phase. Secondly, its cost is much lower than coring. Thirdly, the parameters of oil layer can be obtained from production performance and it is of high accuracy. Fourthly, the formation parameters obtained represents the average value of the area that well affects, which has great representativeness. So well testing has become an important means that reservoir engineer and production engineer understand the reservoir and determine stimulation effect.

The oil-producing capacity should be in line with the lifting ability in well bore, so an optimal working condition is needed, such as an optimal choke size, the optimal gas injection for gas lift and the optimal pumping parameters and so on. All these need to be ascertained by well testing, but it belongs to the research area of production engineering.

As the anti-problem of seepage mechanics in a porous media, well testing mainly determines the formation parameters and skin factor and so on according to the measured production and bottom hole pressure. The initial formation pressure and average reservoir pressure can be obtained according to the shut-in data of short time so as to ascertain the reservoir boundary and determine the effect of reservoir stimulation.

That every actual measured pressure and production do not change with time in the process of well testing is called systematic well testing. While that the production or pressure changes with time is called unsteady state well test. The testing technique of unsteady state well test is easier than the systematic well testing, and its operation is more convenient and also can solve more problems.

Changing the working conditions of production and injection wells. Such as changing the choke of flowing well, the pumping parameters and the injection pressure or injection and so on, after the production attaining a steady state, measuring the production under each working condition and its corresponding bottom hole flowing pressure. Provided the production condition of the surrounding wells do not change more or less during the time of well testing and the production under each working condition really attain steady state, then the production and

心低得多；第三，油层参数由生产动态求得，精确度高；第四，所求地层参数代表井所影响范围的平均值，代表性强。因此试井成为油气藏工程师和采油工程师认识油气藏、判断增产措施效果的重要手段。

油层的生产能力要与井筒内的举升能力相一致，为此要选择一个最佳工作制度，例如最佳油嘴尺寸、最佳气举注气量、最佳抽汲参数等。这些也靠试井确定但属于采油工艺学的研究范围，本课程不讲。

作为渗流力学反问题的试井主要讲根据实测产量、井底压力求地层参数、表皮系数等。根据短时间关井资料可以求原始地层压力和平均地层压力，确定油气藏边界，判断油层措施的效果等。

试井过程中每一实测的压力、产量都不随时间变化的称为稳定试井；产量或压力随时间变化的称为不稳定试井。不稳定试井比稳定试井工艺简单，施工方便，解决的问题更多些。

改变油水井的工作制度，例如改变自喷井的油嘴尺寸、抽油井的抽汲参数、注水井的注水压力或注入量等，待生产达到稳定后，测得每一工作制度下的产量与相应的井底流动压力。只要周围井的生产条件在试井期间大体不变，每一工作制度下的生产真正达到稳定，则在达西定律有效

pressure have a linear relationship in the valid limit of Darcy's law. If the production pressure difference points under every steady state system are drawn on the chart whose x-coordinate is production, and the y-coordinate is pressure difference, a straight line can be obtained, and this line is called index curve. With the increase of pressure difference and production, Darcy's law will be invalid which makes the index curve break away from the straight line and toward the pressure difference coordinate (Figure 2 – 11). Otherwise, the plastic deformation of layers and the degasification in formation, because the bottom hole pressure is lower than saturation pressure, will also make the index curve bend.

的范围内产量与压差呈线性关系。如果将每一稳定制度下的产量、压差点在以产量为横坐标、压差为纵坐标的图上绘出，就得一条直线，该直线称为指示曲线。随着压差和产量的增加，达西定律会失效，结果使指示曲线脱离直线弯向压差轴（图2-11）。另外，井底压力低于饱和压力造成地层内原油脱气以及油层的塑性变形等原因，也会使指示曲线弯曲。

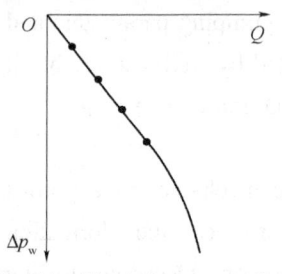

Figure 2 – 11  The index curve of oil well
图2-11  油井指示曲线

As to the straight part of the index curve, the following expression can be obtained:

在指示曲线的直线段，有：

$$Q = J\Delta p \tag{2-38}$$

In the formula above, $J$ is called productivity index, as to an injection well it is called injectivity index. It means the production of oil well or the injection of the injection well under unit producing pressure difference, its unit is m³/(d·MPa). Comparing the formula (2 – 38) and the Dupuit Formula can be seen that:

式中，$J$ 称为产油指数，如果是注水井则称为吸水指数。它是指单位生产压差下油井的产量或注水井的注水量，单位为m³/(d·MPa)。式(2-38)与丘比公式(2-22)相比不难看出：

$$J = \frac{2\pi Kh}{\mu\left(\ln\dfrac{r_e}{r_w} + S\right)} \tag{2-39}$$

Generally, half of the average value of well spacing between wells and its adjacent well is considered as $r_e$, and the skin factor can only be obtained from unsteady state well test or looking up some theoretical curves according to the perforation parameters. Thus, the formation flow coefficient near the well $Kh/\mu$ can be obtained from measured producing index. If the net pay thickness $h$ is obtained according to the

习惯上取本井与周围井距的平均值的一半作为 $r_e$，而表皮系数 $S$ 只有通过不稳定试井或根据射孔参数查某些理论曲线求得。这样就可以由实测的产油指数求出该井附近的地层流动系数 $Kh/\mu$。如果根据测井资

log data, then the formation mobility $K/\mu$ can also be obtained. If the fluid viscosity is known first, then the effective permeability $K$ can be obtained.

If the index curve is bent, the producing index is no longer a constant. It can be seen from Figure 2 – 11 that productivity index is tangent of angle between the straight line section and vertical axis. If the index curve bends, the productivity index is a variable and is the slope of each point on the index curve. With the increase of pressure difference, the producing index decrease constantly.

The key point of systematic well testing is that the production under every working condition must attain steady state, while attaining the steady state needs rather long time, the less permeability is, the longer time reaches steady state, which is the drawback of systematic well testing. However, it is non-replaceable in choosing advisable oil-well working system and needs to be run in actual work.

As to the supply boundary pressure, it is best to use the average reservoir pressure measured when shutting in. If the supply boundary pressure $p_e$ is replaced with average reservoir pressure, the value 1/2 should be deducted in the parentheses of denominator in formula (2 – 39).

## 2.6 Differential equation of single-phase fluid seepage in porous media

In section 2.1 and section 2.2 of this chapter, the seepage theory of planar one-dimensional and radial fluid steady state seepage of single-phase incompressible fluid will be introduced. However, actual reservoir is very complicated, the boundary of reservoir is very irregular, and the well pattern is varied. The effective pay thickness, permeability and burial depth throughout the reservoir are different, and there are numbers of well producing simultaneously in a reservoir and now the seepage law and flow net are definitely very complicated. In order to develop reservoir economically and effectively, it is very important to understand the seepage law under all possible conditions.

## 2.6.1 Continuing equation of seepage of single-phase fluid

The change and movement of all things in nature are regular. All the physical and chemical changes within macroscopy obey the law of energy conservation and law of mass conservation. The course of seepage of the underground fluid also obeys the law of mass conservation. Fluid cannot vanish and also cannot produce in the course of seepage. This law is also the foundation of researching the motion law of fluid.

If an arbitrary confining surface is taken from seepage field, then in a definite time, more (less) fluid mass coming out of this closed surface must be equal to the decreasing (increasing) amount of fluid mass in the confining surface. Apparently, the velocity and density points on the surface are different, and the density and porosity points within the surface are also different. In the view of microscope, within a very tiny domain, around a point the law of mass conservation also stands expressed with mathematical language which means the fluid mass difference flowing in and out around a point should be equal to the change of fluid mass around this point. As to the seepage mechanics, the so-called around a point, the point must include enough solid particles and pore volume. It cannot be infinitely small, or the point will be taken from solid or from fluid which will break away from the limit of fluid seepage mechanics. However, its size also cannot be too large. This is the method that considers the porous media as a continuous media. Provided the relationship between parameters around a point is understood, then it can be expanded into the whole reservoir by adopting advisable method. It is a universal researching method in classical physics.

Understanding the relationship of multiple parameters of a physical phenomenon around a point is setting up differential equation, and resolving concrete problems is solving the differential equation under given conditions or called the integration of differential equation. This is the general method from point to surface then to volume.

In order to understand the relationship among velocity, pressure and the formation and fluid properties of a given point, a small enough confining surface should be taken surrounding this point. Because the surface is small enough, a parallelepiped composed by the surfaces parallel with the coordinate surface is used to represent random confining surface, the error is high order infinitesimal which can be neglected.

Assuming that there is only one kind of fluid in oil layer, as to an arbitrary fluid material point $M(x,y,z)$ in oil layer, consider the point $M$ as center and take an infinitesimal hexahedron whose edges parallel with coordinate plane is respectively d$x$, d$y$ and d$z$ as shown in Figure 2-12. If there is no source or sink in the hexahedron, then as to an arbitrary time $t$, the changing amount (increasing amount) of fluid mass in the pore volume of hexahedron after time d$t$ should be equal to the fluid mass difference flowing in and out the hexahedron within time d$t$.

问题就是在给定的条件下解微分方程,或者称为求微分方程的积分。这也就是由点到面、到体的一般方法。

为了弄清楚某点的速度、压力与地层和流体性质之间的关系,必须围绕这一点取一足够小的封闭曲面。由于曲面足够小,用与坐标面平行的面所构成的平行六面体来代替任意封闭曲面,其误差是高阶小量,可以忽略不计。

假设油层中只有一种流体。对于油层中的任一流体质点$M(x,y,z)$,以$M$点为中心取一个与坐标面平行的长度分别为d$x$、d$y$、d$z$的微元六面体,如图2-12所示。如果六面体内不存在源或汇,那么,对于任一时刻$t$,经过d$t$时间后六面体孔隙体积内的流体质量的变化量(增加量)应该等于d$t$时间内流入与流出六面体的流体质量之差。

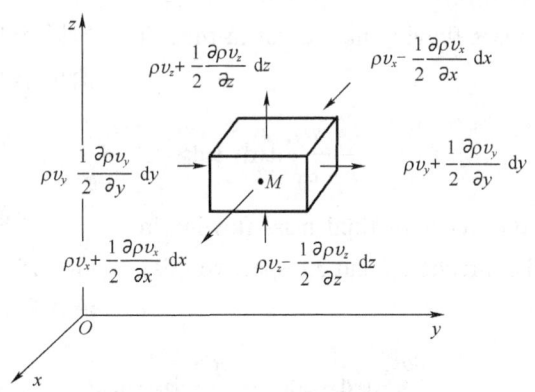

Figure 2-12  The chart of formation infinitesimal
图2-12  地层微元图

Assuming the fluid density is $\rho$ on the location $M$, the seepage velocity is $v$, then the mass velocity is $\rho v$, the components are respectively $\rho v_x$, $\rho v_y$ and $\rho v_z$ on the coordinate $x$, $y$ and $z$, and consider the positive direction of $v_x$, $v_y$ and $v_z$ as the positive direction of coordinate $x$, $y$ and $z$. Because the d$x$, d$y$ and d$z$ of hexahedron is very small, so after the high order infinitesimal is omitted, the variability of

设点$M$处流体的密度为$\rho$,渗流速度为$v$,则质量流速为$\rho v$,在坐标$x$、$y$、$z$上的分量分别为$\rho v_x$、$\rho v_y$和$\rho v_z$,并且认为$v_x$、$v_y$、$v_z$的正方向就是坐标$x$、$y$、$z$的正方向。由于六面体的几何尺寸d$x$、d$y$、d$z$非常小,因此,

mass velocity with distance along the direction $x$ in the infinitesimal body is a constant and is equal to $\dfrac{\partial \rho v_x}{\partial x}$ of the location $M$. The flowing in mass velocity from the back of hexahedron along the direction $x$ is:

$$\rho v_x - \frac{1}{2}\frac{\partial \rho v_x}{\partial x}\mathrm{d}x$$

While the flowing out mass velocity from the front of hexahedron is:

$$\rho v_x + \frac{1}{2}\frac{\partial \rho v_x}{\partial x}\mathrm{d}x$$

The fluid mass flowing in from the back of the time d$t$ is:

$$\left(\rho v_x - \frac{1}{2}\frac{\partial \rho v_x}{\partial x}\mathrm{d}x\right)\mathrm{d}y\mathrm{d}z\mathrm{d}t$$

While the fluid mass flowing out from the front of the time d$t$ is:

$$\left(\rho v_x + \frac{1}{2}\frac{\partial \rho v_x}{\partial x}\mathrm{d}x\right)\mathrm{d}y\mathrm{d}z\mathrm{d}t$$

So the difference of fluid mass flowing in and out in time d$t$ along the direction $x$ is:

$$-\frac{\partial \rho v_x}{\partial x}\mathrm{d}x\mathrm{d}y\mathrm{d}z\mathrm{d}t$$

The same principle, difference of fluid mass flowing in and out in time d$t$ along the direction $y$ and $z$ respectively is:

$$-\frac{\partial \rho v_y}{\partial y}\mathrm{d}x\mathrm{d}y\mathrm{d}z\mathrm{d}t;\ -\frac{\partial \rho v_z}{\partial z}\mathrm{d}x\mathrm{d}y\mathrm{d}z\mathrm{d}t$$

The total difference of fluid mass flowing in and out of in time d$t$ is:

$$-\left(\frac{\partial \rho v_x}{\partial x}+\frac{\partial \rho v_y}{\partial y}+\frac{\partial \rho v_z}{\partial z}\right)\mathrm{d}x\mathrm{d}y\mathrm{d}z\mathrm{d}t$$

Because porosity $\phi$ and the fluid density in pores are relevant with formation pressure, when the formation pressure in hexahedron changes, the fluid mass in it will also

忽略高阶无穷小后,微元体内沿 $x$ 方向的质量速度随距离的变化率为常数,且等于点 $M$ 处的 $\dfrac{\partial \rho v_x}{\partial x}$,沿 $x$ 方向通过六面体后面流入六面体的质量流速为:

而通过六面体前面流出的质量流速为:

在 d$t$ 时间内经后面流入单元体的流体质量为:

而 d$t$ 时间内经前面流出单元体的流体质量为:

所以 d$t$ 时间内沿 $x$ 方向流入与流出单元体的流体质量差为:

同理在 d$t$ 时间内沿 $y$ 方向和 $z$ 方向流入和流出单元体的流体质量差分别为:

在 d$t$ 时间内流入和流出单元体的总的流体质量差为:

由于孔隙度 $\phi$ 及孔隙中的流体密度与地层压力有关,因此,当六面体内地层压力发生变

change. The fluid mass in hexahedron is:

$$\rho\phi dxdydz$$

Assuming the variability $\frac{\partial \rho\phi}{\partial t}dxdydz$ of fluid mass changing in time d$t$ is a constant, the changing amount (increasing amount) of fluid mass in time d$t$ is:

$$\frac{\partial \rho\phi}{\partial t}dxdydzdt$$

According to the law of mass conservation, the increasing amount of fluid mass in hexahedron in time d$t$ should be equal to the difference of fluid mass flowing in and out of the hexahedron in time d$t$ that is:

$$-\left(\frac{\partial \rho v_x}{\partial x}+\frac{\partial \rho v_y}{\partial y}+\frac{\partial \rho v_z}{\partial z}\right)dxdydzdt = \frac{\partial \rho\phi}{\partial t}dxdydzdt$$

The both sides divided by $-dxdydzdt$ simultaneously:

$$\frac{\partial \rho v_x}{\partial x}+\frac{\partial \rho v_y}{\partial y}+\frac{\partial \rho v_z}{\partial z} = -\frac{\partial \rho\phi}{\partial t} \qquad (2-40)$$

The formula (2-40) is the mass conservation equation of single-phase compressible fluid in deformable porous media, and it is also called continuing equation. This equation not only suits for seepage of single-phase fluid, but also suits for seepage of single-phase gas.

Continuing equation can also be expressed with the following vector form:

$$\mathrm{div}\rho\boldsymbol{v} + \frac{\partial \rho\phi}{\partial t} = 0 \qquad (2-41)$$

or

$$\nabla\rho\boldsymbol{v} + \frac{\partial \rho\phi}{\partial t} = 0 \qquad (2-42)$$

Sometimes, when studying the fluid percolate, adopting cylindrical coordinate will be more convenient than using rectangle coordinate. Taking an element under cylindrical coordinate, the continuing equation of single-phase seepage under cylindrical coordinate can be set up using the same way (Appendix A):

$$\frac{1}{r}\frac{\partial}{\partial r}(r\rho v_r) + \frac{1}{r}\frac{\partial}{\partial \theta}(\rho v_\theta) + \frac{\partial}{\partial z}(\rho v_z) = -\frac{\partial(\rho\phi)}{\partial t} \quad (2-43)$$

When two phases of oil and water seepage simultaneously, according to the law of conservation of mass, the changing amount (increasing amount) of the mass of oil and water in time dt should be equal to the mass difference of oil and water flowing in and out of the unit in the same time. The same as deriving continuing equation of single-phase fluid seepage. The continuing equation of two-phase of oil and water can also be derived as:

$$\mathrm{div}\,\rho_o \boldsymbol{v}_o = -\frac{\partial(\rho_o \phi S_o)}{\partial t} \quad (2-44)$$

$$\mathrm{div}\,\rho_w \boldsymbol{v}_w = -\frac{\partial(\rho_w \phi S_w)}{\partial t} \quad (2-45)$$

### 2.6.2 State equation

The oil, gas and water in reservoir lie in the state of high temperature and pressure for long time, the density or volume of oil, gas and water underground and the porosity of oil layer are relevant with the formation temperature and pressure. The formation temperature can be considered as isothermal after a long geological era. So as to the oil and water, the state equation can be derived from the formula (1-2), the definition of compressibility coefficient:

$$C_L = -\frac{1}{V}\frac{\mathrm{d}V}{\mathrm{d}p}$$

As to the liquid with mass $M$, its density and volume are respectively $\rho$ and $V$ when the pressure is $p$, then $M = \rho V$ can be obtained. Substitute the equation into the formula above:

$$C_L = \frac{1}{\rho}\frac{\mathrm{d}\rho}{\mathrm{d}p}$$

Within the range of pressure change, the compressibility coefficient of liquid $C_L$ changes little with pressure, $C_L$ can be approximately considered as a constant. After separating the variables and integrating the formula above, the following expression can be obtained:

$$\rho = \rho_0 e^{C_L(p-p_0)} \quad (2-46)$$

In the formula above, $\rho_0$ is the density when pressure is $p_0$. The formula (2-46) is the isothermal state equation of liquid. If substitute the density of oil and water into the formula (2-46), the corresponding state equation of oil and water can be obtained.

As for the formation, because the relationship between the bulk volume of rock $V$ and pore volume $V_p$ is $V_p = \phi V$ ($V$ is a constant), substitute the expression into formula (1-5) which is the definition of rock's compressibility coefficient, then the following expression can be obtained:

$$C_f = \frac{1}{\phi}\frac{d\phi}{dp}$$

After separating the variables and integrating the formula above, the state equation of elastic formation rock can be obtained:

$$\phi = \phi_0 e^{C_f(p-p_0)} \tag{2-47}$$

As to the seepage of natural gas, its state equation is:

$$\frac{p}{\rho} = \frac{ZRT}{M} \tag{2-48}$$

In the formula above, $Z$ is the compressibility factor (deviation factor) of actual gas under the condition of pressure is $p$ and temperature is $T$, $M$ is the molecular weight of natural gas, $R$ is the universal gas constant.

### 2.6.3 Darcy equation

The seepage velocity of fluid is relevant with pressure gradient when the seepage of fluid obeys the Darcy's law, the relationship between seepage velocity and pressure can be set up using the Darcy equation mentioned before, that is:

$$\begin{cases} v_x = -\dfrac{K}{\mu}\dfrac{\partial p}{\partial x} \\ v_y = -\dfrac{K}{\mu}\dfrac{\partial p}{\partial y} \\ v_z = -\dfrac{K}{\mu}\dfrac{\partial p}{\partial z} \end{cases} \tag{2-49}$$

If it is three-phase seepage of oil, gas and water, the following expression can be obtained:

$$\begin{cases} v_o = K\dfrac{K_{ro}}{\mu_o}\nabla p_o, p_o - p_w = p_{owc} \\ v_w = K\dfrac{K_{rw}}{\mu_w}\nabla p_w, p_o - p_g = p_{ogc} \\ v_g = K\left(\dfrac{K_{rg}}{\mu_g}\nabla p_g + \dfrac{R_s}{B_o}v_o B_g\right) \end{cases} \tag{2-50}$$

In the formula above, $p_o$, $p_w$ and $p_g$ are respectively the pressure of oil, water and gas phase. $B_o$ and $B_g$ is respectively the volume factor of oil and gas. $K_{ro}$, $K_{rw}$ and $K_{rg}$ are respectively the relative permeability of oil, water and gas. $p_{owc}$ and $p_{ogc}$ are respectively the capillary pressure of oil-water and oil-gas. $R_s$ is the dissolved oil and gas ratio.

### 2.6.4 Basic differential equation of single-phase slightly compressible fluid seepage

When researching the seepage of single-phase fluid, the six unknown variables $p$, $\rho$, $\phi$, $v_x$, $v_y$ and $v_z$ are obtained from six equations of the continuing equation, the state equation of fluid and rock, and the Darcy equation, the equation group is closed.

Undoubtedly, it is not convenient to solve the equation group composed of six equations when solving an actual problem. We can follow the elimination method in linear algebra to obtain the equation with only one unknown variable. The concrete way is substituting the state equation formula (2-46), the formula (2-47) and the Darcy equation formula (2-49) into continuing equation formula (2-40), that is:

$$\nabla \left[ \rho_0 e^{C_L(p-p_0)} \frac{K}{\mu} \nabla p \right] = \rho_0 \phi_0 \frac{\partial}{\partial t} e^{(C_L + C_f)(p-p_0)} \quad (2-51)$$

Make $C_t = C_L + C_f$, $C_t$ is called the total compressibility coefficient of formation, the right side of formula (2-51) can be simplified as:

$$\rho_0 \phi_0 \frac{\partial}{\partial t} e^{(C_L + C_f)(p-p_0)} = \rho \phi C_t \frac{\partial p}{\partial t} \quad (2-52)$$

Because the value of $C_t$ is very small and the order is generally $10^{-3}$, so it would be all right to expand the exponential function into Taylor series and take the front two items, that is:

$$e^{C_t(p-p_0)} \approx 1 + C_t(p - p_0)$$

Now it only needs to figure out one item of the left side of the formula (2-51), because:

$$\frac{\partial}{\partial x}\left[\rho_0 e^{C_L(p-p_0)} \frac{K}{\mu} \frac{\partial p}{\partial x}\right] = \frac{K}{\mu} \rho_0 \left(\frac{\partial p}{\partial x}\right) \frac{\partial}{\partial x} e^{C_L(p-p_0)} + \rho \frac{\partial}{\partial x}\left(\frac{K}{\mu} \frac{\partial p}{\partial x}\right)$$

That is:

$$\frac{\partial}{\partial x}\left[\rho_0 e^{C_L(p-p_0)}\frac{K}{\mu}\frac{\partial p}{\partial x}\right] = \frac{K}{\mu}\rho C_L\left(\frac{\partial p}{\partial x}\right)^2 + \rho\frac{\partial}{\partial x}\left(\frac{K}{\mu}\frac{\partial p}{\partial x}\right) \qquad (2-53)$$

While the other two items of formula (2 – 51) respectively are:

而式(2-51)的另外两个分量为：

$$\frac{\partial}{\partial y}\left[\rho_0 e^{C_L(p-p_0)}\frac{K}{\mu}\frac{\partial p}{\partial x}\right] = \frac{K}{\mu}\rho C_L\left(\frac{\partial p}{\partial y}\right)^2 + \rho\frac{\partial}{\partial y}\left(\frac{K}{\mu}\frac{\partial p}{\partial y}\right) \qquad (2-54)$$

$$\frac{\partial}{\partial z}\left[\rho_0 e^{C_L(p-p_0)}\frac{K}{\mu}\frac{\partial p}{\partial z}\right] = \frac{K}{\mu}\rho C_L\left(\frac{\partial p}{\partial z}\right)^2 + \rho\frac{\partial}{\partial z}\left(\frac{K}{\mu}\frac{\partial p}{\partial z}\right) \qquad (2-55)$$

Substitute the formula (2 – 52), formula (2 – 53), formula (2 – 54) and formula (2 – 55) simultaneously into the formula (2 – 51), the following expression can be obtained:

将式(2-52)、式(2-53)、式(2-54)和式(2-55)同时代入式(2-51)中得：

$$\frac{K}{\mu}C_L\left[\left(\frac{\partial p}{\partial x}\right)^2 + \left(\frac{\partial p}{\partial y}\right)^2 + \left(\frac{\partial p}{\partial z}\right)^2\right] + \left[\frac{\partial}{\partial x}\left(\frac{K}{\mu}\frac{\partial p}{\partial x}\right) + \frac{\partial}{\partial y}\left(\frac{K}{\mu}\frac{\partial p}{\partial y}\right) + \frac{\partial}{\partial z}\left(\frac{K}{\mu}\frac{\partial p}{\partial z}\right)\right] = \phi C_t\frac{\partial p}{\partial t} \qquad (2-56a)$$

If it is written as the vector form:

若写成矢量形式则有：

$$C_L\frac{K}{\mu}(\nabla p)^2 + \nabla\left(\frac{K}{\mu}\nabla p\right) = \phi C_t\frac{\partial p}{\partial t} \qquad (2-56b)$$

In most cases, the pressure gradient of seepage $\nabla p$ is very small, and its second power is much smaller. After multiplying a small item $C_L$, it can be omitted completely. So the formula (2 – 40) can be simplified as:

在绝大多数情况下，渗流的压力梯度$\nabla p$很小，其平方项更小，再乘以小量$C_L$后完全可以忽略不计，所以式(2-40)可简化成：

$$\frac{1}{\phi C_t}\left[\frac{\partial}{\partial x}\left(\frac{K}{\mu}\frac{\partial p}{\partial x}\right) + \frac{\partial}{\partial y}\left(\frac{K}{\mu}\frac{\partial p}{\partial y}\right) + \frac{\partial}{\partial z}\left(\frac{K}{\mu}\frac{\partial p}{\partial z}\right)\right] = \frac{\partial p}{\partial t} \qquad (2-57a)$$

Or written as the vector form:

或写成矢量形式：

$$\frac{1}{\phi C_t}\nabla\left(\frac{K}{\mu}\nabla p\right) = \frac{\partial p}{\partial t} \qquad (2-57b)$$

If the formation is homogeneous, then viscosity has nothing with pressure. The course of seepage is isothermal, then:

若地层是均质的，黏度与压力无关，渗流过程是等温的，则有：

$$\frac{K}{\mu\phi C_t}\nabla^2 p = \frac{\partial p}{\partial t} \qquad (2-58)$$

Or expended as:

或展开为：

$$\frac{K}{\mu\phi C_t}\left(\frac{\partial^2 p}{\partial x^2} + \frac{\partial^2 p}{\partial y^2} + \frac{\partial^2 p}{\partial z^2}\right) = \frac{\partial p}{\partial t} \qquad (2-59)$$

The formula (2 – 59) is called diffusion equation or heat conduction equation. It belongs to parabolic equation according to the classification of partial differential equation. The constant factor on the left side is called diffusion coefficient or pressure transmitting coefficient expressed with $\eta$:

式(2-59)称为扩散方程或热传导方程，按二阶偏微分方程分类，属于抛物线形方程。右边的常数因子称为扩散系数或导压系数，以$\eta$表示：

$$\eta = \frac{K}{\mu\phi C_t}$$

If the seepage is steady state, the pressure of arbitrary point will not change with time, then the right side of formula (2-57b) is equal to zero, then the following expression can be obtained:

$$\nabla\left(\frac{K}{\mu}\nabla p\right) = 0 \qquad (2-60)$$

As for the homogeneous formation, the following expression can be obtained:

$$\nabla^2 p = 0$$

Or written as:

$$\frac{\partial^2 p}{\partial x^2} + \frac{\partial^2 p}{\partial y^2} + \frac{\partial^2 p}{\partial z^2} = 0 \qquad (2-61)$$

This is the famous Laplace equation which belongs to ellipse equation according to the classification of second power partial differential equation. Please pay more attention to this: the preconditions of obeying the diffusion equation of homogeneous compressible liquid in compressible formation are: (1) The compressibility of formation and fluid is very small. (2) The pressure gradient is not big in formation. (3) The formation is homogeneous.

The basic differential equation of gas and multiphase seepage will be talked about in corresponding chapters.

After obtaining the unknown function $p(x, y, z, t)$ from the equations above, the velocity of an arbitrary point can be obtained using the Darcy formula (2-49); the fluid density and formation porosity can also be obtained by using formula (2-46) and formula (2-47).

The diffusion equation and its special case—the Laplace equation are suited for any shape of reservoir and well patten under the condition of meeting the assumptions. But if it is used to solve any concrete problem, the concrete conditions of reservoir and wells and the initial state of reservoir must be added, they are the boundary condition and initial condition.

如果渗流是稳定的,任意点的压力不随时间变化,式(2-57b)右边等于零,于是有：

对于均质地层,则有：

或写作：

这就是著名的拉普拉斯(Laplace)方程,在二阶偏微分方程分类上属椭圆形方程。请注意,均质可压缩液体在可压缩地层中渗流服从扩散方程的前提是:第一,液体及地层的压缩性很小;第二,地层中的压力梯度不大;第三,地层是均质的。

气体和多相渗流的基本微分方程将在相应的章节中讨论。

从上述方程式中解出未知函数$p(x, y, z, t)$,就可利用达西方程式(2-49)求出任意一点的速度;利用状态方程式(2-46)、式(2-47)可以求出任意点的液体密度和地层孔隙度了。

扩散方程及其特例——Laplace方程在满足其假设条件下适用于任何形状的油藏和井网。但要用它来解决任何具体问题,必须加上油藏及井的具体条件和油藏的初始状态,这就是边界条件与初始条件。

## 2.6.5 Boundary condition and initial condition of reservoir

The flow of fluid in reservoir is affected by the shape and the boundary properties of reservoir. Whether the boundary pressure remains unchanged or impermeable, the well produces with known production or under stipulated bottom hole pressure and so on. All will affect the solution of seepage issues. The boundary condition and the producing condition on the well bottom in reservoir are both named the boundary condition. The boundary condition of reservoir is usually called outer boundary condition and the producing condition on well bottom are called inner boundary condition. The ordinary boundary conditions of reservoir mainly can be classified as the following two kinds.

1. The first-kind boundary condition

The numeric value of unknown function which can change with time on the boundary is known. For example, there is a boundary $\Gamma$, the pressure of an arbitrary point $M(x,y,z)$ on the boundary can be expressed with the known function $f(x,y,z,t)$, and then the boundary condition of this boundary is:

$$p(x,y,z)\big|_{(x,y,z)\in \Gamma} = f(x,y,z,t)$$

As to the planar radial flow mentioned before, its supply boundary pressure is constant $p_e$, and the bottom hole pressure is constant $p_w$, so the inner and outer boundary condition of radial flow belong to the first-kind boundary condition which can be written as:

$$p\big|_{r=r_e} = p_e; \quad p\big|_{r=r_w} = p_w$$

2. The second-kind boundary condition

If the gradient on the boundary is known of unknown function, or the gradient of unknown function is known on boundary. Such kind of condition is called the second-kind boundary condition. As the closed rectangular reservoir shown in Figure 2-13, the outer boundary is impermeable. Fluid can not pass through the closed boundary, so the normal derivative of pressure on closed boundary is zero (that is the normal flow velocity is zero), so its outer boundary condition is:

## 2.6.5 油藏的边界条件和初始条件

流体在油藏内的流动受油藏形状和油藏边界性质影响。油藏边界上压力是维持不变还是不渗透的,井是以已知产量还是在规定的井底压力下生产等对渗流问题的解都有影响。油藏的边界状况和井底的生产条件统称边界条件。习惯上称油藏边界状况为外边界条件,井底生产条件为内边界条件。常见的油藏边界条件主要有以下两类。

1. 第一类边界条件

未知函数在边界上的数值为已知,这个数值可以随时间 $t$ 而变化。例如,设有一边界 $\Gamma$,对于此边界上的任意一点 $M(x,y,z)$ 的压力值可以用已知函数 $f(x,y,z,t)$ 来表示,则该边界上的边界条件为:

对于前面所讨论的平面径向渗流,其供给边界压力为常数 $p_e$,井底压力为常数 $p_w$,因此,平面径向渗流的内、外边界条件属于第一类边界条件,并可写成:

2. 第二类边界条件

若未知函数在边界上的梯度已知,或边界上未知函数的梯度为已知函数,则称为第二类边界条件。如图 2-13 所示,外边界是不渗透的封闭矩形油藏,由于流体不能穿过封闭边界所以在封闭边界上压力的法向导数为零,即法向渗流速度为零,其外边界条件为:

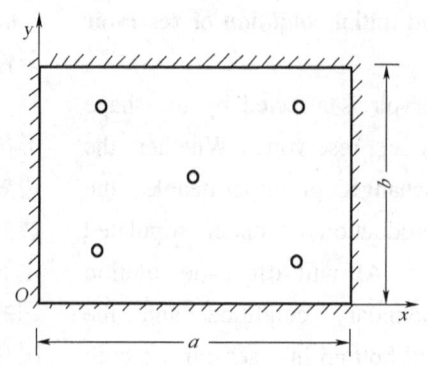

Figure 2 – 13  Closed boundary reservoir
图 2 – 13  封闭边界油气藏

$$\left.\frac{\partial p}{\partial x}\right|_{x=0}=0;\quad \left.\frac{\partial p}{\partial x}\right|_{x=a}=0;\quad \left.\frac{\partial p}{\partial y}\right|_{y=0}=0;\quad \left.\frac{\partial p}{\partial y}\right|_{y=b}=0$$

If the production well $Q$ is known, its inner boundary condition can be written as:

如果已知油井的产量 $Q$,则内边界条件可写作:

$$\left.r\frac{\partial p}{\partial r}\right|_{r=r_w}=\frac{\mu Q}{2\pi Kh}$$

Generally, if the gradient of unknown function on boundary $\Gamma$ is known function $f(x,y,z,t)$, that is:

一般的,如果未知函数在边界 $\Gamma$ 上的梯度为已知函数 $f(x,y,z,t)$,即:

$$\left.\frac{\partial p}{\partial \overline{n}}\right|_{(x,y,z)\in\Gamma}=f(x,y,z,t)$$

Then this kind of boundary condition is called the second-kind boundary condition.

则这类边界条件称为第二类边界条件。

If partial pressure on boundary and the pressure gradient of another part are known, it is called the third-kind boundary condition. For example, the pressure of perforation is equal to the bottom hole pressure, and the radial seepage velocity of the imperforate place is equal to zero.

若边界上的部分压力已知,另一部分压力梯度已知,则称为第三类边界条件。例如,射孔孔眼处压力等于井底压力,无孔处的径向渗流速度等于零。

In reservoir engineering and seepage mechanics, the initial condition that usually meets is $p\mid_{t=0}=p_i$ that the pressure of every point is equal to original formation pressure before development and is equal to the initial formation pressure $p_i$ ($p_i$ can also be the function of coordinate).

油气藏工程和渗流力学中,常遇到的初始条件是 $p\mid_{t=0}=p_i$,即开发前地层各点压力相等,并等于原始地层压力 $p_i$($p_i$ 也可以是坐标的函数)。

## 2.6.6 Basic differential equation and its solutions of single-phase incompressible fluids seepage

### 2.6.6 单相不可压缩液体渗流的基本微分方程及其解

If the state equation and the Darcy equation are substituted into the continuing equation on cylindrical coordinate

如果将状态方程和达西方程代入柱坐标形式的连续性方程中,

form, and assuming the formation is horizontal (that is $v_z = 0$), the circumferential component velocity in homogeneous formation $v_\theta$ is 0, therefore, the corresponding basic differential equation is:

$$\frac{d^2 p}{dr^2} + \frac{1}{r}\frac{dp}{dr} = 0 \qquad (2-62)$$

并假设地层水平(即$v_z = 0$),均质地层中的周向分速度$v_\theta$为0,因此,相应的基本微分方程为:

Derivative replaces partial derivative in formula (2-62) is because the fluid pressure has nothing to do with time $t$ and is only relevant with coordinate $r$ when fluid flow in steady state. The solution of seepage issue can be obtained according to this equation and combined the boundary condition.

式(22-62)用导数代替了偏导数是因为稳定渗流时流体压力与时间$t$无关,而只与坐标$r$有关。根据这一方程并结合边界条件就可以得渗流问题的解。

As to the planar one-dimensional seepage, the seepage law can be described through the following mathematical definite problems:

$$\begin{cases} \dfrac{d^2 p}{dx^2} = 0 \\ p\big|_{x=0} = p_e \\ p\big|_{x=L} = p_w \end{cases}$$

对于平面一维渗流,可以通过以下数学定解问题来描述其渗流规律:

Using the basic knowledge of ordinary differential equation, it is easy to obtain the solution of the problem which is:

$$\begin{cases} p = p_e - \dfrac{p_e - p_w}{L} x \\ p = p_w + \dfrac{p_e - p_w}{L}(L - x) \end{cases}$$

运用常微分方程的基本知识,很容易得到该问题的解为:

The production formula can also be obtained using the Darcy's law which is:

$$Q = vA = -\frac{KA}{\mu}\frac{dp}{dx} = \frac{KA(p_e - p_w)}{\mu L}$$

利用达西定律还可以求其产量公式:

From this it can be seen, the result is consistent with that talked before just in different ways.

由此可见,其结果与前面所讨论的完全一致,只不过是所用方法不同罢了。

As to planar radial fluid seepage, it can be expressed with the following definite problems:

$$\begin{cases} \dfrac{d^2 p}{dr^2} + \dfrac{1}{r}\dfrac{dp}{dr} = 0 \\ p\big|_{r=r_e} = p_e \\ p\big|_{r=r_w} = p_w \end{cases}$$

对于平面径向渗流可以用以下定解问题来描述:

The solution of this problem is:

$$p = \begin{cases} p_e - \dfrac{p_e - p_w}{\mu \ln(r_e/r_w)} \ln(r_e/r) \\ p_w + \dfrac{p_e - p_w}{\mu \ln(r_e/r_w)} \ln(r/r_w) \end{cases}$$

The production $Q$ can be further obtained using the Darcy's law:

$$Q = 2\pi rh \frac{K}{\mu} \frac{dp}{dr} = \frac{2\pi Kh(p_e - p_w)}{\mu \ln(r_e/r_w)}$$

In a word, when researching the seepage law of single-phase incompressible fluid, it must begin with the continuing equation, state equation and the Darcy's law, and derive the basic differential equation, and then carry out the mathematical solution combining with the boundary condition. In the following chapters, the seepage law of single-phase incompressible fluid will be introduced detailedly, and studied with the same method. As to the homogeneous formation of the regularly shaped pattern, the pressure distribution and motion law can be obtained using the method of complex function theory and this problem can also be solved with the method of infinite product. But it is more troublesome. As to the seepage problem of arbitrary shape and non-isopachous non-homogeneous reservoir, numerical integration is the only way to obtain the solution; this is the so-called numerical simulation.

### Exercises

2-1 There is a complete penetrating well penetrating four layers with different $K$ and $h$ in a circular reservoir (Table 2-2), the porosity of each layer is $\phi = 0.2$, $r_e = 2000$m, $r_w = 10$cm, $p_e = 9$MPa, $p_w = 8$MPa, $\mu_0 = 3$mPa·s. Please solve the following problems: (1) The total production of the well $Q$. (2) The average formation permeability $K_p$. (3) Draw the pressure distribution curve and find out the pressure loss from the supply boundary to 10m and 1000m from well. (4) The needed time of fluid moving from the supply boundary to bottom hole.

**Table 2-2  The height and permeability of each layer**
表 2-2  各小层的厚度与渗透率

| Layer | 1 | 2 | 3 | 4 |
|---|---|---|---|---|
| Height, m | 3.0 | 6.0 | 8.0 | 10.0 |
| Permeability, $\mu m^2$ | 0.1 | 0.4 | 0.6 | 1.0 |

2-2  There is a circular reservoir, $r_e = 10$km, $p_e = 10$MPa, $K_1 = 1.5 \mu m^2$, $h = 5$m, $r_w = 0.1$m, $p_w = 9.6$MPa, $\mu = 6$mPa·s. Please solve the following problems: (1) The flow rate $Q_0$. (2) Because of the reason of technology, the permeability within the area of $r_1 = 1$m drops to $K_2 = 0.3 \mu m^2$, please solve the production $Q_1$ and $Q_1/Q_0$, and draw the pressure distribution curve. (3) If the well is acidized, and the permeability within the area of $r_2 = 50$m is increased to $K_3 = 3 \mu m^2$, please solve the present production $Q_2$ and draw the pressure distribution curve.

2-3  If there is a shut-in well 1km far from the production well and its bottom hole pressure is $p_{w1} = 8$MPa, the bottom hole pressure of the production well is $p_{w2} = 4$MPa and the bottom radius $r_w = 10$m. The production well lies in the center of circular formation that its supply radius is $r_e = 10$km. It is known that the Darcy's law stands in the whole formation, please solve the supply boundary pressure $p_e$.

2-4  Assuming that there is a production well in the center of circular formation, the seepage of fluid obeys the Darcy's law, and $r_e = 10$km, $r_w = 10$cm, please solve the following problem: how far is the point from well whose pressure is just right equal to the average value of $p_e$ and $p_w$, that is $p = (p_e + p_w)/2$.

2-5  If pressure is expressed with MPa, viscosity is expressed with mPa·s, the unit of thickness is m, the unit of permeability is $\mu m^2$, the unit of production is $m^3/d$, and change the natural logarithm to common logarithm, please try to rewrite the Dupuit Formula.

2-2  一圆形油气藏 $r_e = 10$km, $p_e = 10$MPa, $K_1 = 1.5 \mu m^2$, $h = 5$m, $r_w = 0.1$m, $p_w = 9.6$MPa, $\mu = 6$mPa·s。(1) 求流量 $Q_0$。(2) 如果由于技术上的原因, 使得井周围半径 $r_1 = 1$m 范围内的渗透率下降到 $K_2 = 0.3 \mu m^2$, 求井的产量 $Q_1$ 和 $Q_1/Q_0$, 并画出压力分布曲线。(3) 对该井进行酸化处理, 使得半径 $r_2 = 50$m 内地层的渗透率提高到 $K_3 = 3 \mu m^2$, 求此时井的产量 $Q_2$, 绘出压力分布曲线。

2-3  如果距生产井 1km 处有一停产井, 井底压力 $p_{w1} = 8$MPa, 生产井的井底压力 $p_{w2} = 4$MPa, 井底半径 $r_w = 10$cm, 生产井位于圆形地层中心, 地层的供给半径 $r_e = 10$km, 已知整个地层内达西定律都成立, 求供给边线压力 $p_e$。

2-4  圆形地层中心有一口生产井, 液体渗流服从达西定律, $r_e = 10$km, $r_w = 10$cm, 求距井多远的点的压力恰好等于 $p_e$ 和 $p_w$ 的平均值, 即 $p = (p_e + p_w)/2$。

2-5  若压力用 MPa 表示, 黏度以 mPa·s 表示, 厚度以 m 为单位, 渗透率以 $\mu m^2$ 为单位, 产量单位为 $m^3/d$, 并将自然对数化为常用对数, 试改写丘比公式。

2 – 6  If seepage obeys the binomial law, please try to derive the seepage law of planar uniflow and radial fluid flow (the relationship of production, formation properties, fluid properties with pressure difference).

2 – 7  Assuming that the formation is homogeneous, isopachous and isotropic, and the seepage of fluid is isothermal and flow towards along radial direction, please try to derive the continuing equation and basic differential equation on cylindrical coordinate.

2 – 8  Assuming that there is a fan-shaped reservoir with included angle $\theta$ as shown in the Figure 2 – 14, there is a production well on the top of the reservoir, and assuming the formation, fluid and production data all are known. Please try to derive the production formula and pressure distribution formula of the oil well.

2 – 9  Assuming that there is a core as shown in Figure 2 – 7, the core, fluid and the height difference of piezometer $\Delta H$ are known, please try to start from the differential form of Darcy's law to prove the flow rate is:

$$Q = \frac{KA\rho g \Delta H}{\mu L}$$

2 – 10  As shown in the Figure 2 – 15, a hemispherical well that its radius is $r_w$ is drilled in the center of homogeneous and incompressible hemispherical formation. Assuming the formation permeability is $K$, fluid viscosity is $\mu$, the radius of supply boundary line is $r_e$, the supply pressure is $p_e$, the bottom pressure is $p_w$, please solve the relationship of production and differential pressure.

2 – 6  若渗流服从二项式定理，试推导平面单向渗流和平面径向渗流的渗流规律（产量、地层性质、流体性质与压力差的关系）。

2 – 7  设地层是均质、等厚、各向同性的，液体作等温径向渗流，试推导柱坐标形式的连续性方程和基本微分方程式。

2 – 8  设有一如图 2 – 14 所示的夹角为 $\theta$ 的扇形油气藏。在该油气藏的顶点有一生产井，设地层、流体和生产数据都已知。试推导油井产量公式和压力分布公式。

2 – 9  设有一岩心如图 2 – 7 所示，岩心、流体及测压管的高差 $\Delta H$ 已知，试从达西定律的微分形式出发，证明流量公式为：

2 – 10  如图 2 – 15 所示的均质、不可压缩半球地层的中心钻了一口半径为 $r_w$ 的半球形井，设地层渗透率为 $K$，液体黏度为 $\mu$，供给边线半径为 $r_e$，供给压力为 $p_e$，井底压力为 $p_w$，求产量与压力差的关系。

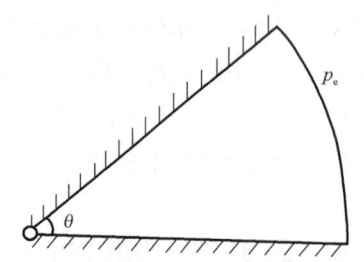

Figure 2 – 14  The sketch of fan-shaped reservoir
图 2 – 14  扇形油气藏

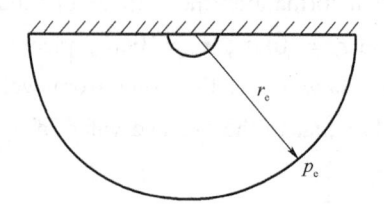

Figure 2 – 15  The sketch of half-round reservoir
图 2 – 15  半球形油气藏

2 – 11  The measured average formation pressure of an imperfect well when shut in is $\bar{p} = 25.60 \text{MPa}$. It is known that the supply radius is $r_e = 250\text{m}$, the bottom radius is $r_w =$

2 – 11  某一不完善油井关井测得的平均地层压力为 $\bar{p} = 25.60\text{MPa}$，已知供给半径 $r_e$

0.1m, the fluid viscosity is $\mu = 4\text{mPa}\cdot\text{s}$, the formation effective thickness is $h = 8\text{m}$. Now the well is conducted the systematic well testing, and the measured production and pressure are shown in Table 2-3, please try to solve the formation permeability.

250m，井底半径 $r_w = 0.1\text{m}$，流体黏度 $\mu = 4\text{ mPa}\cdot\text{s}$，地层有效厚度 $h = 8\text{m}$。现对该井进行稳定试井，实测产量和压力见表2-3，试求地层的渗透率。

**Table 2-3  The measured production and pressure**
**表2-3  实测产量和压力**

| $Q$, m³/d | 35.58 | 62.26 | 94.88 | 127.48 |
|---|---|---|---|---|
| $p_w$, MPa | 24.4 | 23.5 | 22.4 | 21.3 |

2-12  If the changing law of formation permeability with formation pressure is $K = K_0 e^{a(p-p_0)}$, the relationship of fluid viscosity and pressure is $\mu = \mu_0 e^{b(p-p_0)}$, the supply radius is $r_e$, the supply pressure is $p_e$, the radius of well is $r_w$, the bottom pressure is $p_w$, the formation effective thickness is $h$, please solve the production formula of radial fluid seepage.

2-12  若地层渗透率随地层压力的变化规律为 $K = K_0 e^{a(p-p_0)}$，流体黏度与压力的关系为 $\mu = \mu_0 e^{b(p-p_0)}$，供给半径为 $r_e$，供给压力为 $p_e$，井半径为 $r_w$，井底压力为 $p_w$，地层有效厚度为 $h$，求液体作平面径向稳定渗流的产量公式。

2-13  The well testing on a well shows: the well is damaged, the skin factor is $S = 3.8$, the formation permeability is $K = 0.4\mu\text{m}^2$, the oil layer effective pay thickness is $h = 6\text{m}$, the fluid viscosity is $\mu = 4.5\text{mPa}\cdot\text{s}$, and the oil volume factor is $B_o = 1.2$, the bottom radius is $r_w = 10\text{cm}$. Please solve the following problems:

(1) The reduced radius of the well.

(2) What is the additional resistance caused by skin effect when the well producing with the production of $32\text{m}^3/\text{d}$?

2-13  对某井进行试井的结果表明：该井受到伤害，表皮系数 $S = 3.8$，地层渗透率 $K = 0.4\mu\text{m}^2$，油层有效厚度 $h = 6\text{m}$，流体黏度 $\mu = 4.5\text{mPa}\cdot\text{s}$，原油体积系数 $B_o = 1.2$，井底半径 $r_w = 10\text{cm}$。
(1) 该井的折算半径为多少？
(2) 若该井以 $32\text{m}^3/\text{d}$ 的产量生产时，由表皮效应引起的附加阻力为多少？

2-14  Please prove that the continuing equation of single-phase fluid (2-57) can also be written as the form below ($B$ is volume factor).

2-14  证明单相流体的连续性方程式(2-40)可以写成如下形式($B$为体积系数)：

$$\frac{\partial}{\partial x}\left(\frac{v_x}{B}\right) + \frac{\partial}{\partial y}\left(\frac{v_y}{B}\right) + \frac{\partial}{\partial z}\left(\frac{v_z}{B}\right) = -\frac{\partial}{\partial t}\left(\frac{\phi}{B}\right)$$

Chapter 2 Study Guide
第二章学习指南

# 3　Interference theory of wells under rigid water drive

刚性水压驱动下的油井干扰理论

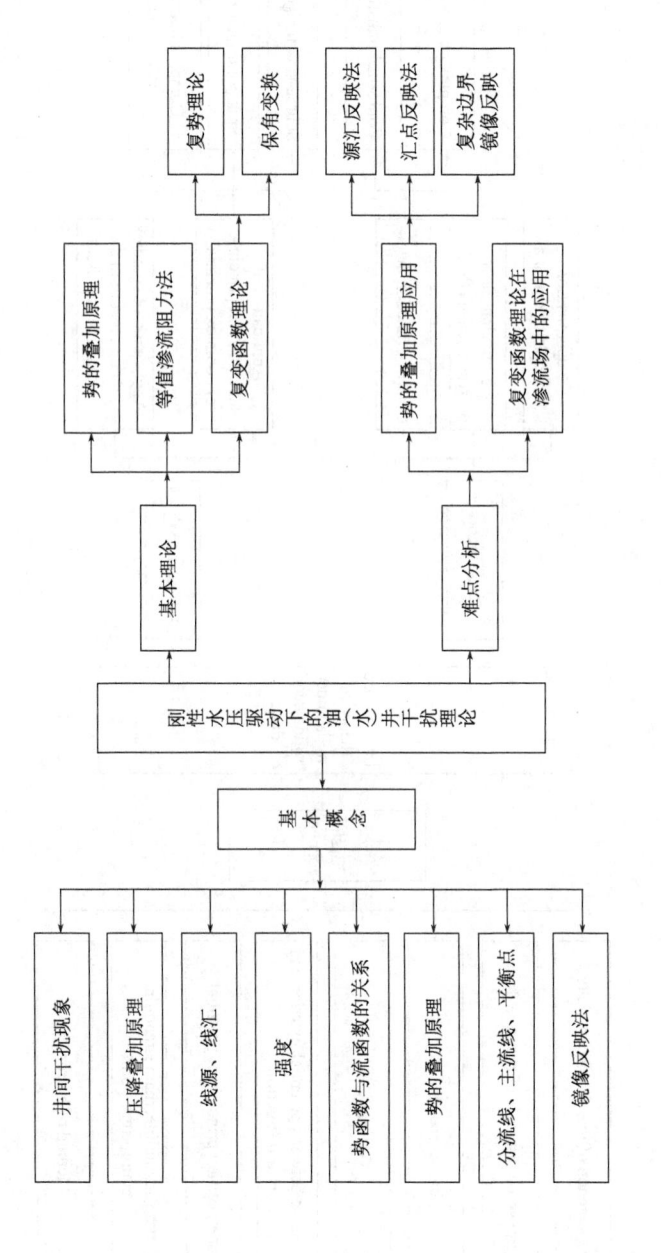

# 第三章知识图谱

```
刚性水压驱动下的油(水)井干扰理论
├── 基本理论
│   ├── 势的叠加原理 → 复势理论
│   ├── 等值渗流阻力方法 → 保角变换
│   └── 复变函数理论
└── 难点分析
    ├── 势的叠加原理应用
    │   ├── 源汇反映法
    │   ├── 汇点反映法
    │   └── 复杂边界镜像反映
    └── 复变函数理论在渗流场中的应用

基本概念
├── 井间干扰现象
├── 压降叠加原理
├── 线源、线汇
├── 强度
├── 势函数与流函数的关系
├── 势的叠加原理
├── 分流线、主流线、平衡点
└── 镜像反映法
```

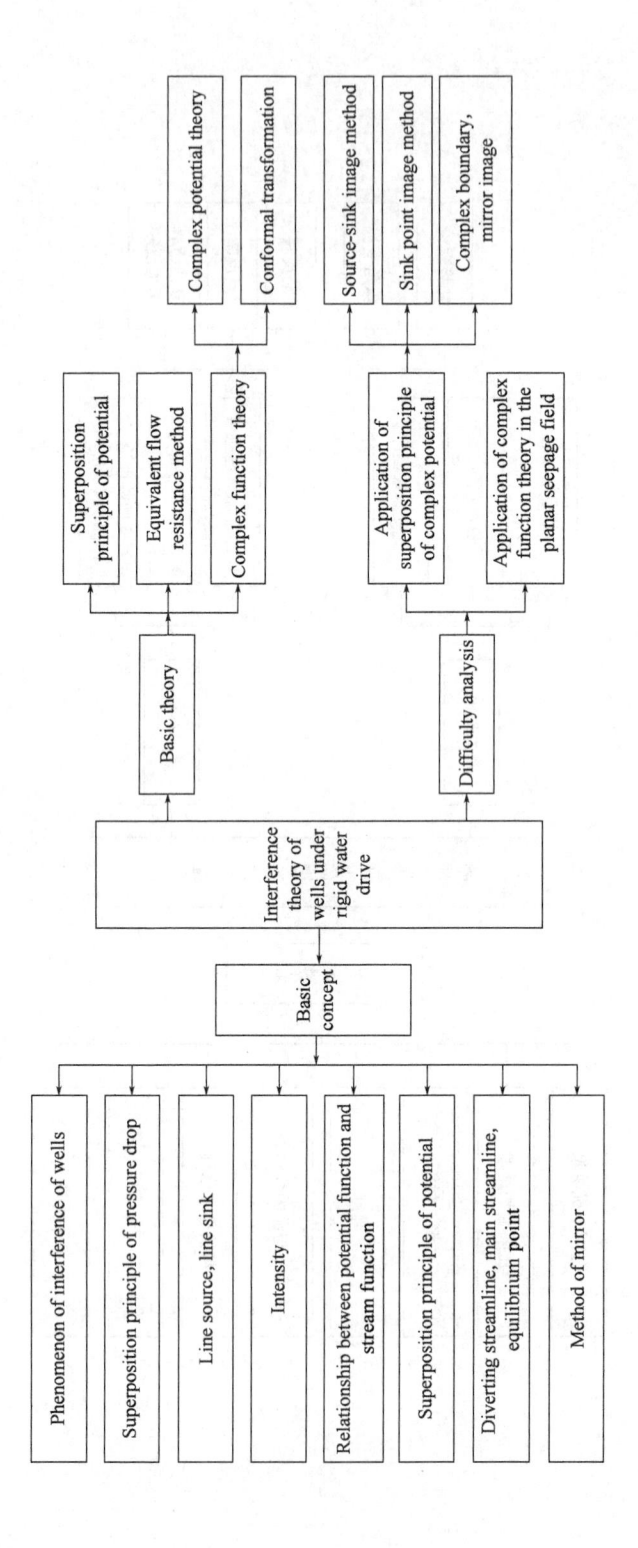

# Chapter 3 Knowledge Graph

**Interference theory of wells under rigid water drive**

- Basic concept
  - Phenomenon of interference of wells
  - Superposition principle of pressure drop
  - Line source, line sink
  - Intensity
  - Relationship between potential function and stream function
  - Superposition principle of potential
  - Diverting streamline, main streamline, equilibrium point
  - Method of mirror

- Basic theory
  - Superposition principle of potential
  - Equivalent flow resistance method
  - Complex function theory
    - Complex potential theory
    - Conformal transformation

- Difficulty analysis
  - Application of superposition principle of complex potential
    - Source-sink image method
    - Sink point image method
    - Complex boundary, mirror image
  - Application of complex function theory in the planar seepage field

— 86 —

The theory of one dimensional and racial seepage of single-phase incompressible fluid is introduced in the last chapter. However, it is not a single well producing in the production process for most reservoirs but many production wells and injection wells work simultaneously, and with the further development of reservoir, some of the wells are put into production or shut down constantly. On the other hand, the working system of the wells put into production will also be changed frequently in the period of development. The initial pressure distribution will be destroyed because of the change of the working system of wells and drilling new wells, which also makes the whole seepage field change, and naturally will affect the production of adjacent wells. The phenomenon of mutual effect between wells is called interference of wells. Also, the shape of actual reservoir is very irregular. Some of the boundaries are full opening, Some are closed and some are partly closed, and the impermeable boundary of pinchout line and closed fault often exist in the reservoir. These irregular boundaries also have influence on the production of wells.

In this chapter, some typical interference between wells and boundary-well of homogeneous incompressible fluid in incompressible formation steady state seepage will be introduced.

## 3.1 Phenomenon of interference between wells

In order to know the phenomenon of interference of wells, the two production wells in Figure 3-1 (Dynamic graph 3-1) are shown as an example to be analyzed. In the figure $O—O'$ is the initial formation pressure line. Assuming that the formation and fluid are homogeneous and incompressible, the geometry of formation is much bigger than well spacing. So when the well $A_1$ produces independently with the production $Q_1$, the pressure of every point in formation will drop. Its pressure distribution is shown as the dashed curve Ⅰ. When the well $A_2$ produces independently with the production $Q_2$, its pressure distribution is shown as the dashed line Ⅱ。 Even if the production of well $A_1$ is $Q_1=0$, the pressure drop $AB$ in the place of well

上一章介绍了单相不可压缩液体的平面一维和平面径向稳定渗流理论。然而，绝大多数的油藏在生产过程中不是一口井单独生产，而是许多生产井和注水井同时工作，并且，随着油藏的进一步开发，不断有井投入使用或关闭。已投入生产的井在整个开发期间也会频繁地改变工作制度。油水井工作制度的变化及新井的投产会使原来的压力分布状态遭到破坏，引起整个渗流场发生变化，自然会影响邻井的产量，这种井间相互影响的现象称为井间干扰。另外，实际油藏的形状也是很不规则的，其边界有的是敞开的，有的是封闭的或局部封闭的，而且，油藏内部常有尖灭线和封闭性断层等不渗透边界存在，这些不规则的边界对井的生产也会产生影响。

本章将介绍均质不可压缩液体在不可压缩地层中作稳定渗流的几种典型的井与井、边界与井的干扰问题。

## 3.1 井间干扰现象

为了认识井间干扰现象，以图3-1（动态图3-1）中两口生产井为例进行分析。图中$O—O'$为原始地层压力线。设地层及液体是均质、不可压缩的，地层的几何尺寸比井距大得多。那么，当$A_1$井以产量$Q_1$单独生产时，地层中每一点的压力都要下降，其压力分布如虚线Ⅰ所示。当$A_2$井以产量$Q_2$单独生产时，在地层中造成的压力分布如虚线Ⅱ所示。即使$A_1$井的产量$Q_1=0$，由于$A_2$井的生产

$A_1$ will also emerge because of the production of well $A_2$. When the well $A_1$ and well $A_2$ produce simultaneously and maintain the bottom pressure $C$ and $C'$ unchanged because of the interference of well $A_2$, the pressure drop value caused in the place of $A_1$ reduces $AB$ than that of working independently, and the producing pressure difference changes to be $BC$, so the corresponding production will certainly decrease. If the production $Q_1$ of well $A_1$ is maintained, the bottom pressure should be reduced from the point $C$ to point $E$ and make $CE = AB$. The interference of well $A_1$ to well $A_2$ is also the same. Apparently, when the well $A_1$ and well $A_2$ produce simultaneously respectively with production $Q_1$ and $Q_2$, the pressure distribution curve will no longer be the dashed line Ⅰ and Ⅱ. At this time, the pressure drop of every point in formation should be equal to the algebraic sum of pressure drop on this point when well $A_1$ and well $A_2$ produce independently; the pressure distribution is shown as block curve Ⅲ.

As for the water flooding reservoir, there are not only production wells but also injection wells. The production wells make the formation pressure drop while the injection wells make the pressure buildup (negative pressure drop). As shown in Figure 3-2 (Dynamic graph 3-2), the pressure distribution curve of well $A_1$ producing with production $Q_1$ in formation is shown as the dash curve Ⅰ, the pressure distribution curve of well $A_2$ when working with the injection $Q_2$ in formation is shown as the dash curve Ⅱ. When the well $A_1$ and $A_2$ working simultaneously with the production $Q_1$ and injection $Q_2$, the total pressure drop of an arbitrary point in formation should be equal to the difference of the pressure drop on this point. When well $A_1$ works independently and the build-up pressure on this point when well $A_2$ works independently, in other words, it is the algebraic sum of pressure drop on this point when the two wells work independently. The difference is that the injection well causes negative pressure drop, while the pressure distribution in formation is shown as the block curve Ⅲ.

在 $A_1$ 井处也会造成数值等于 $AB$ 的压力降。当 $A_1$ 井和 $A_2$ 井同时生产并维持井底压力 $C$ 和 $C'$ 不变时，$A_2$ 井的干扰在 $A_1$ 井处引起的压降比 $A_1$ 井单独工作时减少了 $AB$，生产压差变为 $BC$，相应的产量必然降低。若要维持 $A_1$ 井的产量 $Q_1$ 不变，则井底压力应从 $C$ 点降到 $E$ 点，且使 $CE = AB$。$A_1$ 井对 $A_2$ 井的干扰也是如此。显然，$A_1$ 井和 $A_2$ 井同时分别以产量 $Q_1$ 和 $Q_2$ 生产时，油层中的压力分布不再如虚线Ⅰ、Ⅱ所示。此时，地层中任一点处的压降应等于 $A_1$ 井和 $A_2$ 井单独工作在该点造成的压降的代数和，压力分布如实线Ⅲ所示。

对于注水开发的油田，不仅有生产井，而且还有注水井，生产井导致地层压力下降，而注水井引起压力回升（负压降）。如图 3-2（动态图 3-2）所示，$A_1$ 井以产量 $Q_1$ 单独生产时在地层中造成的压力分布如虚线Ⅰ所示，$A_2$ 井以注入量 $Q_2$ 单独工作时引起的压力分布如虚线Ⅱ所示。当 $A_1$ 井和 $A_2$ 井同时分别以产量 $Q_1$ 和注入量 $Q_2$ 工作时，地层中任一点的总压降等于 $A_1$ 井单独工作在该点造成的压降与 $A_2$ 井单独工作在该点引起的压力升之差，也可以说是由两井单独工作在该点造成的压降的代数和，所不同的是注水井产生负压降，而地层中的压力分布如实线Ⅲ所示。

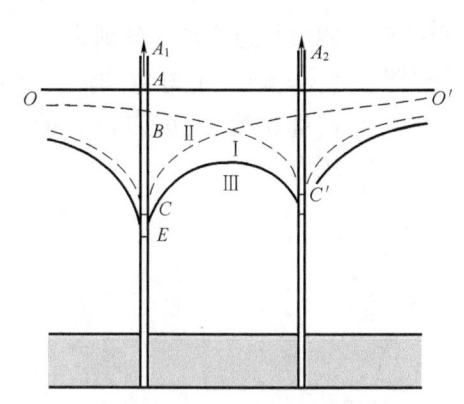

Figure 3 – 1   The interference of
two production wells
图 3 – 1   两口生产井的干扰

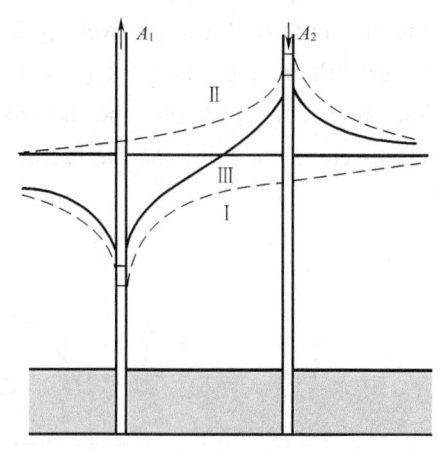

Figure 3 – 2   The interference of
a production well and an injection well
图 3 – 2   一口生产井和一口注水井的干扰

Dynamic graph 3-1  The interference
of two production wells
动态图3-1   两口生产
井的干扰

Dynamic graph 3-2  The interference of
a production well and an injection well
动态图3-2   一口生产
井和一口注水井的干扰

In a word, as to two or more wells work simultaneously in a large formation, the following superposition principle of pressure drop exists: when many wells (production wells and injection wells) work simultaneously, the pressure drop on an arbitrary point in formation should be equal to the pressure-drop algebraic sum of each well on this point when it works independently with its own unchanged production.

## 3.2　Superposition principle of potential

For many wells working simultaneously, it needs to know the pressure distribution or seepage field in formation, so it should start from describing of differential equation of fluid seepage.

For example, assuming that there is a kind of incompressible fluid flowing in the horizontal homogeneous, isopachous and incompressible formation, and the supply radius is $r_e$. There are four wells working simultaneously (Figure 3 – 3), the bottom pressure of the first well and the

综上所述,对于很大地层上的两口或两口以上的井同时生产,存在以下压降叠加原理:多井(生产井和注水井)同时工作时,地层中任一点处的压降应等于各井以各自不变的产量单独工作时在该点处造成的压降的代数和。

## 3.2　势的叠加原理

对于多井同时工作,要了解地层内的压力分布或渗流场,应从解描述液体渗流规律的微分方程组出发。

例如,设水平均质、等厚、不可压缩、供给半径为 $r_e$ 的地层中只有一种不可压缩的液体在流动。地层中有四口井同时工作(图 3 – 3),第一、第二口井的井

— 89 —

second well is respectively $p_{w1}$ and $p_{w2}$, the production of the third and the fourth well is respectively $Q_3$ and $Q_4$, the radius of the four well is respectively $r_{wi}(i = 1,\ldots,4)$, please solve the production of the first and the second well and the bottom pressure of the third and the fourth well.

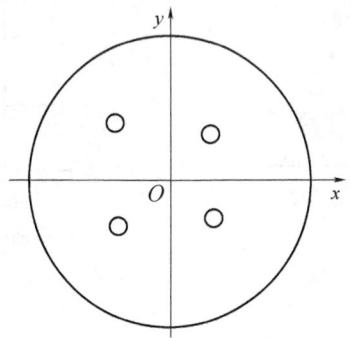

Figure 3-3  The well distribution graph
图 3-3  井的分布示意图

Set up the coordinate system as shown in Figure 3-3 and assume the coordinates of the four wells are respectively $(x_i, y_i)$ ($i = 1, \ldots, 4$). Because formation and fluid are incompressible, so the seepage obeys the Laplace equation, then the problem of multi interference of wells can be expressed with the following mathematical definite problem combining the boundary condition.

$$\nabla^2 p = 0$$

$$p \big|_{x^2+y^2 = r_e^2} = p_e$$

$$p \big|_{c_1} = p_{w1}$$

$$p \big|_{c_2} = p_{w2}$$

$$-\frac{Kh}{\mu}\oint_{c_3} \frac{\partial p}{\partial \boldsymbol{n}} \mathrm{d}s = Q_3$$

$$-\frac{Kh}{\mu}\oint_{c_4} \frac{\partial p}{\partial \boldsymbol{n}} \mathrm{d}s = Q_4$$

In the expression above, $c_i$ is the boundary of $(x-x_i)^2 + (y-y_i)^2 = r_{wi}^2$ ($i = 1,\ldots, 4$); $n$ is the normal vector of corresponding boundary; $ds$ is the infinitesimal of corresponding boundary and its direction is the outer normal direction of boundary. It is very troublesome to solve the Laplace's equation under the boundary condition above. Actually, the well radius is very small relative to the size of formation, the well can be completely considered as a geometry line (a point on planar), and thus the influence of inner boundary condition can be canceled. If the formation is finite, the pressure distribution caused by well location is different. In order to cancel the influence of outer boundary, the formation is assumed as infinite. The geometric condition of wells considered as a line in infinite formation is the same. While considering the formation as infinite and considering the well in it as a line is far away from the reality. It can be seen that the actual finite formation is only a special case in infinite formation, while borehole wall is general an isobar.

It is the commonly used method in science and technology that draws out the concrete influence of inner and outer boundaries and various concrete problem solutions are included in abstract solutions. Right abstraction can much more reflect the nature of the problems.

The line absorbing fluid from formation is called line sink; the line supplying fluid into formation is called line source. The line sink is equivalent to production well and the line source is equivalent to injection well. The production or injection in oil layer of unit effective thickness $q$ is called the intensity of line sink (source). As to the horizontal isopachous formation, it is enough to talk about one of its horizontal cross section and now the line sink (source) changes to be point sink (source). The line sink (source) and point sink (source) will not be distinguished here after. In this chapter, in order to introduce briefly, the formation height is generally assumed as $h = 1\text{m}$.

Assuming that homogeneous incompressible fluid percolates in homogeneous incompressible formation. Please solve the pressure distribution caused by a point sink with intensity $q$ in infinite formation.

Take the point sink as the zero point on the polar-coordinate system, because the formation and fluid are all homogeneous, so the Laplace's equation is:

$$\frac{1}{r}\frac{\mathrm{d}}{\mathrm{d}r}\left(r\frac{\mathrm{d}p}{\mathrm{d}r}\right) = 0 \tag{3-1}$$

The solution of the formula above can be easily obtained as:

$$p = C_1 \ln r + C_2 \tag{3-2}$$

In the formula above, $C_1$、$C_2$ is all arbitrary constant. taking the derivative of both sides with respect to $r$, the following expression can be obtained:

$$\frac{\mathrm{d}p}{\mathrm{d}r} = C_1 \frac{1}{r} \tag{3-3}$$

It is known that the intensity of the point sink is $q$, and then the following expression can be obtained:

$$\lim_{r \to 0} \frac{2\pi r K}{\mu} \frac{\mathrm{d}p}{\mathrm{d}r} = \frac{2\pi K C_1}{\mu} \lim_{r \to 0} r \cdot \frac{1}{r} = q$$

So:

$$C_1 = \frac{q\mu}{2\pi K}$$

Substitute the value of $C_1$ into the formula (3-2), the pressure distribution caused by a point sink with the intensity $q$ on the planar infinite formation is:

$$p = \frac{q\mu}{2\pi K}\ln r + C_2 \tag{3-4}$$

Apparently, the factor $\frac{\mu}{K}$ always appears in the pressure distribution formula, in order to keep in line with the theory of potential in mathematical physics and also for the purpose of writing conveniently, the concept of potential is introduced here:

$$\Phi = \frac{K}{\mu}p + C_3 \tag{3-5}$$

In the formula above: $\Phi$ is called potential. $C_3$ is an arbitrary constant. So the Darcy's law can be written as the expression below with the concept of potential:

$$\bar{v} = -\mathrm{grad}\Phi \qquad (3-6)$$

As for the radial fluid flow, it can be obtained:

$$v = -\frac{\mathrm{d}\Phi}{\mathrm{d}r} \qquad (3-7)$$

Combine the formula (3-4) and formula (3-5), it can be obtained:

$$\Phi = \frac{q}{2\pi}\ln r + C \qquad (3-8)$$

对于平面径向渗流就有：

联立式(3-4)和式(3-5)可得：

Apparently, C is an arbitrary constant which needs to be determined according to the outer boundary condition. Because the negative gradient is equal to seepage velocity, so potential is also called velocity potential. The nature of potential is the same as pressure which is a scalar. The gradient of potential which is also the velocity field can be obtained if the distribution of potential is known, so the solution of problem can also be obtained.

If there are lots of point sinks and point sources working simultaneously in the homogeneous isopachous incompressible infinite formation, we will naturally think about that the potential of an arbitrary point in formation should be equal to the algebraic sum of potential on this point caused by every point sink and point source when working independently, and this is the superposition principle of potential. Assume that the potential of $n$ point sinks (source) is $q_i(i=1,\ldots,n)$, the distance of an arbitrary point $M$ to every point sink (source) is respectively $r_i(i=1,\ldots,n)$, then the potential of point $M$ is:

显然 C 是任意常数，需要根据外边界条件来确定。因为势的负梯度等于渗流速度，所以势也称为速度势。势的本质与压力一样，是标量，知道了势的分布，就可以求出势的梯度，即速度场，这样就得出了问题的解。

若均质、等厚、不可压缩、无限大地层上有许多个点源、点汇同时工作，自然就会想到地层上任一点的势应该等于每个点源、点汇单独工作时在该点所引起的势的代数和，这就是势的叠加原理。设 n 个点汇(源)的强度为 $q_i(i=1,\cdots,n)$，地层中任一点 M 距各个点汇(源)的距离分别为 $r_i(i=1,2,\cdots,n)$，则 M 点的势为：

$$\Phi = \frac{1}{2\pi}\sum_{i=1}^{n} \pm q_i \ln r_i + C \qquad (3-9)$$

A negative value should be taken in front of the point source $q_i$. As for point sink, positive value should be taken in front.

The theoretical foundation is: the Laplace's equation is linear; the linear combination of particular solution to the linear equation is just the solution to equation. It is not difficult to see the formula (3-9) meets the Laplace's equation. So how to guarantee the solution obtained by such way is the same with that of resolving the Laplace's equation directly? The theorem of existence and uniqueness

对于点源 $q_i$ 的前面应取负值，而对点汇 $q_i$ 前取正值。

这样做的理论根据是：拉氏方程是线性的，线性方程的特解的线性组合还是方程的解，不难看出式(3-9)满足拉氏方程。那么，怎么能保证这样求出的解与直接解拉氏方程得到的解一样呢？拉氏方程的存在与唯一

of the Laplace's equation tell us: to the Laplace's equation meeting definite boundary condition there must be one and just only one solution, only one constant is permitted to two solutions. It is not a question as to the solution obtained according to the superposition principle to meet the Laplace's equation, the only thing to do is just substituting the boundary condition into the formula (3-9) again and obtain the value of undetermined constant $C$, the problem will be solved completely. Apparently, it is much easier to settle the interference of wells with superposition principle than solving the Laplace's equation directly.

**Example 3-1** In formation that is very big compared with well spacing there are three wells, which is known that the pressure on supply boundary $p_e$ is a constant, the formation is horizontal and its height is $h$; the permeability is $K$; the bottom pressure of the first well is $p_{w1}$; the bottom pressure of the second well is $p_{w2}$, and the intensity of the third well is $q_3$ as shown in Figure 3-4. The radius of the three wells is equal and is $r_w$, please solve the intensity of the first and second well $q_1$ and $q_2$, and the bottom pressure of the third well $p_{w3}$.

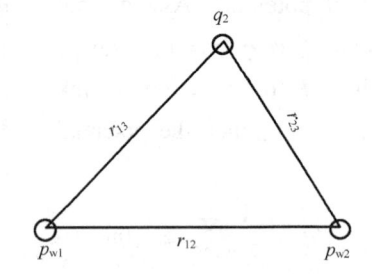

Figure 3-4  The location of the three wells
图 3-4  三口井的井位图

Solution: According to the formula (3-9), the potential of an arbitrary point $M$ in formation is:

$$\Phi = \frac{1}{2\pi}(q_1 \ln r_1 + q_2 \ln r_2 + q_3 \ln r_3) + C \qquad (3-10)$$

Substitute $\Phi = \frac{K}{\mu} p + C_3$ into the formula above, the following expression can be obtained:

$$\frac{K}{\mu} p = \frac{1}{2\pi}(q_1 \ln r_1 + q_2 \ln r_2 + q_3 \ln r_3) + C \qquad (3-11)$$

In the formula above: $C$ is a new arbitrary constant. If the point $M$ is chosen on the first well wall, and assume the distance between the first well and the second well is $r_{12}$; the distance between the first well and the third well is $r_{13}$; the distance between the second well and the third well is $r_{23}$, then following expression can be obtained:

$$\frac{K}{\mu}p_{w1} = \frac{1}{2\pi}(q_1 \ln r_w + q_2 \ln r_{12} + q_3 \ln r_{13}) + C \qquad (3-12)$$

Choose the point $M$ respectively on the well wall of the second and the third well, the following expression can be obtained:

$$\frac{K}{\mu}p_{w2} = \frac{1}{2\pi}(q_1 \ln r_{12} + q_2 \ln r_w + q_3 \ln r_{23}) + C \qquad (3-13)$$

$$\frac{K}{\mu}p_{w3} = \frac{1}{2\pi}(q_1 \ln r_{13} + q_2 \ln r_{23} + q_3 \ln r_w) + C \qquad (3-14)$$

Three equations are obtained now. $q_1$, $q_2$, $p_{w3}$ and $C$ are four unknown variables. The unique solution can not be obtained from three equations with four unknown variables. The condition that the outer boundary of atmospheric pressure is far from well must be considered now, and the point $M$ is chosen on the outer boundary, because:

$$r_{1M} \approx r_{2M} \approx r_{3M} \approx r_e$$

So the following expression can be obtained:

$$\frac{K}{\mu}p_e = \frac{1}{2\pi}(q_1 + q_2 + q_3)\ln r_e + C \qquad (3-15)$$

There are four equations from formula (3-12) to formula (3-15). Naturally four unknown variables can be obtained from the equations. The complex process of solving the Laplace's equation can be changed into the simple process to solve the simultaneous linear algebra equation with the superposition principle. However, there are two problems needed to be thought about. The first point is if the formation is not large, how about a small fault block oil reservoir? The second point is whether the well wall really can be considered as constant pressure surface?

As for many point sinks (source) working simultaneously, the velocity of arbitrary point in formation can adopt the principle of vector addition of velocity, that is: the seepage velocity $v$ of an arbitrary point $M$ is equal to the vector sum of seepage velocity $v_i$ ($i = 1, 2, \ldots, n$) caused by every point

上式中的 $C$ 为新的任意常数。若把 $M$ 点选在第一口井的井壁上,并设第一口井与第二口井的距离为 $r_{12}$,第一口井与第三口井的距离为 $r_{13}$,第二口井与第三口井的距离为 $r_{23}$,则有:

再分别将 $M$ 点选在第二、第三口井的井壁上,结果得:

至此,得到了三个方程式,其中 $q_1$、$q_2$、$p_{w3}$ 和 $C$ 为四个未知量,三个方程四个未知量是得不到唯一解的。此时必须考虑常压外边界距井很远这一条件,把 $M$ 点选在外边界上,由于:

于是又得:

从方程(3-12)到方程(3-15)共四个方程式,从中自然可以求出四个未知量。利用叠加原理可以把解拉氏方程的复杂过程转化为解线性代数方程的简单过程。不过其中有两个问题值得思考,其一是若地层不是很大,如小断块油藏怎么办?其二是井壁果真能看作等压面吗?

许多点汇(源)同时工作时,地层任意点处的速度,可以采用速度的矢量合成这一原理,即:任意点 $M$ 处的渗流速度 $v$ 等于各点汇(源)单独工作时在

sink (source) on this point when working independently as shown in Figure 3-5 and the formula is as follow:

$$v = \sum_{i=1}^{n} v_i \qquad (3-16)$$

Of course, $v$ can also be solved according to the Darcy's law, that is:

$$v = -\nabla \Phi$$

Actually, the seepage velocity component in the direction of $x$ and $y$ of $v$ can be solved firstly:

$$v_x = -\frac{\partial \Phi}{\partial x}; \quad v_y = -\frac{\partial \Phi}{\partial y}$$

And then the following expression can be obtained:

$$|v| = \sqrt{\left(\frac{\partial \Phi}{\partial x}\right)^2 + \left(\frac{\partial \Phi}{\partial y}\right)^2}; \quad \tan\theta = \frac{\frac{\partial \Phi}{\partial y}}{\frac{\partial \Phi}{\partial x}}$$

In the formula above, $\theta$ is the included angle between $v$ and $x$ axis as shown in Figure 3-6.

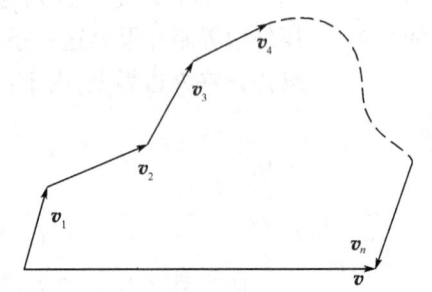

Figure 3-5  The vector sum of velocity
图 3-5  速度的矢量和

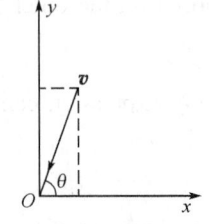

Figure 3-6  The included angle between velocity and $x$ axis
图 3-6  速度与 $x$ 轴的夹角

## 3.3 Method of mirror

It is always from the simple to the complex to research problems. However, the general law can usually be obtained from solving the simple problems. In order to solve the problem of interference of wells, we also start from the simplest interference problem, from which we can not only understand the essence of interference problem but also obtain two important methods to solve interference problem. The simplest interference problems are one source and one sink and two sinks interference, while one of the important methods to deal with interference problem is the method of mirror image. This is because the superposition principle of potential is set up on the assumption of infinite formation, but the actual reservoir is finite and some wells may be very near the boundary and these boundaries may be supplied boundary or impermeable boundary. Because the existence of boundary will make the fluid seepage change, thereby affects the production of oil well. As to these problems, it is hoped to change these problems to be multi interference of wells problem in infinite formation, and solve it with the superposition principle of potential.

### 3.3.1 Interference of uniform-intensity one source and one sink—the source-sink image method

Assuming that there is a uniform-intensity one source and one sink respectively on the two points, the distance between the two points far away with each other is $2a$ in infinite formation as shown in Figure 3-7. Take the connection between source and sink as axis $x$ and take the vertical bisection of the connection as axis $y$. The point $(a, 0)$ is the sink, while the point $(-a, 0)$ is the source.

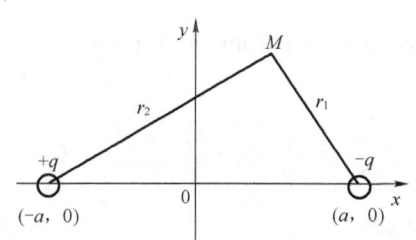

Figure 3-7  The interference of one source and one sink

According to the superposition principle of potential, the potential of an arbitrary point $M(x,y)$ in formation is:

$$\Phi = \frac{q}{2\pi}\ln\frac{r_1}{r_2} + C \qquad (3-17)$$

In the formula above, $r_1$ and $r_2$ are respectively the distance from the point $M$ to sink and source, that is:

$$r_1 = \sqrt{(x-a)^2 + y^2}$$
$$r_2 = \sqrt{(x+a)^2 + y^2}$$

If it is known that the bottom pressure of source and sink are respectively $p_j$ and $p_w$, the well radius is $r_w$, if the point $M$ is respectively chosen on the bore wall of source and sink, at the point source wall, $r_1 \approx 2a$, $r_2 = r_w$, so:

$$\Phi_j = \frac{q}{2\pi}\ln\frac{2a}{r_w} + C \tag{3-18}$$

As to the point sink wall, $r_1 = r_w$, $r_2 \approx 2a$, so:

$$\Phi_w = \frac{q}{2\pi}\ln\frac{r_w}{2a} + C \tag{3-19}$$

Combine the formula (3 – 18) and the formula (3 – 19), the produced (injected) fluid intensity formula of one source and one sink is:

$$q = \frac{\pi(\Phi_j - \Phi_w)}{\ln(2a/r_w)} = \frac{\pi K(p_j - p_w)}{\mu\ln(2a/r_w)} \tag{3-20}$$

Combine the three equations of the formula (3 – 17) ~ formula (3 – 19), the potential distribution formula can be obtained:

$$\Phi(x,y) = \begin{cases} \Phi_j - \dfrac{q}{2\pi}\ln\dfrac{2ar_2}{r_1 r_w} \\ \Phi_w + \dfrac{q}{2\pi}\ln\dfrac{2ar_1}{r_2 r_w} \end{cases} \tag{3-21}$$

The isobar equation can be obtained from the formula (3 – 17) which is:

$$\frac{r_1}{r_2} = \frac{\sqrt{(x-a)^2 + y^2}}{\sqrt{(x+a)^2 + y^2}} = C$$

Collating the formula above we get:

$$\left(x - \frac{1+C^2}{1-C^2}a\right)^2 + y^2 = \left(\frac{2aC}{1-C^2}\right)^2 \tag{3-22}$$

When $C \neq 1$, what the formula (3 – 22) shows is a circle with the center $\left(\dfrac{1+C^2}{1-C^2}a, 0\right)$ and the radius $\dfrac{2aC}{1-C^2}$. And because of $\left|\dfrac{1+C^2}{1-C^2}\right| > 1$, the coordinate of centre is in the left side of the point source and right of the point sink. Strictly speaking, the bore wall is that the point sink (source) is considered as the centre not an equipotential surface. But if the value of $C$ is big enough, $\left|\dfrac{1+C^2}{1-C^2}\right|$ is close to 1, so the error is very small to consider them as equipotential surface. It can be known from the formula (3 – 17). That is $C = 1$

($r_1 = r_2$) also an equipotential line which is also the axis $y$, this is because straight line is circular arc with infinite radius.

According to the principle that streamlines are vertical with an equipotential line from which the streamline equation can be obtained:

$$x^2 + \left(y + \frac{C}{2}\right)^2 = a^2 + \frac{C^2}{4} \qquad (3-23)$$

The equation above is a circle with the centre $(0, \frac{-C}{2})$ and the radius $\sqrt{a^2 + \frac{C^2}{4}}$ and all circles are through the point $(\pm a, 0)$. Actually, all streamlines start from the point source and reach the point sink along the circular arc. So the axis $x$ is composed of three streamlines, one streamline is from the point source to infinite far, another line is from the right infinite far to the point sink, the third line is from the point source to point sink. The Figure 3-8 (Dynamic graph 3-3) is the interference figure of one point source and one point sink.

As shown in Figure 3-9, if there is a point sink with intensity $q$ from the supply boundary line with the distance $a$, consider the supply boundary line as the axis $y$ and the line which is through the point sink and vertical with the supply boundary line as the axis $x$, the seepage law of fluid on the semi-infinite plane which is $x \geq 0$ can be expressed with the following mathematical equation:

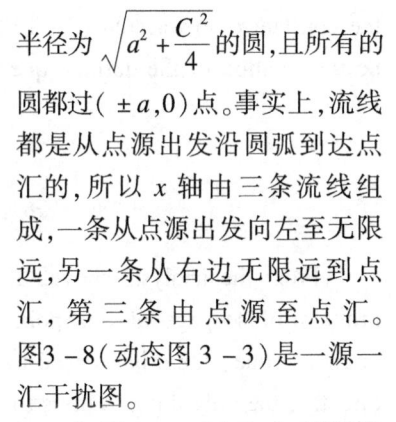

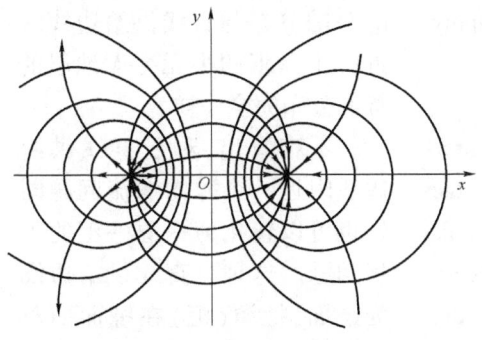

Figure 3-8  The interference of one source and one sink

图 3-8  一源一汇干扰图

Dynamic graph 3-3  Streamline diagram of uniform-intensity one source and one sink in infinite formation

动态图3-3  无穷大地层等强度一源一汇流线图

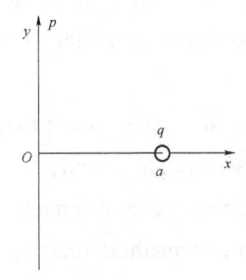

Figure 3-9  A well by the side of the straight supply boundary line

图 3-9  直线供给边线一口井

$$\begin{cases} \dfrac{\partial^2 \Phi}{\partial x^2} + \dfrac{\partial^2 \Phi}{\partial y^2} = 0 \\ q = \oint \nabla \Phi \mid_c \cdot \mathrm{d}s \\ \Phi \mid_{x=0} = \Phi_e \end{cases}$$

In the expression above: $C$ shows a circle which is $(x-a)^2 + y^2 = r_w^2$. $\mathrm{d}s$ is the infinitesimal length of circle.

Because the formula (3-17) meets the Laplace's equation, it will be the solution of this question if meeting the proposed boundary condition. Because the interference of one source and one sink has the property of $\Phi \mid_{x=0} = C$, only the constant $C$ in formula (3-17) is equal to $\Phi_e$, it would be the solution of the definite question above, that is:

$$\Phi = \frac{q}{2\pi} \ln \frac{r_1}{r_2} + \Phi_e \qquad (x \geq 0)$$

From the analysis above, it can be known: as to the problem of seepage of one point sink near straight supply boundary line, there is no need to solve the definite question directly, the only thing to do is to assume that point sink $(a, 0)$, there is a uniform-intensity source at the symmetric place about the straight supply boundary line $(-a, 0)$. Thus the question is turned to be the interference of one source and one sink. When the potential distribution in formation under the interference of one source and one sink meets the condition of straight-line supply boundary, the potential distribution of semi-infinite plane when $x \geq 0$ is completely the same as that of one point sink near straight supply boundary line of the semi-infinite plane.

It is easy to see that the seepage problem of one source near straight supply boundary can also be turned to the problem of interference of one source and one sink. This kind of method that considers the straight supply boundary line as a mirror which reflects a uniform-intensity sink (source) with the inverse sign of source (sink) at the symmetric place about mirror, thereby we change the seepage problem of semi-infinite formation with constant pressure boundary into the multi interference of wells in infinite formation. It is called the source-sink image method which is also called contrary sign image. The well reflected is called image well.

With the source-sink reflection method, after reflecting the problem of a production well near the straight supply boundary line to be the interference of one source and one sink in infinite formation, the potential of an arbitrary point $M(x,y)$ in formation is:

$$\Phi = \frac{q}{2\pi}\ln\frac{r_1}{r_2} + C \quad (x \geq 0) \quad (3-24)$$

Because the potential on the straight supply boundary line must meet the boundary condition $\Phi|_{x=0} = \Phi_e$, and when $x=0$, $r_1 = r_2$, the following equation can be obtained:

$$C = \Phi_e$$

Substitute the expression above into the formula (3-24), the following equation can be obtained:

$$\Phi = \frac{q}{2\pi}\ln\frac{r_1}{r_2} + \Phi_e \quad (3-25)$$

This is the potential distribution formula of a point sink near the straight-line constant pressure supply boundary line, if the point $M$ is chosen on the wall of the point sink, then $r_1 = r_w$ and $r_2 \approx 2a$, so:

$$\Phi_w = \frac{q}{2\pi}\ln\frac{r_w}{2a} + \Phi_e$$

The liquid producing intensity of a point sink near the constant pressure supply boundary of straight line can be obtained after collating the formula above:

$$q = \frac{2\pi(\Phi_e - \Phi_w)}{\ln(2a/r_w)} = \frac{2\pi K(p_e - p_w)}{\mu\ln(2a/r_w)} \quad (3-26)$$

It is very hard to know exactly the supply radius of actual oil well; it has been told before that the shape of supply area has little effect on the Dupuit formula. Assuming that the supply boundary line is a straight line, but it is mistaken as a circle, how much error will it cause? Ratio of the production $Q_1$ with that of circular reservoir $Q_c$ with supply radius $a$:

$$\frac{Q_1}{Q_c} = \frac{\lg \dfrac{a}{r_w}}{\lg \dfrac{a}{r_w} + \lg 2}$$

It can be seen from the comparison that the production calculated by the circular reservoir is higher, but the derivation is not big, because the only difference of the two expressions is that there is lg2 in the nominator, and the degree of $a/r_w$ is usually bigger than $10^3$. For example, when the radius of oil well $r_w$ = 10cm, $a$ = 200m, the derivation is 9.16%. This result is very valuable, because the general supply line can neither the straight line nor a circle, it is mostly closed and irregular curve, thus, the error between the production calculating with the Dupuit formula and the real production will be much smaller, but there will be a great difference between the potential distribution of the two.

Actually, the source-sink reflection method can not only be used to the supply boundary of straight line but also to the circular supply boundary. All the equipotential lines of the one source and one sink working simultaneously on infinite formation are eccentric circles whose center is on the extension line of the connection of source and sink, and every equipotential line can be considered as the supply boundary. If one of the equipotential lines on the right half-planar is considered as the supply boundary line in Figure 3–8, then the seepage situation of fluid in the equipotential line is the seepage problem: There is a eccentric well within the circular supply boundary, and its seepage filed is the same as that of one source and one sink in planar infinite formation. So the seepage problem of eccentric well within the circular supply boundary is changed to the interference of one source and one sink in infinite formation.

As shown in Figure 3–10, suppose that the radius of circular supply boundary line is $r_e$ and the pressure on it is $p_e$, the radius of well is $r_w$, the distance from the well centre to the boudary centre (eccentricity) is $d$, the bottom pressure is $p_w$, the fluid producing intensity is $q$, the height

经过比较可以看出，按圆形油藏计算的产量偏高，但是，偏差不大，因为两式仅在分母中相差一个lg2，而一般$\dfrac{a}{r_w}$至少是$10^3$级。例如，当油井半径$r_w$ = 10cm，$a$ = 200m 时，偏差为9.16%。这个结论很有价值，因为一般供给边线既不会是直线也不会是圆，大都是封闭的不规则的曲线，这样，用丘比公式计算的产量与真实产量之间的误差还会更小，然而，两者的势的分布却会有很大差异。

实际上，不仅对直线供给边可以使用源汇反映法，就是对圆形供给边线也可以应用源汇反映法。平面无穷大地层上一源一汇同时工作时所有的等势线都是圆心位于源汇点连线的延长线上的偏心圆，每条等势线都可当作供给边线。若将图3–8的右半平面的某一条等势线看作供给边线，则该等势线内液体的渗流情况就是圆形供给边线内一口偏心井的渗流问题，而其渗流场与平面无穷大地层上一源一汇的渗流场一样，于是就可以把圆形供给边线向偏心井的渗流问题化成平面无限地层上一源一汇的干扰问题。

如图3–10所示，设圆形供给边线的半径为$r_e$，其上的压力为$p_e$，井的半径为$r_w$，井中心离地层中心的距离（偏心距）为$d$，井底压力为$p_w$，产液强度为$q$，

of formation is $h$, the permeability is $K$, the fluid viscosity is $\mu$, please find the seepage law of fluid flowing to well bottom.

地层厚度为 $h$,渗透率为 $K$,流体黏度为 $\mu$,求液体流向井底的渗流规律。

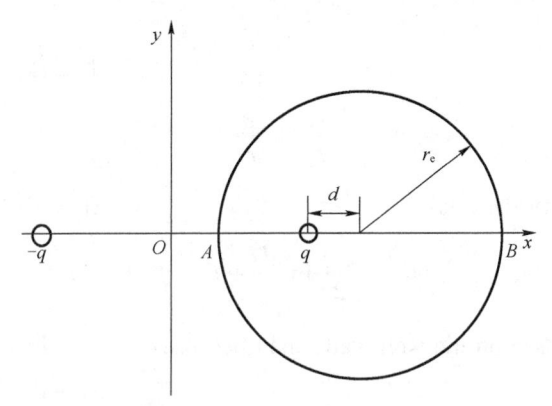

Figure 3 – 10  An eccentric well in circular formation
图 3 – 10  圆形地层中一口偏心井

First of all, change the actual supply boundary into an equipotential line in infinite formation. In concrete, an imaginary uniform-intensity point source needs to be added on the extension line of diameter of circular supply boundary, the diameter must be through the point sink, and the distance is $2a$ from the real well to the point source that reflects the finite circular formation to be infinite. Because the circle with the radius $r_e$ is an equipotential line, so the ratio of the distance from an arbitrary point $M$ on the equipotential line to the point sink $r_1$ and the distance to the point source $r_2$ must be constant, that is:

首先要将实际供给边线变成无穷大地层上一条等势线,这就要求在通过点汇的供给边界的直径的延长线上距真井 $2a$ 处加上一个假想的等强度的点源,把有限的圆形地层反映成无限大地层。由于半径为 $r_e$ 的圆是等势线,所以其上任意一点 $M$ 到点汇的距离 $r_1$ 和到点源的距离 $r_2$ 的比值为一常数,即:

$$\frac{r_1}{r_2} = C$$

For simplicity, the point $M$ is chosen on the two intersection points of the diameter passing through the point sink and the supply boundary line $A$ and $B$ as shown in Figure 3 – 10, and consider the connection of point source and point sink as the axis $x$, the vertical bisector of the connection as the axis $y$, then on the point $A$, the following expression can be obtained:

为了简单起见,把 $M$ 点选在过点汇的直径与供给边线的两个交点 $A$、$B$ 上,如图 3 – 10 所示,并以点源与点汇的连线为 $x$ 轴,以源汇连线的垂直平分线为 $y$ 轴,则在点 $A$ 处有:

$$r_1 = r_e - d; \quad r_2 = 2a + d - r_e$$

On the point $B$, the following expression can be obtained:

在点 $B$ 处有:

$$r_1 = r_e + d; \quad r_2 = 2a + d + r_e$$

So, the following expression can be obtained:

因此有:

$$\frac{r_1}{r_2} = \frac{r_e - d}{2a + d - r_e} = \frac{r_e + d}{2a + d + r_e}$$

The solution of the above equation is:

$$2a = \frac{r_e^2 - d^2}{d} \qquad (3-27)$$

Then:

$$\frac{r_1}{r_2} = \frac{d}{r_e} \qquad (3-28)$$

So on the supply boundary there is:

$$\Phi_e = \frac{q}{2\pi} \ln \frac{d}{r_e} + C \qquad (3-29)$$

If the point $M$ is chosen on the well wall, then because:

$$r_1 = r_w; \quad r_2 \approx 2a = \frac{r_e^2 - d^2}{d}$$

The potential on the hole wall is:

$$\Phi_w = \frac{q}{2\pi} \ln \frac{dr_w}{r_e^2 - d^2} + C \qquad (3-30)$$

The liquid production per unit thickness can be obtained by combining the formula (3-29) and the formula (3-30), that is:

$$q = \frac{2\pi(\Phi_e - \Phi_w)}{\ln \frac{r_e^2 - d^2}{r_e r_w}} = \frac{2\pi K(p_e - p_w)}{\mu \ln \frac{r_e}{r_w}\left(1 - \frac{d^2}{r_e^2}\right)} \qquad (3-31)$$

It is easy to see in the formula above, when $d \ll r_e$, there is little difference with the Dupuit formula, even if when $d = \frac{r_e}{2}$ and $\frac{r_e}{r_w} > 10^3$, the error of the production formula of the eccentric well and the Dupuit formula is just 4.1%. However, the pressure distribution is very different from that of a well in the center of circular formation.

If there are many wells within the circular supply boundary line, as to every well, the mirror well with same number and adverse sign can be obtained by source-sink reflection to circular constant pressure supply boundary, and deal with it with the superposition principle.

## 3.3.2 The interference of two sinks equal in intensity—sink point image method

Assuming that there are two point sinks in a homogeneous, isopachous and infinite formation, and the distance between the two sinks is $2a$ as shown in Figure 3-11, consider the connection of the two point sinks as axis $x$, and the vertical bisector of the connection as the axis $y$, the coordinate of the two-sinks is respectively $(a,0)$ and $(-a,0)$.

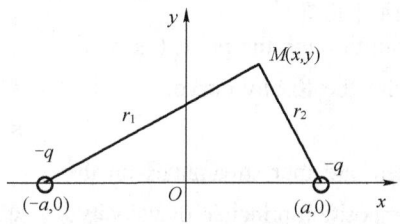

Figure 3-11  The interference of uniform-intensity two sinks

According to the superposition principle, the potential of an arbitrary point $M(x,y)$ in formation is:

$$\Phi = \frac{q}{2\pi}\ln r_1 r_2 + C = \frac{q}{4\pi}\ln[(x-a)^2 + y^2][(x+a)^2 + y^2] + C \quad (3-32)$$

It can be known from the formula that the equipotential line equation is:

$$r_1 r_2 = C$$

That is:

$$(x^2 + y^2)^2 + 2a^2(y^2 - x^2) + a^4 = C^2$$

This family of curves is called the Kozny curve. Apparently, this is a curve symmetric about axis $x$ and $y$. When $C < a^2$, the shape of the curve is two egg-shaped circles surrounding the point sinks, and both pointed ends point to the origin; when $C = a^2$, the Kozny curve changes to be the Bernoulli lemniscate $\rho^2 = 2a^2\cos 2\theta$; when $C > a^2$, the shape of curve is a flat circle surrounding the point sink. The bigger the value of $C$ is, the closer the curve is to the circle centered at the origin (Figure 3-12、Dynamic graph 3-4).

According to the property that the streamline is vertical with the equipotential line, it is not hard to obtain the streamline family equation as follows:

$$x^2 - y^2 - 2C_1 xy - a^2 = 0$$

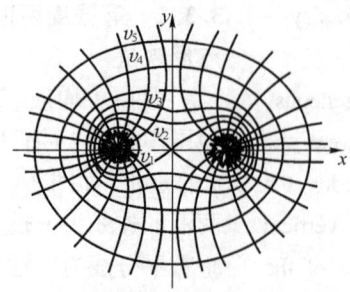

Figure 3 – 12  The interference of two point sinks
图 3 – 12  两汇干扰图

Dynamic graph 3-4  Streamline diagram of the interference of two sinks in infinite formation
动态图3-4  无穷大地层两汇干扰流线图

This is a family of hyperbola through the point ($\pm a, 0$) and each of which is vertical with the Kozny curve.

It is not hard to see that there are four streamlines on the axis $x$ according to the vector composition principle of velocity, two of which flow from the origin to the two point sinks, and two of which flow from the positive and negative infinite to the two point sinks. There are also two streamlines flowing from the positive and negative infinite to the origin on the axis $y$ as shown in Figure 3 – 13. This is because the distance from the arbitrary point $M(0, y)$ on axis $y$ to the two point sinks is the same, so the absolute value of velocity caused on point $M$ when the two point sinks working independently is equal, and the following expression can be obtained:

这是一簇过($\pm a, 0$)点的双曲线,其中每一条流线都与Kozny曲线正交。

根据速度的矢量合成原理不难看出 $x$ 轴上有四条流线,从原点流向两点汇的两条,从正负无限远流到点汇的两条。$y$ 轴上也有两条流线分别从正负无限远流向原点,如图 3 – 13 所示。这是因为 $y$ 轴上任一点 $M(0, y)$ 与两点汇的距离一样,所以两个点汇单独工作时在 $M$ 点引起的速度的绝对值相等,且有

$$|v_1| = |v_2| = \frac{q}{2\pi\sqrt{y^2 + a^2}}$$

However, the direction of the two velocities above is different. Decompose $v_1$ and $v_2$ to $v_{x1}$, $v_{y1}$ and $v_{x2}$, $v_{y2}$, apparently, $v_{x1}$ and $v_{x2}$ have the same absolute value and adverse direction which results in the component of resultant velocity in the $x$ direction is zero and only the component velocity in the $y$-direction is left, so axis $y$ is streamline.

If the point $M$ is chosen on the well wall, then $r_1 = r_w$ and $r_2 \approx 2a$ or $r_1 \approx 2a$ and $r_2 = r_w$, so the following expression can be obtained:

但二者方向不一样。将 $v_1$、$v_2$ 分别分解成 $v_{x1}$、$v_{y1}$ 和 $v_{x2}$、$v_{y2}$,显然 $v_{x1}$ 与 $v_{x2}$ 大小相等方向相反,结果合成速度在 $x$ 方向上的分量为零,只有 $y$ 方向上的分速度,所以 $y$ 轴是流线。

如果将 $M$ 点选在井壁上,则有 $r_1 = r_w$, $r_2 \approx 2a$,或 $r_1 \approx 2a$, $r_2 = r_w$,于是有:

$$\Phi_w = \frac{q}{2\pi}\ln 2ar_w + C \tag{3 – 33}$$

In order to eliminate the arbitrary constant $C$, the outer boundary condition must be considered. According to the analysis before, at the place far away from the two point sinks, the equipotential line is similar to the circle with the

为了消除任意常数 $C$,必须考虑外边界条件。根据前面分析,在离两点汇很远的地方,等势线近似于一个以原点为中心

center at origin. Assume the radius of the circle is $r_e$ ($r_e \gg 2a$), then $r_1 \approx r_2 \approx r_e$, the following expression can be obtained:

$$\Phi_e = \frac{q}{2\pi}\ln r_e^2 + C \tag{3-34}$$

The intensity of two sinks interference can be obtained by combining the formula (3-33) and the formula (3-34):

$$q = \frac{2\pi(\Phi_e - \Phi_w)}{\ln\dfrac{r_e^2}{2ar_w}} = \frac{2\pi K(p_e - p_w)}{\mu\ln\dfrac{r_e^2}{2ar_w}} \tag{3-35}$$

It can be known from the formula (3-35), under the condition that the bottom pressure stays unchanged, the intensity of the two sinks working simultaneously is smaller than that of one well working independently, and it is relevant with the well spacing. The smaller the well spacing $2a$ is, the smaller well spacing the bigger decrement rate is. The formula of potential distribution in corresponding formation is:

$$\Phi = \Phi_w + \frac{q}{2\pi}\ln\frac{r_1 r_2}{2ar_w} = \Phi_w + \frac{q}{2\pi}\ln\frac{\sqrt{[(x-a)^2 + y^2][(x+a)^2 + y^2]}}{2ar_w} \tag{3-36}$$

As shown in Figure 3-14, assuming that there is a point sink with intensity $q$ near the linear impermeable boundary (fault) with the distance $a$, consider the linear fault as the axis $y$ and the line passing through the point sink and vertical with axis $y$ as the axis $x$. The seepage law of fluid on semi-infinite planar, that is $x \geqslant 0$, can be described with mathematical equation as the expression below:

的圆，设此圆半径为 $r_e$（$r_e \gg 2a$），则 $r_1 \approx r_2 \approx r_e$，于是有：

联立式（3-33）与式（3-34）可得等强度两汇干扰的产液强度公式：

由式（3-35）可知，在井底压力不变的条件下，两汇同时工作时的强度比单井工作时要小，且与井距有关，井距 $2a$ 越小，减小的幅度越大。相应的地层中势的分布公式为：

如图 3-14 所示，设距离直线不渗透边界（断层）为 $a$ 的某点处有一强度为 $q$ 的点汇，以直线断层为 $y$ 轴，以通过点汇且与 $y$ 轴垂直的直线为 $x$ 轴。液体在 $x \geqslant 0$ 的半无限大平面上的渗流规律可以用以下数学方程来描述：

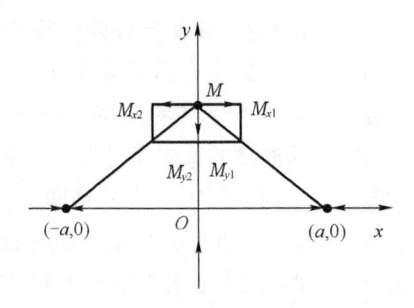

Figure 3-13　The vector composition of velocity
图 3-13　速度的矢量合成

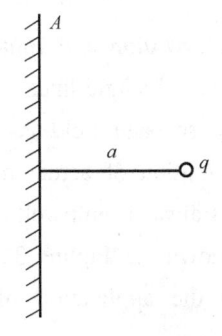

Figure 3-14　An oil well near linear impermeable boundary
图 3-14　直线不渗透边界附近一口井

$$\begin{cases} \dfrac{\partial^2 \Phi}{\partial x^2} + \dfrac{\partial^2 \Phi}{\partial y^2} = 0 \\ q = \oint \nabla \Phi \big|_c \cdot \mathrm{d}s \\ \dfrac{\partial \Phi}{\partial x}\bigg|_{x=0} = 0 \quad (x \geq 0) \end{cases}$$

In the formula above, $c$ shows the cylindrical surface of $(x-a)^2 + y^2 = r_w^2$. Apparently, the formula (3-32) meets the Laplace's equation and the boundary condition of the problems above, so it is the solution to the definite question.

It can be found from the analysis above that if there is a point sink near the linear impermeable boundary, there is no need to solve its definite solution problem. The right way is: consider the linear impermeable boundary as a mirror which reflects a uniform-intensity point sink at the symmetric place of point sink about the mirror. Thereby, the seepage problem of point sink near the linear impermeable boundary is turned to the interference problem of two point sinks in infinite formation. This kind of method considers the impermeable boundary as a mirror and reflects a uniform-intensity point source (sink) at the symmetric place of point source (sink) about the mirror. That the problem of impermeable boundary put into the problem of two sources (sinks) interference on the symmetric place in infinite formation is called the point sink reflection method which is also called homo-sign reflection. And both the source-sink reflection and the point sink reflection are generally called mirror image.

### 3.3.3 Popularization and application of mirror image

The affect of single linear boundary on production of oil well and the seepage field is introduced above, while the boundary condition of actual reservoir is more complicated than this and always multiwell working simultaneously. The reservoirs shown in Figure 3-15 are all like this, so popularizing the application of mirror image is of great importance.

(a)Two vertical inpermeable boundaries
(a)两条垂直不渗透边界

(b)Impermeable boundary with 45° included angle
(b)45°夹角不渗透边界

(c)Parallel impermeable boundaries
(c)平行不渗透边界

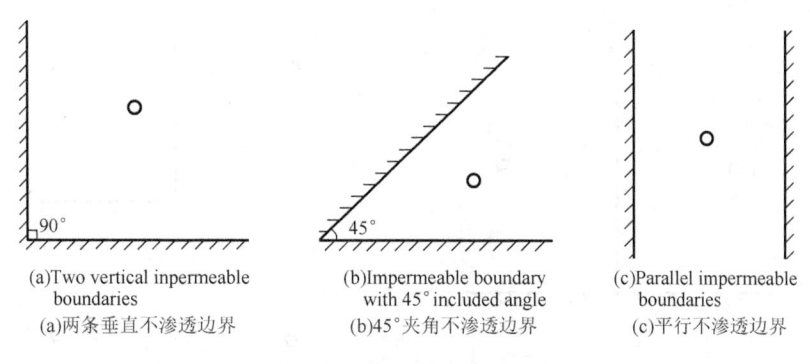

Figure 3 – 15　Reservoirs with impermeable boundary
图 3 – 15　不渗透边界油气藏示意图

As for the mirror image of complicated boundaries, the first point needed to notice is: the well number located on the both sides of boundary and its extension line (including the real well and the image well) must be equal. If the boundary is of constant pressure, the real well and the image well are of contrary sign. If the boundary is impermeable, the real well and the image well are of the same sign. The reason of keeping the well number equal of the both sides of boundary is the invariability of boundary condition. Second, the formation and the other boundaries on one side of image are all needed to reflect to the other side of image symmetrically that is reflecting the partial infinite formation to be the total planar infinite formation.

According to the mirror image principles, as to the reservoir that there is a point sink with intensity $q$ in infinite formation with two vertical impermeable boundaries as shown in Figure 3 – 15 (a), firstly consider the vertical impermeable boundary as mirror and reflect it to be the situation of uniform-intensity two point sinks near linear impermeable boundary in semi-infinite formation as shown in Figure 3 – 16 (a). And further proceed the point sink reflection considering the horizontal impermeable boundary as mirror as shown in Figure 3 – 16 (a), then the problem is changed to be the four point sinks interference in infinite formation as in Figure 3 – 16 (b). And the problem can be solved using the superposition principle of potential at last.

对于复杂边界的镜像反映,首先要注意边界及其延长线两侧的井数(包括真井及镜像井)必须相等。如果边界是定压的,则镜像井与真井异号。如果边界是不渗透的,则镜像井与真井同号。之所以要保持边界两侧井数相等,就是边界条件的不变性。其次,镜面一侧的地层和其他边界也都要对称地反映到镜面的另一侧去,即要把局部的无限地层映射成全平面无限大地层。

根据这些镜像反映原则,对于图 3 – 15(a)所示的有两条垂直不渗透边界的无限大地层中有一强度为 $q$ 的点汇的油藏,首先以其中的垂直不渗透边界为镜面,将之反映成直线不渗透边界附近等强度两汇的半无穷大地层的情况,如图 3 – 16(a)所示。进一步以图 3 – 16(a)中的水平不渗透边界为镜面进行汇点反映,则将原问题转化成了无穷大地层中四个等强度点汇的干扰问题,如图 3 – 16(b)所示。再利用势的叠加原理,该问题就迎刃而解了。

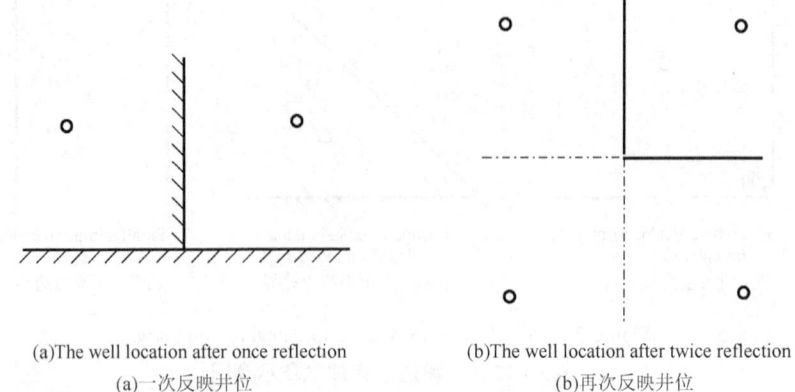

(a)The well location after once reflection
(a)一次反映井位

(b)The well location after twice reflection
(b)再次反映井位

Figure 3 – 16　The mirror image of reservoir with two vertical impermeable boundaries
图 3 – 16　两条垂直不渗透边界油气藏镜像反映图

The Figure 3 – 15 (b) shows the well located on the diagonal line of two faults with included angle 45°, with the same reflecting course, the situation is changed to be the interference of 8 uniform-intensity point sinks in infinite formation as shown in Figure 3 – 17. As to the situation shown in Figure 3 – 15 (c) that a point sink keeps in the center of two parallel linear impermeable boundaries, after mirror image, the linear impermeable boundary can maintain as streamline only when the well number of both sides of the boundary must be equal. That is to say the flow velocity direction of an arbitrary fluid mass point on boundary is parallel with faults, the original problem is reflected to be the interference of infinite well chain in infinite formation as shown in Figure 3 – 18.

图 3 – 15(b)表示的是井位于交角为 45°的两条断层的分角线上,经过同样的镜像反映过程而转化成了图 3 – 17 所示的无穷大地层中 8 个等强度点汇的干扰问题。而对于图 3 – 15(c)所示的两条互相平行的直线不渗透边界中间有一点汇的情况,经过镜像反映后,每条边界两侧的井数必须相等才能保持直线不渗透边界为分流线,即在边界上任一液体质点的渗流速度的方向与断层平行,这样原问题就反映成了无限地层上的无穷大井链的干扰问题,如图 3 – 18 所示。

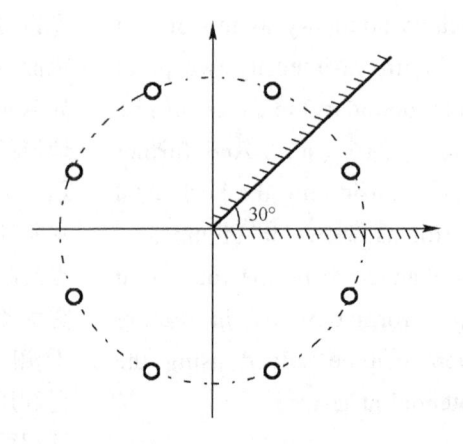

Figure 3 – 17　The mirror image of reservoir with impermeable boundary with 30° included angle
图 3 – 17　夹角为 30°的不渗透边界油气藏镜像反映图

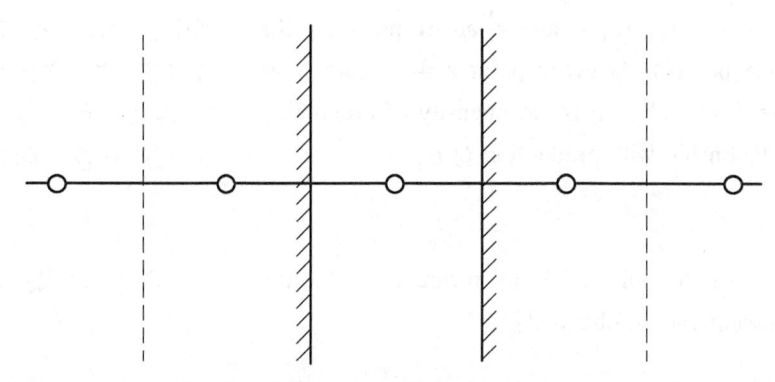

Figure 3–18  The mirror image of two parallel linear impermeable boundaries
图 3–18  两条平行直线不渗透边界油气藏镜像反映图

**Example 3–2** As shown in Figure 3–19(a), OA is the linear supply boundary line, the pressure on it is $p_e$ = 12MPa, and bottom pressure is $p_w$ = 9MPa, the well radius is $r_w$ = 10cm, the distance from well to the supply boundary line is $a$ = 300m and to the impermeable boundary OB is $b$ = 400m, the formation height is $h$ = 3m, the permeability is $K$ = 0.5μm², the fluid viscosity is $\mu$ = 4mPa·s. Please try to derive the production formula of oil well and solve the production.

**例 3–2**  如图 3–19(a)所示,OA 为直线供给边线,其上的压力 $p_e$ = 12MPa,井压力 $p_w$ = 9MPa,井半径 $r_w$ = 10cm,井距供给边线的距离 $a$ = 300m,距离不渗透边界 OB 的距离 $b$ = 400m,地层厚度 $h$ = 3m,渗透率 $K$ = 0.5μm²,液体黏度为 $\mu$ = 4mPa·s,试推导出油井的产量公式,并求出油井的产量。

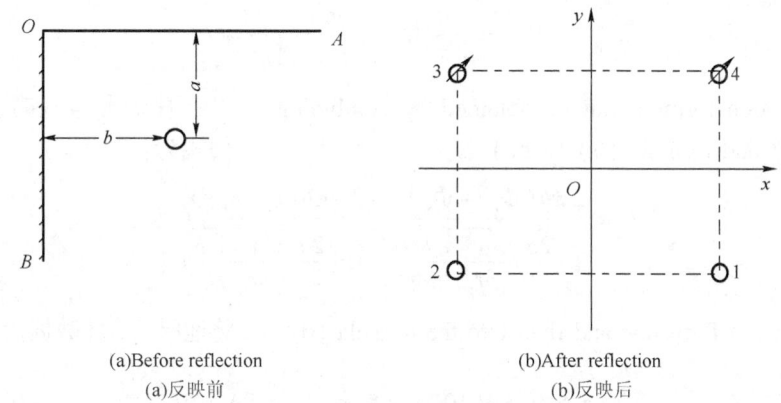

(a)Before reflection　　　　　(b)After reflection
(a)反映前　　　　　　　　　(b)反映后

Figure 3–19  The mirror reflection of two vertical linear supply boundary and closed boundary
图 3–19  垂直的两条直线供给边界与封闭边界油藏镜像反映图

**Solution**: According to the principle of mirror reflection, reflect the reservoir to be the problem of uniform-intensity two sources and two sinks in infinite formation as shown in Figure 3–19(b). And according to the superposition principle of potential, the potential of an arbitrary point $M(x,y)$ in formation is:

**解**:根据镜像反映原理,将该油藏反映成如图 3–19(b)所示的无穷大地层上等强度两源两汇的干扰问题。再根据势的叠加原理,地层中任一点 $M(x,y)$ 处的势为:

$$\Phi = \frac{q}{2\pi}\ln\frac{r_1 r_2}{r_3 r_4} + C \tag{1}$$

In the formula above: $r_1$, $r_2$, $r_3$ and $r_4$ are respectively the distance from the point $M$ to every point sink (source) as shown in Figure 3-19(b). $q$ is the intensity of production well and the relationship with production $Q$ is:

$$q = \frac{Q}{h} \tag{2}$$

Substitute the formula (2) to formula (1), the following expression can be obtained:

$$\Phi = \frac{Q}{2\pi h}\ln\frac{r_1 r_2}{r_3 r_4} + C \tag{3}$$

If the point $M$ is chosen on the linear supply boundary line, then $r_1 = r_4$ and $r_2 = r_3$, substitute these into formula (3), the following expression can be obtained:

$$\Phi_e = C \tag{4}$$

If the point $M$ is chosen on the well wall of production well, then $r_1 = r_w$, $r_2 \approx 2b$, $r_4 \approx 2a$ and $r_3 \approx 2\sqrt{a^2 + b^2}$, substitute these into formula (3), the following expression can be obtained:

$$\Phi_w = \frac{Q}{2\pi h}\ln\frac{r_w b}{2a\sqrt{a^2 + b^2}} + C \tag{5}$$

The production formula can be obtained by combining the formula (4) and formula (5), which is:

$$Q = \frac{2\pi h(\Phi_e - \Phi_w)}{\ln\dfrac{2a\sqrt{a^2+b^2}}{br_w}} = \frac{2\pi Kh(p_e - p_w)}{\mu\ln\dfrac{2a\sqrt{a^2+b^2}}{br_w}} \tag{6}$$

Substitute the data of formation and fluid into the formula (6):

$$Q = \frac{2\pi \times 0.5 \times 10^{-12} \times 3 \times (12-9) \times 10^6}{4 \times 10^{-3}\ln\dfrac{2 \times 300 \times 500}{400 \times 0.1}}$$

$$= 7.922 \times 10^{-4}(\text{m}^3/\text{s}) = 68.45(\text{m}^3/\text{d})$$

### 3.3.4 Equivalent flow resistance method

Starting from the seepage field and according to the principle of interference of wells, Borisov of Soviet Union obtained an approximate calculation which has clear physical meaning and enough accuracy, and describing the seepage process of interference of wells with the figure of equivalent flow resistance, and solve it by the Kirchhoff's law.

1. The approximate calculation formula of linear supply boundary

1) The seepage formula

As shown in Figure 3-20, according to the source-sinkimage method, a production well is set in the centre of parallel impermeable boundary which is also $L$ from the linear supply boundary; this is equivalent to a row of wells arranged parallel the linear supply boundary in semi-infinite formation with the well spacing $2a$.

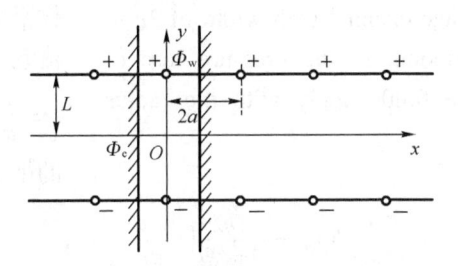

Figure 3-20 The reflection of linear well line
图 3-20 直线井列的反映

The production formula of single well is:

$$Q = \frac{2\pi Kh(p_e - p_w)}{\mu \ln \dfrac{a}{\pi r_w} 2\mathrm{sh}\dfrac{\pi L}{a}} \quad (3-37)$$

Because of $2\mathrm{sh}\dfrac{\pi L}{a} = \mathrm{e}^{\frac{\pi L}{a}} - \mathrm{e}^{-\frac{\pi L}{a}}$, $L \gg a$ and $\dfrac{L}{a} \gg 1$, so $\mathrm{e}^{\frac{\pi L}{a}} \gg \mathrm{e}^{-\frac{\pi L}{a}}$, so $\mathrm{e}^{-\frac{\pi L}{a}}$ can be omitted compared with $\mathrm{e}^{\frac{\pi L}{a}}$. For example, $L = a$, $\mathrm{e}^{\pi} = 23.14$, $\mathrm{e}^{-\pi} = 0.04321$, and the production formula of single well is:

$$Q = \frac{2\pi Kh(p_e - p_w)}{\mu\left(\dfrac{\pi L}{a} + \ln\dfrac{a}{\pi r_w}\right)} \quad (3-38)$$

2) The physical meaning of seepage formula

The formula (3-38) can be reformed bellow:

基尔霍夫定律求解。

1. 直线供给边界的近似计算公式

1) 渗流公式

如图 3-20 所示,根据源汇反映法,在两互相平行的不渗透边界中间距直线供给边缘 $L$ 处布一口生产井,相当于在半无限大地层上距直线供给边缘平行布置一排井,井距为 $2a$。

单井的产液量公式为:

因为 $2\mathrm{sh}\dfrac{\pi L}{a} = \mathrm{e}^{\frac{\pi L}{a}} - \mathrm{e}^{-\frac{\pi L}{a}}$,通常 $L \gg a$,$\dfrac{L}{a} \gg 1$,因此 $\mathrm{e}^{\frac{\pi L}{a}} \gg \mathrm{e}^{-\frac{\pi L}{a}}$,所以与 $\mathrm{e}^{\frac{\pi L}{a}}$ 相比 $\mathrm{e}^{-\frac{\pi L}{a}}$ 可以忽略不计。例如 $L = a$,$\mathrm{e}^{\pi} = 23.14$,$\mathrm{e}^{-\pi} = 0.04321$。此时单井的产液量公式为:

2) 渗流公式的物理意义

式(3-38)变形为:

$$p_e - p_w = \frac{Q\mu}{2\pi Kh}\left(\frac{\pi L}{a} + \ln\frac{a}{\pi r_w}\right) = \frac{\mu L}{Kh}\frac{Q}{2a} + \frac{Q\mu}{2\pi Kh}\ln\frac{a}{\pi r_w} \qquad (3-39)$$

The first right item of the formula (3-39) $\frac{\mu L}{Kh}\frac{Q}{2a}$ is the pressure loss when seepage liquid flows from the supply boundary to the linear drainage channel with the length of $L$ and the width of $2a$:

式(3-39)右端第一项 $\frac{\mu L}{Kh}\frac{Q}{2a}$ 是渗流液体从供给边缘流经距离为 $L$、宽度为 $2a$ 的直线排液道中的压力损失：

$$\Delta p_L = \frac{\mu L}{Kh}\frac{Q}{2a}$$

The second right item of the formula (3-39) is fluid flowing again along the drainage channel with width of $2a$ to the well. Its resistance is amount to the pressure loss of radial fluid seepage from the fluid supply with the radius $r = \frac{a}{\pi}$ to radius $r_w$:

式(3-39)右端第二项是液体沿宽度为 $2a$ 的排液道再向井流动，其阻力相当于从半径为 $r = \frac{a}{\pi}$ 的供液边缘到半径为 $r_w$ 的平面径向渗流的压力损耗：

$$\Delta p_w = \frac{Q\mu}{2\pi Kh}\ln\frac{a}{\pi r_w}$$

Of course, the actual stream line won't bend at the right angle, it is just a kind of proximity. Thus, pressure $\Delta p = p_e - p_w$ is composed of two items:

实际流线当然不会拐直角弯，这只是一种近似。这样，压力 $\Delta p = p_e - p_w$ 由两部分损耗组成：

$$\Delta p = p_e - p_w = \Delta p_L + \Delta p_w$$

When there are $n$ wells producing simultaneously:

有 $n$ 口井生产时：

$$nq = Q, q = \frac{Q}{n}, 2an = B$$

In the formula above, $B$ is the width of formation.

式中，$B$ 为地层宽度。

The formula (3-39) can be written as:

式(3-39)可以写为：

$$p_e - p_w = \frac{\mu L}{BKh}Q + \frac{\mu Q}{2n\pi Kh}\ln\frac{a/\pi}{r_w} \qquad (3-40)$$

The $\frac{\mu L}{BKh}$ in formula (3-40) is just the resistance term from the supply boundary to the drainage channel located in the well array.

式(3-40)中的 $\frac{\mu L}{BKh}$ 正是从供给边缘达到位于井排处的排液道的阻力项。

$$R_1 = \frac{\mu L}{BKh}$$

The resistance $\frac{\mu}{2n\pi Kh}\ln\frac{a/\pi}{r_w}$ from drainage channel to well bottom is equivalent to the resistance term from supply boundary with

从排液道流向井底的阻力 $\frac{\mu}{2n\pi Kh}\ln\frac{a/\pi}{r_w}$ 相当于从半径为 $r =$

radius $r = a/\pi$ to the planer radial flow with radius $r_w$ in seepage.

$\frac{a}{\pi}$的供给边缘到半径为 $r_w$ 的平面径向渗流系统中的阻力项。

$$R_2 = \frac{\mu}{2n\pi Kh}\ln\frac{a/\pi}{r_w}$$

The seepage system can be expressed with series circuit as shown in Figure 3 - 21. So the following expression can be obtained:

可将渗流系统以串联的等值图表示,如图 3 - 21 所示。于是有:

$$p_e - p_w = R_1 Q + R_2 Q$$

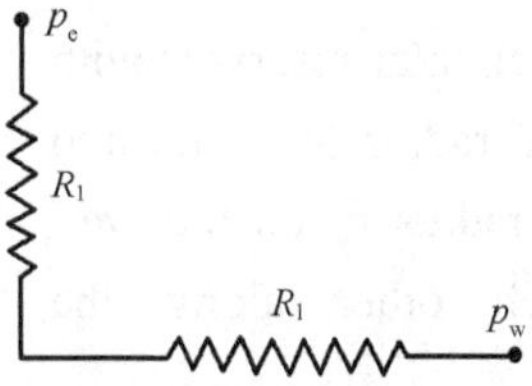


Figure 3 - 21 The equivalent seepage figure of linear well line

图 3 - 21 直线井列的等值渗流图

3) The similarity with the Ohm's Law

3)与欧姆定律的相似性

Write the formula (3 - 40) to be the form of Ohm's Law:

将式(3 - 40)写成欧姆定律的形式:

$$Q = \frac{p_e - p_w}{R_1 + R_2} = \frac{p_e - p_w}{R} \qquad (3-41)$$

Where: $R_1$—the outer resistance of seepage;

$R_2$—the inner resistance of seepage;

$R$—the flow resistance.

式中:$R_1$ = 渗流外阻;

$R_2$ = 渗流内阻;

$R$ = 渗流阻力。

The flow resistance $R_1$ and $R_2$ are considered as the series circuit electric resistance of Ohm's Law. It is more or less equal to the assumed condition of resistance in the two continous seepage, and $R_1$ is called outer resistance of seepage and $R_2$ is called inner resistance of seepage. Inner resistance is the resistance in the drainage channel. $n$ in the resistance item $R_2$ is the number of producing well. The inner resistance of every well has the parallel relationship in equivalent circuit, and the equipotent electric resistance of $n$ parallel electric resistances is:

渗流阻力 $R_1$ 和 $R_2$ 视作欧姆定律的串联电阻。这是与两个连续渗流过程中所受的阻力的假想情况是相当的,并且称 $R_1$ 为渗流外阻、$R_2$ 为渗流内阻。内阻就是排液道内的阻力。$R_2$ 阻力项中 $n$ 是生产井的井数,按等值电路每口井的内阻是并联关系。$n$ 个并联电阻的等效电阻为:

$$\frac{1}{R} = \sum_{i=1}^{n}\frac{1}{R_i}$$

When $n$ equal electric resistances are paralleling, $\frac{1}{R} = \frac{n}{r}$, so $R = \frac{r}{n}$, and $r$ is the inner resistance of one well, so the following expression can be obtained:

当 $n$ 个相等电阻并联时,$\frac{1}{R} = \frac{n}{r}$,所以 $R = \frac{r}{n}$,$r$ 为一口井的内阻,所以有:

$$R_2 = \frac{\mu}{2n\pi Kh}\ln\frac{a/\pi}{r_w}$$

This kind of similarity between seepage of linear well line and current circuit makes the seepage course described by electric circuit, and seepage problem can be solved with the Kirchhoff's law after drawing out the seepage picture.

2. Approximate calculation formula of circular supply boundary

1) The seepage formula

As shown in Figure 3 – 22, in the circular reservoir with radius $r_e$, the annular well line with radius $r$ is located symmetrically. There are $n$ wells with radius $r_w$ on the well line and the well spacing of each other along the circumference is $2a$.

这种直线井排渗流与电路中电流的相似性,使得可以用电路图来描述渗流过程,绘出电路式的渗流图后,利用基尔霍夫定律求解渗流问题。

2. 圆形供给边界的近似计算式

1) 渗流公式

如图 3 – 22 所示,在 $r_e$ 为半径的圆形油气藏上,对称布置半径为 $r$ 的环状井排,井排上有 $n$ 口半径为 $r_w$,沿圆周井距离 $2a$ 的井。

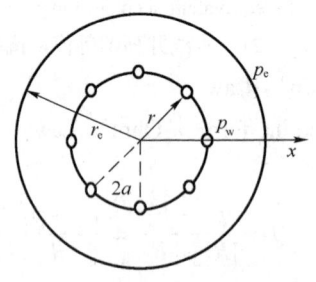

Figure 3 – 22 The annular well line

图 3 – 22 环状井排

The liquid production intensity $q'$ formula of single well is:

单井的产液强度 $q'$ 公式为

$$q' = \frac{2\pi K(p_e - p_w)}{\mu\ln\frac{r_e^n}{nr_w r^{n-1}}\left(1 - \frac{r^{2n}}{r_e^{2n}}\right)}$$

Generally, well number $n > 4$ and $r/r_e < 1$, so $(r/r_e)^{2n} \ll 1$ can be omitted, so the following expression can be obtained:

一般井数 $n > 4$,且 $r/r_e < 1$,所以 $(r/r_e)^{2n} \ll 1$ 可以忽略不计,因而:

$$q' = \frac{2\pi K(p_e - p_w)}{\mu\ln\frac{r_e^n}{nr_w r^{n-1}}} = \frac{2\pi K(p_e - p_w)}{\mu\left(n\ln\frac{r_e}{r} + \ln\frac{r}{nr_w}\right)} \qquad (3-42)$$

The liquid production intensity well line when $n$ production wells work simultaneously is:

有 $n$ 口生产井的产液强度为:

$$q = nq'$$
$$q = n \cdot \frac{2\pi K(p_e - p_w)}{\mu\left(n\ln\frac{r_e}{r} + \ln\frac{r}{nr_w}\right)}$$

From the length of well line $2\pi r = n \cdot 2a$, $\dfrac{r}{n} = \dfrac{a}{\pi}$ can be obtained, substitute it into the formula above, the following expression can be obtained:

$$q = n \cdot \frac{2\pi K(p_e - p_w)}{\mu\left(n\ln\dfrac{r_e}{r} + \ln\dfrac{a/\pi}{r_w}\right)} \qquad (3-43)$$

2) The physical analysis of seepage formula

The pressure difference is:

$$p_e - p_w = \frac{\mu}{2\pi K}\ln\frac{r_e}{r}q + \frac{\mu}{2n\pi K}\ln\frac{a/\pi}{r_w}q \qquad (3-44)$$

$\left(\dfrac{\mu}{2\pi K}\ln\dfrac{r_e}{r}\right)q$ which the first item on the right side of the formula (3-44) is just the pressure loss along planar radial direction of fluid with production $q$ flowing from the supply boundary to the circular drainage channel with radius $r$.

$$\Delta p_L = \frac{\mu}{2\pi K}\ln\frac{r_e}{r}q$$

In the formula above, $\dfrac{\mu}{2\pi K}\ln\dfrac{r_e}{r}$ is the resistance undergone flowing from the circular supply boundary with radius $r_e$ to well line with radius $r$. It is the outer resistance.

$$R_1 = \frac{\mu}{2\pi K}\ln\frac{r_e}{r}$$

$\dfrac{\mu}{2n\pi K}\ln\dfrac{a/\pi}{r_w}q$ which the second item on the right side of the formula (3-44) shows the resistance of fluid flowing from the drainage channel to well. On numeric value, it is amount to the pressure loss of planar radial seepage from the circular supply boundary with radius $\dfrac{a}{\pi}$ to the well with radius $r_w$.

In the formula, $\dfrac{\mu}{2n\pi K}\ln\dfrac{a/\pi}{r_w}$ shows the resistance of seepage in the circular drainage channel that is inner resistance. The same with the form of linear well line, it is also the result of $n$ wells in parallel.

按井排环长 $2\pi r = n \cdot 2a$，得 $\dfrac{r}{n} = \dfrac{a}{\pi}$，代入上式得：

2) 渗流公式物理分析

压力差为：

式（3-44）右端第一项 $\left(\dfrac{\mu}{2\pi K}\ln\dfrac{r_e}{r}\right)q$ 正是产量为 $q$ 的液体从供给边缘到半径为 $r$ 的圆形排液道的平面径向压力损耗。

式中，$\dfrac{\mu}{2\pi K}\ln\dfrac{r_e}{r}$ 是从半径为 $r_e$ 的圆形供给地层渗流到半径为 $r$ 的井排所受的阻力，即外阻。

式（3-44）右端第二项中的 $\dfrac{\mu}{2n\pi K}\ln\dfrac{a/\pi}{r_w}q$ 表示排液道向井流动的阻力，它在数值上相当于沿半径为 $\dfrac{a}{\pi}$ 圆形供给边界向半径为 $r_w$ 的井作平面径向渗流时的压力损耗。式中，$\dfrac{\mu}{2n\pi K}\cdot\ln\dfrac{a/\pi}{r_w}$ 表示在圆形排液道内渗流时所受的阻力，即内阻。形式与直线井排一样，也是 $n$ 口井并联的结果。

$$R_2 = \frac{\mu}{2n\pi K}\ln\frac{a/\pi}{r_w}$$

So:

$$p_e - p_w = R_1 q + R_2 q$$

The figure of equivalent flow resistance can also be drawn for the same reason.

The content mentioned above is a row of wells. It can be assumed that when there are rows of wells, the thing that needs to be done is adding the corresponding equivalent seepage resistance of every row. Because the seepage law is similar to the Ohm's law, the seepage issue can be dealt with the way of solving the direct current circuit that is the Kirchhoff's law.

## 3.4 Application of complex function theory in the planar seepage field

The problem of multi interference of wells with complex boundary solved with mirror image and superposition principle of potential are talked about in the former section. In this section, the application of complex function theory to the planar seepage field will be introduced. With this theory, not only the steady state seepage field can be described easily, but also the production of oil well of more complicated reservoir can be obtained.

### 3.4.1 Foundation of complex function

The definition of a complex number $z$ is:

$$z = x + iy$$

In the formula above: $x$ and $y$ are respectively arbitrary real number. $i = \sqrt{-1}$. The formula above is also called the algebraic expression of complex number, $x$ and $y$ are respectively called the real part and imaginary part of complex number which are respectively marked with $\operatorname{Re}(z)$ and $\operatorname{Im}(z)$.

If the polar coordinate is used, the complex number $z$ can also be shown as the expression below which are:

$$z = r(\cos\theta + i\sin\theta) = re^{i\theta}$$
$$r = \sqrt{x^2 + y^2}; \quad \theta = \arg z$$

In the expression above: $r$ is called the module of $z$ which is also marked as $|z|$. And $\theta$ is the argument of $z$ which means the intersection angle of vector $z$ with axis $x$.

When the complex variable $z$ changes continuously on a given area $B$ of complex plane, if the value of complex variable $W$ changes with that of $z$, the $W$ is called the function of complex variable $z$ on area $B$, and $z$ is called the argument of $W$. The function is marked as:

$$W = f(z)$$

It is regulated that the domain $B$ is single connected, that is to say, the area cannot be divided into small domains which are noncommunicating. If the domain $B$ does not include the points on boundary but the points inside, then the domain is called open. While if the domain $B$ not only includes the points inside, but also the points on boundary, the domain is called closed domain.

When the real part and imaginary part of complex function $f(z)$ are respectively marked as $u(x,y)$ and $v(x,y)$, that is:

$$f(z) = u(x,y) + iv(x,y)$$

Complex function can be solved into a pair of real variable function. Therefore, many definitions, formulas and laws in real variable function can be transplanted directly into the theory of complex function.

For example, the definition of the complex function $f(z)$ which continues on the point $z_0 = x_0 + y_0 i$ is: when $z \to z_0$, $f(z) \to f(z_0)$.

It can be solved into a pair of binary real variable function $u(x,y)$ and $v(x,y)$ continues on the point $(x_0, y_0)$, that is:

$$\begin{cases} x \to x_0 \\ y \to y_0 \end{cases} \Rightarrow \begin{cases} u(x,y) \to u(x_0, y_0) \\ v(x,y) \to v(x_0, y_0) \end{cases}$$

If $\Delta z$ approaches 0 along any path, the proportionality $\dfrac{f(z + \Delta z) - f(z)}{\Delta z}$ always approaches the same one limit, and the limit is called the derivative of function $f(z)$ on point $z$. Apparently, the function is derivable on point $z$, which must meet the condition that the function is continuous on this point simultaneously.

The necessary condition that complex function $f(z)$ is derivable meets the Cauchy-Riemann equations, that is:

$$\begin{cases} \dfrac{\partial u}{\partial x} = \dfrac{\partial v}{\partial y} \\ \dfrac{\partial u}{\partial y} = -\dfrac{\partial v}{\partial x} \end{cases}$$

On the contrary, $u(x,y)$ and $v(x,y)$ have first-order continuous derivative and meet the Cauchy-Riemann equation, it is a sufficient condition for the derivability of complex functions. The derivation law of complex function is the same with that of real variable function which has the following rules:

$$\begin{cases} \dfrac{d}{dz}(W_1 + W_2) = \dfrac{dW_1}{dz} + \dfrac{dW_2}{dz} \\ \dfrac{d}{dz}(W_1 W_2) = \dfrac{dW_1}{dz}W_2 + W_1\dfrac{dW_2}{dz} \\ \dfrac{d}{dz}\left(\dfrac{W_1}{W_2}\right) = \dfrac{W_1' W_2 - W_1 W_2'}{W_2^2} \\ \dfrac{d}{dz}F(W) = \dfrac{dF}{dW}\dfrac{dW}{dz} \end{cases}$$

The complex function which is derivable everywhere on a given domain is called an analytic function on the domain. The analytic function has an intimate relationship with the planar seepage field. It can be known from the Cauchy-Riemann equation that the two real variable function of the analytic function $u(x,y)$ and $v(x,y)$ have the following properties:

$$\dfrac{\partial u}{\partial x}\dfrac{\partial v}{\partial x} + \dfrac{\partial u}{\partial y}\dfrac{\partial v}{\partial y} = 0; \quad \dfrac{\partial^2 u}{\partial x^2} + \dfrac{\partial^2 u}{\partial y^2} = 0; \quad \dfrac{\partial^2 v}{\partial x^2} + \dfrac{\partial^2 v}{\partial y^2} = 0$$

The first formula shows that the two real function $u(x,y)$ and $v(x,y)$ are orthogonal each other. The second and the third formulas show that the function $u(x,y)$ and $v(x,y)$ all meet the Laplace's equation. In other words, they are all harmonic function. Because they are the real part and imaginary part of one complex function, it is also called conjugate harmonic function.

### 3.4.2 Complex potential of planar seepage field

The potential of planar seepage field is defined in the former section which has the following properties:

$$\frac{\partial^2 \Phi}{\partial x^2} + \frac{\partial^2 \Phi}{\partial y^2} = 0 \; ; \; v_x = -\frac{\partial \Phi}{\partial x} \; ; \; v_y = -\frac{\partial \Phi}{\partial y}$$

Generally, the line formed by connecting points with same potential is called equipotential line which is also constant pressure line.

In the course of fluid seepage, in order to describe the moving direction of a series of mass points, this kind of space curve can be drawn at a given moment. The velocity direction of every fluid mass point on this curve is tangent to this curve and this curve is called streamline. Take an arbitrary point on this streamline, and the seepage velocity on this point is $v$, the two components on axis $x$ and $y$ are respectively $v_x$ and $v_y$, according to the similarity relationship of triangle, there is:

$$\frac{\mathrm{d}x}{v_x} = \frac{\mathrm{d}y}{v_y}$$

Or

$$v_y \mathrm{d}x - v_x \mathrm{d}y = 0 \qquad (3-45)$$

The formula above is the differential equation of streamline. Assuming that the function $P(x,y)$ and $Q(x,y)$ have first-order continuous partial derivative in single connected domain $G$, then the sufficient and necessary condition of $P(x,y)\mathrm{d}x + Q(x,y)\mathrm{d}y$ is the total differential of a given function $\Psi(x,y)$ in $G$ is:

$$\frac{\partial P}{\partial y} = \frac{\partial Q}{\partial x} \qquad (3-46)$$

Now $v_y$ and $-v_x$ are verified whether meet the formula (3-46), according to the Darcy's law and the properties of potential function meeting the Laplace's equation, the following expression can be obtained:

$$\frac{\partial v_y}{\partial y} = -\frac{\partial^2 \Phi}{\partial y^2} \; ; \; -\frac{\partial v_x}{\partial x} = \frac{\partial^2 \Phi}{\partial x^2}$$

So, $v_y$ and $-v_x$ meet the formula (3-46), then $v_y \mathrm{d}x - v_x \mathrm{d}y$ is the total differential of function $\Psi(x,y)$, that is:

$$\mathrm{d}\Psi = v_y \mathrm{d}x - v_x \mathrm{d}y \qquad (3-47)$$

$\Psi(x,y)$ is called stream function. Substitute the streamline formula (3-45) into the formula (3-47), the following expression can be obtained:

— 121 —

$$\mathrm{d}\Psi = 0$$

That is:

即:

$$\Psi(x,y) = C.$$

It shows that the curve composed of points with the same stream function in planar seepage field is streamline. As to the two points $A$ and $B$ with different stream function values as shown in Figure 3 - 23, the difference of stream function value is the flow rate passing through the two points $A$ and $B$.

这说明平面渗流场中流函数相等的各点所组成的曲线就是流线。而对于图 3 - 23 所示的流函数值不同的 $A$、$B$ 两点,其流函数值的差就是穿过 $A$、$B$ 两点之间的流量。

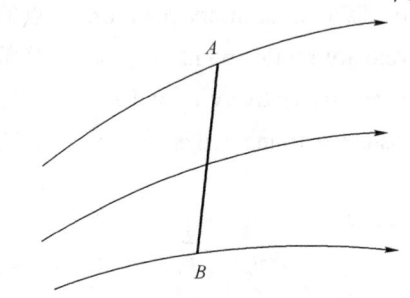

Figure 3 - 23  The schematic map of streamline
图 3 - 23  流线示意图

The total differential of stream function can be written as:

流函数的全微分可以写作:

$$\mathrm{d}\Psi = \frac{\partial \Psi}{\partial x}\mathrm{d}x + \frac{\partial \Psi}{\partial y}\mathrm{d}y \qquad (3-48)$$

By combining the formula (3 - 47) and formula (3 - 48), the following expression can be obtained:

联立式(3 - 47)和式(3 - 48)得:

$$\begin{cases} v_x = -\dfrac{\partial \Psi}{\partial y} \\ v_y = \dfrac{\partial \Psi}{\partial x} \end{cases}$$

Substitute $v_x = -\dfrac{\partial \Phi}{\partial x}$ and $v_y = -\dfrac{\partial \Phi}{\partial y}$ into the expression above, the following expression can be obtained, that is:

将 $v_x = -\dfrac{\partial \Phi}{\partial x}$ 和 $v_y = -\dfrac{\partial \Phi}{\partial y}$ 代入上式得:

$$\begin{cases} \dfrac{\partial \Phi}{\partial x} = \dfrac{\partial \Psi}{\partial y} \\ \dfrac{\partial \Phi}{\partial y} = -\dfrac{\partial \Psi}{\partial x} \end{cases} \qquad (3-49)$$

Because the potential function $\Phi$ meets the Laplace's equation, it is easy to be obtained from the formula (3 - 49), that is:

由于势函数 $\Phi$ 满足拉氏方程,因此由式(3 - 49)易得:

$$\frac{\partial^2 \Psi}{\partial x^2} + \frac{\partial^2 \Psi}{\partial y^2} = 0$$

Because the seepage velocity is continuous in the whole seepage field, therefore, the potential function and stream function have first order continuous partial derivative,

由于渗流速度在整个渗流场中是连续的,因此,势函数和流函数具有一阶连续的偏导数,

and the formula (3-49) shows that they meet the condition of Cauchy-Riemann, and the following expression is defined as:

$$W(z) = \Phi(x,y) + i\Psi(x,y) \qquad (3-50)$$

Apparently, $W(z)$ is an analytic function. This kind of analytic function whose real part is composed of potential function and its imaginary part is composed of stream function, Which is called complex potential function of planar seepage field which is also called complex potential for short. According to the property of analytic function, the following expression can be obtained:

$$\frac{\partial \Phi}{\partial x} \cdot \frac{\partial \Psi}{\partial x} + \frac{\partial \Phi}{\partial y} \cdot \frac{\partial \Psi}{\partial y} = 0$$

The expression above shows that potential function and stream function are a pair of orthogonal functions.

If the complex potential of a given planar seepage field is known and it is written as the form of formula (3-50), the characteristic function of the seepage field—potential function and stream function can be obtained and also the equation of an equipotential line and streamline, so as to draw the figure of seepage field.

If the formula (3-50) is differentiated, the following expression can be obtained:

$$dW(z) = d\Phi(x,y) + id\Psi(x,y) = \left(\frac{\partial \Phi}{\partial x} + i\frac{\partial \Psi}{\partial x}\right)dx + \left(\frac{\partial \Phi}{\partial y} + i\frac{\partial \Psi}{\partial y}\right)dy$$

Take the Cauchy-Riemann equation into account, the expression above can be further written as:

$$dW(z) = \left(\frac{\partial \Phi}{\partial x} - i\frac{\partial \Phi}{\partial y}\right)dx + i\left(\frac{\partial \Phi}{\partial x} + i\frac{\partial \Phi}{\partial y}\right)dy = \left(\frac{\partial \Phi}{\partial x} - i\frac{\partial \Phi}{\partial y}\right)(dx + idy) = -(v_x - iv_y)dz$$

So the following expression can be obtained:

$$\left|\frac{dW(z)}{dz}\right| = |-(v_x - iv_y)| = \sqrt{v_x^2 + v_y^2} = |v|$$

Hence, only the complex potential of planar seepage field is known, the seepage velocity of an arbitrary point in formation can be obtained. Some examples which research the planar seepage field with complex potential theory will be given below.

**Example 3-3** Assuming that the complex potential of planar seepage field is:

$$W(z) = AZ + C$$

In the formula above: $A$ is real constant. $C$ is complex constant.

Make $C = C_1 + iC_2$, then the complex potential can be written as:

$$W(z) = Ax + C_1 + i(Ay + C_2)$$

In the formula above, $C_1$ and $C_2$ are all real constant. Apparently, this complex function $W(z)$ is an analytic function. The two characteristic functions of the planar seepage field—potential function and stream function respectively are:

$$\Phi(x,y) = Ax + C_1; \quad \Psi(x,y) = Ay + C_2$$

It can be seen that the equipotential lines are family of straight lines which are vertical to the axis $x$, while the streamlines are family of straight lines which are parallel to the axis $x$. The direction of streamline can be ascertained from the following equations:

$$v_x = -\frac{\partial \Phi}{\partial x} = -A; \quad v_y = -\frac{\partial \Phi}{\partial y} = 0$$

When $A > 0$, the direction of streamline is contrary with that of $x$. When $A < 0$, the direction of streamline is in line with that of $x$. The Figure 3-24 is the seepage field figure when $A < 0$. While what the complex potential describes is the planar one-dimensional seepage field.

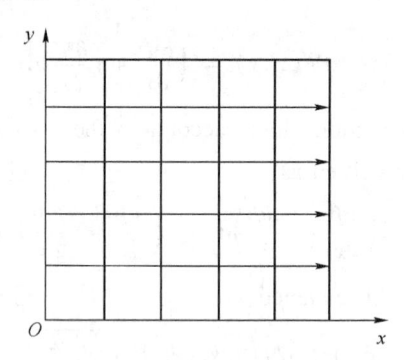

Figure 3-24 The figure of planar one-dimensional seepage field
图 3-24 平面一维渗流场图

The seepage velocity of an arbitrary point in formation is:

$$|v| = \left|\frac{dW}{dz}\right| = |A|$$

While the direction of seepage velocity is decided by $v_x$ and $v_y$.

**Example 3-4** Assuming that the complex potential of a planar seepage field is:

$$W(z) = A\ln z + C$$

In the expression above: $A$ is a real constant. $C$ is a complex constant. Show the complex variable $z$ with polar coordinates which is $z = re^{i\theta}$, then the complex potential above can be written as:

式中, $A$ 为实常数; $C$ 为复常数。将复数 $z$ 用极坐标表示, 即 $z = re^{i\theta}$, 则该复势可写成:

$$W(z) = A\ln r + C_1 + i(A\theta + C_2)$$

In the expression above, $C_1$ and $C_2$ are real constants. And the potential function and stream function of the planar seepage field respectively are:

式中, $C_1$、$C_2$ 为实常数。该平面渗流场的势函数和流函数分别为:

$$\Phi(r,\theta) = A\ln r + C_1; \ \Psi(r,\theta) = A\theta + C_2$$

It can be seen that the equipotential lines are family of concentric circles whose centre is at the origin and the streamlines are family of rays pointing to the origin. And the direction of streamlines can be determined by the following equations:

可见, 等势线是一簇以原点为圆心的同心圆, 而流线则是一簇过原点的射线。流线的方向由下列方程决定:

$$v_r = -\frac{\partial \Phi}{\partial r} = -\frac{A}{r}; \ v_\theta = 0$$

Because there is only radial seepage, what the complex potential describes is the planar radical seepage. When $A > 0$, the streamlines point to the origin which belong to point sink as shown in Figure 3-25. When $A < 0$, the streamlines start from the origin and point to the infinite which belongs to point source. The seepage velocity of an arbitrary point in formation is:

由于只有径向渗流, 因此这个复势描述的是平面径向渗流。当 $A > 0$ 时, 流线指向原点, 属于点汇, 如图 3-25 所示。当 $A < 0$, 流线由原点出发指向无穷远, 属于点源。地层中任一点的渗流速度为:

$$|v| = \left|\frac{dW(z)}{dz}\right| = \left|\frac{A}{z}\right| = \frac{|A|}{r}$$

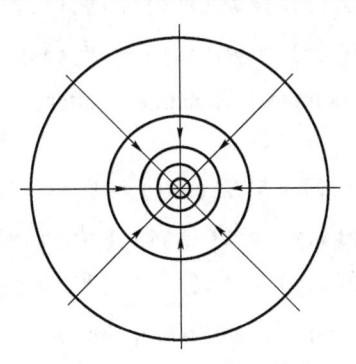

Figure 3-25 The figure of planar radial fluid flow field
图 3-25 平面径向渗流场图

As to an arbitrary point sink with intensity $q$ in formation ($A > 0$), there is the following relationship:

对于地层中有一强度为 $q$ 的点汇 ($A > 0$) 存在以下关系:

$$|v| = \frac{A}{r} = \frac{q}{2\pi r}$$

It is easy to obtain that $A = \frac{q}{2\pi}$, so the complex potential of planar radial seepage field is:

$$W(z) = \frac{q}{2\pi}\ln z + C$$

And the corresponding potential function and stream function respectively are:

$$\Phi(r,\theta) = \frac{q}{2\pi}\ln r + C_1; \quad \Psi(r,\theta) = \frac{q}{2\pi} + C_2$$

In the expression above, $C_1$ and $C_2$ are real constants that need to be decided by the outer boundary condition.

Assuming that the complex potential of a planar radial seepage field is:

$$W(z) = A\ln(z - z_0) + C$$

Make $z_1 = z - z_0$ and do coordinate transformation, then the following expression can be obtained:

$$W(z_1) = A\ln z_1 + C$$

What the complex potential describes is the problem of radial fluid flow which is a point source (sink) on point $z_0$.

### 3.4.3 Superposition principle of complex potential

If there are two point sinks simultaneously on planar one-dimensional seepage field, the complex potentials of planar seepage field when they are working independently respectively are:

$$W_1(z) = \Phi_1(x,y) + i\Psi_1(x,y)$$
$$W_2(z) = \Phi_2(x,y) + i\Psi_2(x,y)$$

The newly formed complex potential after superposition is:

$$W(z) = \Phi_1(x,y) + \Phi_2(x,y) + i[\Psi_1(x,y) + \Psi_2(x,y)] = \Phi(x,y) + i\Psi(x,y)$$
$$\Phi(x,y) = \Phi_1(x,y) + \Phi_2(x,y)$$
$$\Psi(x,y) = \Psi_1(x,y) + \Psi_2(x,y)$$

Because $\Phi_1$ and $\Psi_1$, $\Phi_2$ and $\Psi_2$ are two pairs of conjugate harmonic functions and all meet the Cauchy-Riemann equation, the new formed potential function $\Phi$ and stream function $\Psi$ after superposition are also a pair of conjugate harmonic function which all meet the Cauchy-Riemann equation. While the complex potential $W(z)$ can describe the problem of planar seepage field of two point sinks.

When there are many point sources (sinks) in formation, the algebraic sum of complex potential formed by every point source (sink) when working independently is the complex potential of many sources and sinks working simultaneously, which is called the superposition principle of complex potential.

When there are $n$ sources (sinks) with the intensity $q_i$ ($i = 1, 2, \ldots, n$) and their locations on complex planar respectively are $z_1, z_2, \ldots, z_n$, according to the superposition principle, the complex potential that $n$ wells work simultaneously can be obtained as follows:

$$W(z) = \frac{1}{2\pi} \sum_{i=1}^{n} \pm q_i \ln(z - z_i) + C \qquad (3-51)$$

The corresponding stream function and potential function respectively are:

$$\Phi(z) = \frac{1}{2\pi} \sum_{i=1}^{n} \pm q_i \ln|z - z_i| + C_1 \qquad (3-52)$$

$$\Psi(z) = \frac{1}{2\pi} \sum_{i=1}^{n} \pm q_i \arg(z - z_i) + C_2 \qquad (3-53)$$

As to the point source, the negative is taken in front of $q_i$ and the positive is taken in front of point sink. The following example is used to show the application of the superposition principle of complex potential.

**Example 3-5** There are one source and one sink of uniform intensity in infinite formation.

Assuming the distance of point sink $A$ and point source $B$ is $2a$, the coordinate system is set up as shown in Figure 3-26, then complex potential of one source and one sink interference is:

$$W(z) = \frac{q}{2\pi} \ln \frac{z-a}{z+a} + C \qquad (3-54)$$

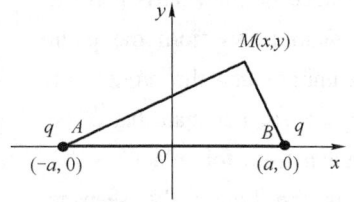

Figure 3-26 The figure of seepage field of uniform-intensity two sinks

图 3-26 等强度两汇渗流均图

If $z - a = r_1 e^{i\theta_1}$ and $z + a = r_2 e^{i\theta_2}$, so:

$$W(z) = \frac{q}{2\pi}\ln\frac{r_1}{r_2} + C_1 + i\left[\frac{q}{2\pi}(\theta_1 - \theta_2) + C_2\right]$$

And its potential function and stream function respectively are: 其势函数和流函数分别为:

$$\Phi = \frac{q}{2\pi}\ln\frac{r_1}{r_2} + C_1; \quad \Psi = \frac{q}{2\pi}(\theta_1 - \theta_2) + C_2$$

It can be seen that the equation of equipotential lines is $\frac{r_1}{r_2} = C$. If the rectangular coordinate is adopted, the equation of equipotential lines is:

由此可见,等势线方程为 $\frac{r_1}{r_2} = C$。如果采用直角坐标,则等势线方程为:

$$\left(x - \frac{1+C^2}{1-C^2}a\right)^2 + y^2 = \left(\frac{2aC}{1-C^2}\right)^2 \tag{3-55}$$

And the equation of streamlines is: 而流线方程为:

$$\theta_1 - \theta_2 = C \tag{3-56}$$

Because $\theta_1 = \arg(z-a)$, $\theta_2 = \arg(z+a)$, after substituting the two expressions above into the formula (3-56), taking tangent to both sides of the equation, using the properties of trigonometric function and simplifying, the following expression can be obtained:

由于 $\theta_1 = \arg(z-a)$, $\theta_2 = \arg(z+a)$, 将上述两式代入式(3-56)后两边取正切并利用三角函数性质,再化简整理得:

$$x^2 + \left(y + \frac{C}{2}\right)^2 = a^2 + \frac{C^2}{4} \tag{3-57}$$

The formula (3-55) and formula (3-57) are respectively totally consistent with the formula (3-22) and formula (3-23), and the figure of seepage field as shown in Figure 3-8.

式(3-55)、式(3-57)分别与式(3-22)、式(3-23)完全一致,其渗流场图如图3-8所示。

The same as the former section, if the inner and outer boundary condition are known, the formula of liquid recovery strength and the formula of potential distribution can be further obtained, which will not be repeated here.

与上节一样,若已知内外边界条件,则还可以进一步得到产液强度公式和势的分布公式,在此不再重复。

According to the formula (3-54), the seepage velocity at an arbitrary point in formation is:

根据式(3-54),可得地层中任一点的渗流速度为:

$$|v| = \left|\frac{dW(z)}{dz}\right| = \frac{q}{2\pi}\left|\frac{1}{z-a} - \frac{1}{z+a}\right| = \frac{qa}{\pi r_1 r_2} \tag{3-58}$$

Because of steady seepage, the trace of fluid mass point is consistent with that of streamline, fluid flows from the point source to point sink along each streamline, and the product on the connection of source and sink $r_1r_2$ is smaller than that of on other streamlines $r_1r_2$. It can be known from formula (3-58) that the smaller the product $r_1r_2$ is, the bigger the seepage velocity is. So the fluid mass point moves fastest along the connection of source and sink and the line segment is called the main streamline. The farther the streamline from the axis $x$ is,

由于稳定渗流液体质点的运动轨迹与流线是一致的,所以液体沿各流线由点源流向点汇,而源汇连线上的乘积 $r_1r_2$ 比其他流线上的乘积 $r_1r_2$ 要小,由式(3-58)可知:$r_1r_2$ 越小,渗流速度越大。所以,液体质点沿源汇连线这条流线运动得最快,这条线段称

the slower the fluid mass point moves along it. If the passing locations of fluid on every point of well wall of point source passing through same time is connected in curves shown in Figure 3-27, it can be seen from the curves that when the fluid flows along the main streamline from the point source well to point sink well, the fluid flows along the other streamlines from the point source well has still not reached the production well and most of which is still far from the production well, and this kind of phenomenon is called the tonguing phenomenon.

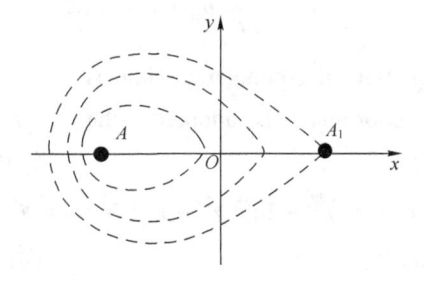

Figure 3-27  The tonguing phenomenon
图 3-27  舌进现象

Because of the existence of the tonguing phenomenon, when water flooding development is adopted, the area the water sweeps through is limited when there is water in oil well. Assuming that at time $t$, the injection water reaches the well wall of point sink along a given streamline, the included angle of main streamline and the tangent line at point $A$ of the streamline is $\theta$, and $2\theta$ is called the water flooded angle.

Similar as the former section, as to the reservoir with linear constant pressure supply boundary or circular constant pressure supply boundary line, the method of mirror image should be applied first and then the superposition principle of complex potential.

**Example 3-6**  There is interference of uniform-intensity two sinks in infinite formation.

Assuming that the distance of point sink $A$ and $B$ is $2a$, and the coordinate system is set up as shown in Figure 3-26. Then the complex potential of interference of two sinks is:

$$W(z) = \frac{q}{2\pi}\ln(z-a)(z+a) + C$$

If $z - a = r_1 e^{i\theta_1}$ and $z + a = r_2 e^{i\theta_2}$, the following expression can be obtained:

$$W(z) = \frac{q}{2\pi}\ln r_1 r_2 + C_1 + i\left[\frac{q}{2\pi}(\theta_1 + \theta_2) + C_2\right]$$

The corresponding stream function and potential function respectively are:

$$\Phi = \frac{q}{2\pi}\ln r_1 r_2 + C_1$$

$$\Psi = \frac{q}{2\pi}(\theta_1 + \theta_2) + C_2$$

It can be seen that the equation of equipotential line is $r_1 r_2 = C$, if the rectangular coordinate is adopted, the equation of equipotential line is:

$$(x^2 + y^2)^2 + 2a^2(y^2 - x^2) + a^4 = C^2$$

While the equation of streamline is:

$$\theta_1 + \theta_2 = C$$

After substituting the two expressions $\theta_1 = \arg(z-a)$ and $\theta_2 = \arg(z+a)$ into the expression above, and taking tangent to both sides of the equation and simplifying it, the following expression can be obtained:

$$x^2 - y^2 - 2Cxy - a^2 = 0$$

According to the complex potential, the seepage velocity at an arbitrary point in formation can also be obtained, which is:

$$|v| = \left|\frac{dW(z)}{dz}\right| = \frac{q}{2\pi}\left|\frac{1}{z-a} + \frac{1}{z+a}\right| = \frac{qr}{\pi r_1 r_2}$$

In the formula above, $r$ is the distance from an arbitrary point in formation to the origin of coordinate. It can be seen from the formula above, the seepage velocity at the origin in formation is 0, which is called the equilibrium point. Generally, the equilibrium point will consequently appear between two sinks near which the seepage velocity is very small and nearly does not move, so the dead oil area is formed near the equilibrium point, as shown in Figure 3-28. By the same reason, as to a well near fault, a dead oil area will be formed near the fault directly pointing to the oil

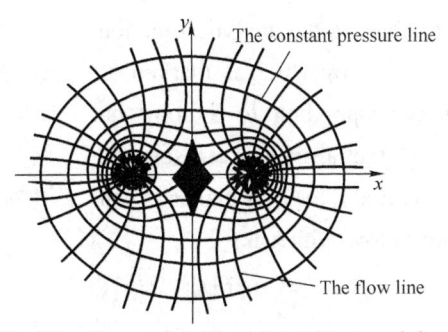

Figure 3 – 28  The figure of uniform-intensity two sinks seepage field
图 3 – 28  等强度两汇渗流场图

well. If the production of two sinks is not equal, the location of the equilibrium point will change which is generally partial to the sink with less production. The intensity of well A and B respectively are $q_1$ and $q_2$, then the seepage velocity of an arbitrary point on the connection of two wells is:

$$|v| = \frac{1}{2\pi}\left|\frac{q_1}{x-a} - \frac{q_2}{x+a}\right|$$

Make $|v| = 0$, then the location of the equilibrium point can be obtained as follows:

$$x = \frac{a(q_2 - q_1)}{q_1 + q_2}$$

The same as the former section, if the inner and outer boundary conditions are known, the formula of two sinks of uniform intensity liquid production and the formula of potential distribution can be further obtained.

When meeting reservoirs with complicated boundary, the method of mirror image should be applied to turn the original problem to the problem of multi interference of wells in infinite formation first, and then using the superposition principle of complex potential to solve problem.

### 3.4.4  Application of conformal mapping in planar seepage field

The problem of planar seepage field is researched with the superposition principle of potential (complex potential) above. While as to the more complicated problem (the infinite well chain for example), it can be solved by proper translation to turn the problem to be more simple problem of seepage firstly, and then the solution to the complicated seepage problem can be obtained easily with simple seepage problem known or adopting the superposition principle of potential (complex potential).

汇的产量不相等,平衡点的位置会发生变化,一般偏向于产量小的点汇一边,A、B 两井的强度分别为 $q_1$、$q_2$,则井连线上任一点的渗流速度值为:

令 $|v| = 0$,则可得平衡点的位置为:

与上节相同,若已知内外边界条件,还可以根据势函数的表达式进一步求出等强度两汇的产液强度公式和势的分布公式。

当遇到复杂边界的油气藏时,首先应该运用镜像反映法将原问题转化成无穷大地层的多井干扰问题,然后再用复势理论加以解决。

### 3.4.4  保角映射在平面渗流场中的应用

以上讨论了运用势(复势)叠加原理来研究平面渗流场问题。而对于更复杂的(如无限井链等)问题,则可以首先通过适当的变换,将其变成简单的渗流问题,然后再用已知的简单的渗流问题的解或采用势(复势)的叠加原理,就可以很容易地得到复杂的渗流问题的解。

Assuming that complex variable $z$ is the analytic function of another complex variable $w = \xi + i\eta$ which is marked as $z(w)$, a point on plane $w$ is corresponding to a point on planar $z$. If $f(z)$ is the complex potential of a given seepage filed on plane $z$, the complex potential $f_1(w)$ on planar $w$ can be obtained from the expression below which is:

$$f(z) = f[z(w)] = f_1(w) \tag{3-59}$$

$f_1(w)$ is the complex potential of another seepage filed on plane $w$. If $f_1(w)$ is known, the seepage filed unknown on the plane $w$ can be obtained. The function $z(w)$ implements this transformation, the flow net on plane $z$ is changed to be the flow net on plane $w$ and the equipotential lines and streamlines are still perpendicular each other after the transformation. Of course, the shape and their geometry of the two flow nets will not be the same. Apparently, the transformation should be one-one correspondent and conformal. It would be all right if $z(w)$ can be proved as analytic function.

$$z(w) = z(\xi + i\eta) = x(\xi, \eta) + iy(\xi, \eta) \tag{3-60}$$

Provided $z(w)$ is a single-valued function, the transformation is one-one correspondence. Now the conformability of transformation will be proven which means the streamlines and equipotential lines are still orthogonal each other after the transformation.

According to the definition of analytic function, on the corresponding point of planar $z$ and $w$ there is:

$$\frac{dz}{dw} = \lim_{\Delta w \to 0} \frac{\Delta z}{\Delta w} = \lim_{\substack{\Delta \xi \to 0 \\ \Delta \eta \to 0}} \frac{\Delta x + i\Delta y}{\Delta \xi + i\Delta \eta} \tag{3-61}$$

The limits are always existing and equal when $\Delta w$ approach zero no matter what direction. If $dz_1, dz_2, dz_3, \ldots$ can be obtained along different direction starting from the point $z_0$, there will undoubtedly be corresponding $dw_1, dw_2, dw_3, \ldots$ on plane $w$ and the following expression can also be obtained:

$$\frac{dz_1}{dw_1} = \frac{dz_2}{dw_2} = \frac{dz_3}{dw_3} = \cdots \tag{3-62}$$

The expression above can also be written as:

$$\frac{dz_1}{dz_2} = \frac{dw_1}{dw_2};\ \frac{dz_2}{dz_3} = \frac{dw_2}{dw_3};\ \frac{dz_1}{dz_3} = \frac{dw_1}{dw_3},\cdots$$

It is known that the argument of the quotient of complex number is different from the arguments of numerator and nominator, that is:

$$\arg(dz_1 - dz_2) = \arg(dw_1 - dw_2)$$
$$\arg(dz_2 - dz_3) = \arg(dw_2 - dw_3)$$
$$\arg(dz_1 - dz_3) = \arg(dw_1 - dw_3)$$

That is the intersection angle of arbitrary two lines from an arbitrary point unchanged after transformation, so it is called conformal transformation. But the shape of streamline and equipotential line will change after transformation. And the change of geometry size after transformation is:

$$dz = f'(w)dw \tag{3-63}$$

If the radius of well on plane $z$ is $r_w$, because $r_w$ is very small so it can be considered as $dz$ which is changed to be $dw$ on plane $w$.

已知复数的商的辐角等于分子的辐角与分母辐角之差,即:

即从任一点出发任意两条线的交角变换后不变,所以称为保角变换。但变换后流线与等势线的形状会变。变换后几何尺寸的变化为:

若 $z$ 平面上井半径为 $r_w$,由于 $r_w$ 很小所以认为是 $dz$,则变为 $w$ 平面上的 $dw$。

$$r_w = \left|\frac{dz}{dw}\right|\rho_w;\ \rho_w = \left|\frac{dw}{dz}\right|r_w \tag{3-64}$$

What it needs to do now is to prove the intensity of source (sink) keeps unchanged after transformation.

As shown in Figure 3-29(a), assuming that we take an arbitrary closed curve $c$ surrounding an arbitrary source (sink) on plane $z$, and then take a little segment $ds$ at an arbitrary point $z_0$ on this curve, while the unit normal vector of this curve is $\boldsymbol{n}$, the intensity of well on plane $z$ is:

现在需要证明变换后源(汇)的强度保持不变。

图 3-29(a) 所示,设在平面 $z$ 上围绕任一点源(汇)任取一条封闭曲线 $c$,在该曲线上任一点 $z_0$ 处取该曲线的一微小线段 $ds$,而该曲线的单位法向向量为 $\boldsymbol{n}$,则在 $z$ 平面上井的强度为:

$$Q = \oint_c \left|\frac{\partial p}{\partial n}\right| ds \tag{3-65}$$

(a) Before transformation
(a) 变换前

(b) After transformation
(b) 变换后

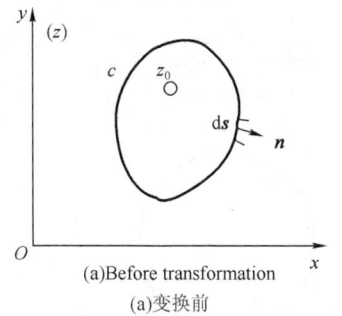

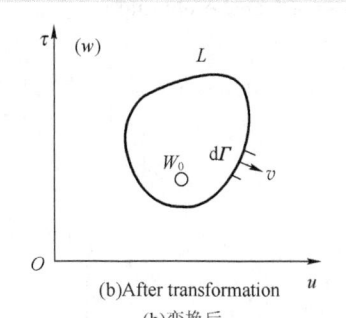

Figure 3-29 The conformal transformation (reflection)
图 3-29 保角变换(映射)

After the conformal transformation $w = f(z)$, the closed curve $c$, point $z_0$, vector $\boldsymbol{n}$ and the segment $ds$ is respectively

经保角映射 $w = f(z)$ 后, $z$ 平面上的封闭曲线 $c$、点 $z_0$、向量 $\boldsymbol{n}$

reflected to be the closed curve $L$, point $w_0$, vector $v$ and the segment $\mathrm{d}\Gamma$ on plane $w$ and $v \perp \mathrm{d}\Gamma$ as shown in Figure 3-29(b). So the intensity of well on plane $w$ is:

$$Q = \oint_L \left|\frac{\partial \Phi}{\partial v}\right| \mathrm{d}\Gamma$$

According to the properties of conformal transformation, it can be known that:

$$\mathrm{d}\Gamma = |f'(z_0)| \mathrm{d}s$$
$$\mathrm{d}v = |f'(z_0)| \mathrm{d}n$$

Substitute the expression above into the intensity formula on plane $w$, the following expression can be obtained:

$$Q = \oint_L \left|\frac{\partial \Phi}{\partial v}\right| \mathrm{d}\Gamma = \oint_c \left|\frac{\partial \Phi}{\partial n}\right| \frac{1}{|f'(z_0)|} |f'(z_0)| \mathrm{d}s = \oint_c \left|\frac{\partial \Phi}{\partial n}\right| \mathrm{d}s$$

The unchangeability of the production of oil well can be proved after the conformal transformation. The following problem of several kinds of planar seepage will be researched through the conformal transformation.

1. A well near linear supply boundary line

Assuming that there is a point sink with intensity $q$ near a linear constant pressure supply boundary line with the distance $a$, and the coordinate is set up as shown in Figure 3-30(a). Take the transformation $w = \dfrac{z - \mathrm{i}a}{z + \mathrm{i}a}$, and then the point sink on plane $z$ is transformed to be the origin on plane $w$. While as to an arbitrary point $z = x + \mathrm{i}y$ on plane $z$ when $y > 0$, the following expression can be obtained:

$$|w(z)| = \left|\frac{x + \mathrm{i}(y - a)}{x + \mathrm{i}(y + a)}\right| = \sqrt{\frac{x^2 + (y - a)^2}{x^2 + (y + a)^2}} < 1$$

(a) Before transformation  (b) After transformation
(a) 变换前  (b) 变换后

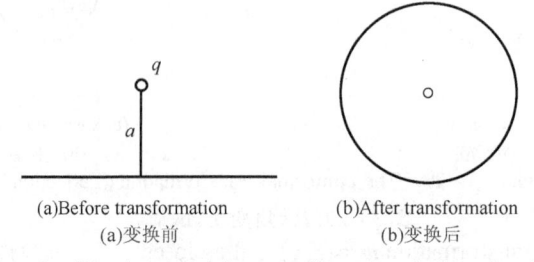

Figure 3-30　A point sink near a linear supply boundary line
图 3-30　直线供给边线旁一点汇

It can be seen that all the points on the upper half planar of complex plane $z$ after mapping all drop in the area of a unit circle $\rho < 1$ on plane $w$. While as to the real axis of complex plane $z$, the following expression can be obtained after mapping:

$$w = \frac{x - ia}{x + ia} = \frac{(x - ia)^2}{x^2 + a^2} = e^{i2\theta}$$

In the expression above, $\theta = \arg(x - ia)$. When $x$ changes from $-\infty$ to $\infty$ and $2\theta$ changes from $-2\pi$ to $0$, the axis $x$ is changed to be a circumference of $\rho = 1$ on plane $w$. Thus the seepage problem of a well near linear supply boundary line on plane $z$ is transformed to be that of a well in the centre of circular formation after conformal transformation and the solution to the latter is known, which is:

$$q = \frac{2\pi(\Phi_e - \Phi_w)}{\ln\frac{1}{\rho_w}}$$

Because $\rho_w = \left|\frac{dw}{dz}\right|_{z=ia}$, $r_w = \frac{r_w}{2a}$, the formula of liquid production per unit thickness of a point sink near linear supply boundary line is:

$$q = \frac{2\pi(\Phi_e - \Phi_w)}{\ln\frac{2a}{r_w}} = \frac{2\pi K(p_e - p_w)}{\mu \ln\frac{2a}{r_w}}$$

The potential distribution formula of potential on plane $w$ is:

$$\Phi = \Phi_e - \frac{\Phi_e - \Phi_w}{\ln\frac{1}{\rho_w}}\ln\frac{1}{\rho}$$

Substitute $\rho = \left|\frac{z - ia}{z + ia}\right| = \sqrt{\frac{x^2 + (y - a)^2}{x^2 + (y + a)^2}}$ into the formula above, the following expression can be obtained:

$$\Phi = \Phi_e - \frac{\Phi_e - \Phi_w}{2\ln\frac{2a}{r_w}}\ln\frac{x^2 + (y - a)^2}{x^2 + (y + a)^2}$$

It can be seen that the result obtained from conformal transformation is completely the same as the result obtained from the superposition of potential.

2. An eccentric well in circular formation

Assuming that there is an eccentric well with eccentricity

$d$ in circular formation, the pressure on circular supply boundary line is $p_e$, and the pressure on well wall is $p_w$, and the coordinate is set up as shown in Figure 3-31. Assuming that the location of oil well on this complex plane is $z_0$, then $|z_0| = d$, the radius of oil well $r_w$, do the transformation then the following expression can be obtained:

$$w = \frac{r_e(z - z_0)}{r_e^2 - \bar{z}_0 z}$$

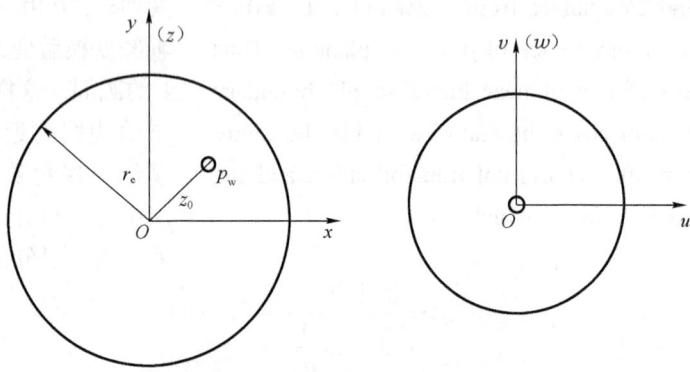

Figure 3-31　The eccentric well
图 3-31　偏心井

When $z = z_0$, $w(z_0) = 0$, that is: after the transformation the oil well lies in the origin of the new complex plane. While as to an arbitrary point $z = r_e e^{i\theta}$ on supply boundary line, it will change to be the following expression after transformation:

$$|w(z)| = \left| \frac{r_e(r_e e^{i\theta} - z_0)}{r_e(r_e - z_0 e^{i\theta})} \right| = 1$$

After transformation, the supply boundary line with radius $r_e$ on plane $z$ is changed to be a circumference with radius 1 on plane $w$, so the formula of intensity is:

$$q = \frac{2\pi(\Phi_e - \Phi_w)}{\ln \dfrac{1}{\rho_w}}$$

While the relationship of the radius of wells on two complex planes are:

$$\rho_2 = \left| \frac{dw}{dz} \right|_{z=z_0} r_w = \frac{r_e r_w}{r_e^2 - d^2}$$

Substitute the formulas above into the equation of liquid producing intensity, the following expression can be obtained:

$$q = \frac{2\pi(\Phi_e - \Phi_w)}{\ln\dfrac{r_e^2 - d^2}{r_e r_w}} = \frac{2\pi K(p_e - p_w)}{\mu \ln\dfrac{r_e}{r_w}\left(1 - \dfrac{d^2}{r_e^2}\right)}$$

### 3. A circular well-drain in circular formation

Assuming that there is a circular well-drain in circular formation with supply radius $r_e$ as shown in Figure 3-32 (a). $n$ wells are evenly distributed on the circumference with radius $r$ and the intensity is all $q$. A coordinate is set up which makes the $n$ wells distribute evenly on-both sides of axis $x$. So the location of every well on plane $z$ is $z_0 = re^{i\frac{2m+1}{n}\pi}$ ($m = 0, 1, \ldots, n-1$).

### 3. 圆形地层中有一圆形井排

设供给半径为 $r_e$ 的圆形地层中有一圆形井排如图 3-32(a) 所示。$n$ 口井均匀地分布在半径为 $r$ 的圆周上，且强度均为 $q$。建立坐标使 $n$ 口井均匀地分布在 $x$ 轴的两侧，因此 $z$ 平面上各井的位置为 $z_0 = re^{i\frac{2m+1}{n}\pi}$ ($m = 0, 1, \cdots, n-1$)。

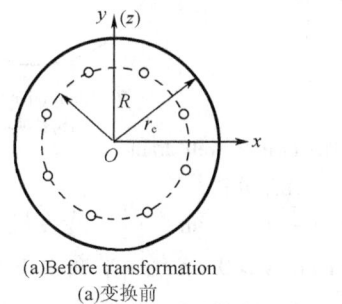

(a) Before transformation　　(b) After transformation
(a) 变换前　　(b) 变换后

Figure 3-32　The circular well-drain
图 3-32　圆形井排

Do the transformation $\omega = z^n$, then the fan-shaped area when $\theta$ changes from $\dfrac{2m}{n}\pi$ to $\dfrac{2(m+1)}{n}\pi$ on plane $z$ will be changed to be a circular reservoir with an eccentric well after transformation as shown in Figure 3-32 (b). The supply radius is $\rho_e = r_e^n$, eccentricity is $d = |z_0^n| = r^n$, while the radius of oil well is:

作变换 $w = z^n$，则 $z$ 平面上 $\theta$ 由 $\dfrac{2m}{n}\pi$ 到 $\dfrac{2(m+1)}{n}\pi$ 的扇形区域经映射后变成有一偏心井的圆形油藏，如图 3-32(b) 所示。供给半径 $\rho_e = r_e^n$，偏心距为 $d = |z_0^n| = r^n$，而油井半径为：

$$\rho_w = \left|\frac{dw}{dz}\right|_{z=z_0} r_w = nr^{n-1}r_w$$

Substitute $\rho_e$, $\rho_w$ and eccentricity $d$ simultaneously into the formula of eccentric well, the following expression can be obtained:

将 $\rho_e$、$\rho_w$ 和偏心距 $d$ 同时代入偏心井的公式中得：

$$q = \frac{2\pi K(p_e - p_w)}{\mu \ln\dfrac{r_e^n}{nr^{n-1}r_w}\left(1 - \dfrac{r^{2n}}{r_e^{2n}}\right)}$$

This is the calculation formula of the intensity of liquid production of single well of circular well-drain.

这就是圆形井排单井产液强度的计算公式。

### 4. An infinite well-drain near a linear supply boundary line

Assuming that there is an infinite well-drain near a

### 4. 直线供给边线附近有一无限井排

设距离直线供给边线 $L$ 有

— 137 —

linear supply boundary line $L$ with the distance $2a$ as shown in Figure 3-33 (a). Set up the following coordinates to make the location of every well on plane $z$ to be $z_0 = \pm 2na + iL$ and $n = 0, \pm 1, \pm 2, \ldots$

一相距为 $2a$ 的无穷大井链,如图3-33(a)所示。建立如下坐标使各井在 $z$ 平面上的位置分别为 $z_0 = \pm 2na + iL$,其中 $n = 0, \pm 1, \pm 2, \cdots$

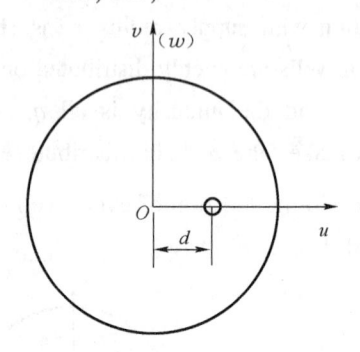

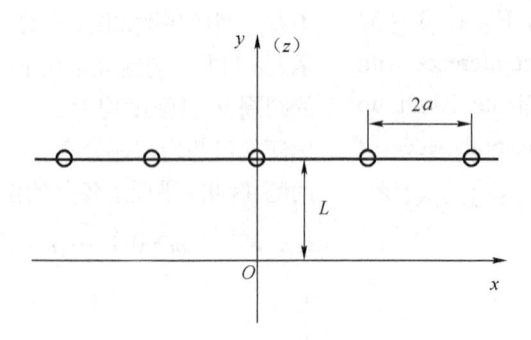

(a) Before transformation
(a) 变换前

(b) After transformation
(b) 变换后

Figure 3-33  The infinite well-drain
图 3-33  无限井排

Making the transformation $w = e^{\frac{i\pi z}{a}} = e^{-\frac{\pi y}{a}} e^{\frac{i\pi x}{a}}$, an arbitrary banding area $(2n-1)a \leq x \leq (2n+1)a, y \geq 0$ on plane $z$ is changed to be a circular reservoir with an eccentric well on plane $w$ after transformation, which is the line segment of $y = 0$ mapped to be a circumference with radius 1, while the area of $y > 0$ is transformed with in the area of the unit circle, and the eccentricity of oil well is:

令变换 $w = e^{\frac{i\pi z}{a}} = e^{-\frac{\pi y}{a}} e^{\frac{i\pi x}{a}}$,对于 $z$ 平面上的任一个 $(2n-1)a \leq x \leq (2n+1)a、y \geq 0$ 的带状区域经变换后变成 $w$ 平面上的含有一口偏心井的圆形油气藏,即 $y = 0$ 的线段映射成半径为 1 的圆周,而 $y > 0$ 的部分映射在该单位圆内,而油井的偏心距为:

$$d = |w(z_0)| = e^{-\frac{\pi L}{a}}$$

The well radius of eccentric well is:

偏心井的井半径为:

$$\rho_w = \left|\frac{dw}{dz}\right|_{x=x_0} r_w = \frac{\pi r_w}{a} e^{-\frac{\pi L}{a}}$$

Substitute $\rho_e$, $d$ and $\rho_w$ simultaneously into the formula of liquid producing intensity of eccentric well, the following expression can be obtained:

将 $\rho_e$、$d$ 和 $\rho_w$ 同时代入偏心井的产液强度公式中得:

$$q = \frac{2\pi K(p_e - p_w)}{\mu \ln\left[\frac{a}{\pi r_w} e^{\frac{\pi L}{a}}(1 - e^{-\frac{2\pi L}{a}})\right]} = \frac{2\pi K(p_e - p_w)}{\mu \ln\left(\frac{2a}{\pi r_w} \operatorname{sh}\frac{\pi L}{a}\right)}$$

This is the calculation formula of liquid production per unit thickness of single well of an infinite well array near a linear supply boundary.

这就是直线井排附近无穷大井链的单井产液强度的计算公式。

5. Zhukovsky transformation

Nowadays, hydraulic fracturing has become a generally

5. 儒阔夫斯基变换

目前水力压裂已成为普遍

adopted stimulation measure, and almost all reservoirs are using this treatment. Generally, the fractures made by the treatment are all vertical to oil layers and the height is the same as that of formation. If the permeability of fractures is much higher than that of formation, it can be considered that the permeability of fractures is infinite, means that the fracture is a constant pressure surface. Then the seepage is actually a planar problem, means that the seepage field along an arbitrary cross section which is parallel with oil layer is the same as shown in Figure 3 – 34. If the supply boundary is assumed to be an ellipse and the vertical fracture to be a straight line, the problem is much easier to solve. Only the length of fracture is not too big, the shape of supply boundary has little influence on the production, then the Zhukovsky transformation can be adopted.

采用的增产措施,几乎所有的油气藏都在压裂。一般情况下,压出的都是垂直于层面、高度等于地层厚度的裂缝。如果裂缝渗透率远高于地层渗透率,可以认为,裂缝渗透率无限大,即裂缝是等压面。这样渗流实际上是平面问题,即沿平行于层面的任一截面渗流场都是一样的,如图3 – 34所示。如果把供给边缘设成椭圆,垂直缝设成一直线,问题容易得多,只要裂缝长度不是太大,供给边缘的形状对产量的影响很小,于是可采用儒阔夫斯基变换。

$$w(z) = \frac{L}{2}\left(z + \frac{1}{z}\right)$$

(a)Before transformation  (b)After transformation
(a)变换前  (b)变换后

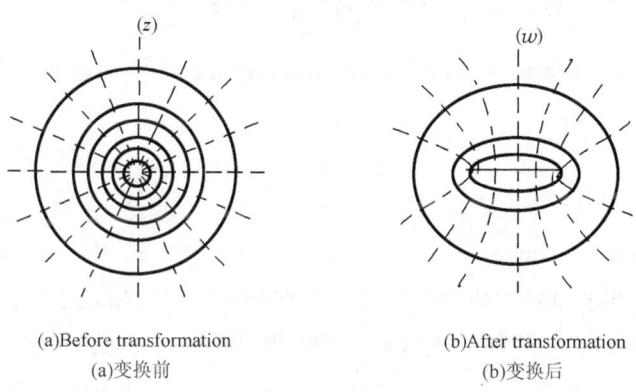

Figure 3 – 34  The figure of Zhukovsky transformation
图 3 – 34  儒阔夫斯基变换图

Its real part and imaginary part respectively are:

它的实部和虚部分别是:

$$\xi = \frac{L}{2}\left(\rho + \frac{1}{\rho}\right)\cos\varphi$$

$$\eta = \frac{L}{2}\left(\rho - \frac{1}{\rho}\right)\sin\varphi$$

The concentric circle family $|z| = \rho_0$ on plane $z$ is changed to be the following on plane $w$:

$z$ 平面上的同心圆簇 $|z| = \rho_0$ 变为 $w$ 平面上的:

$$\xi = \frac{L}{2}\left(\rho_0 + \frac{1}{\rho_0}\right)\cos\varphi$$

$$\eta = \frac{L}{2}\left(\rho_0 - \frac{1}{\rho_0}\right)\sin\varphi$$

After canceling the parameter $\varphi$, the expression below can be obtained:

消去参数 $\varphi$,得:

$$\frac{\xi^2}{a^2}+\frac{\eta^2}{b^2}=1$$

In the expression above: $a=\frac{L}{2}\left(\rho_0+\frac{1}{\rho_0}\right)$, $b=\frac{L}{2}\left(\rho_0-\frac{1}{\rho_0}\right)$, this is the ellipse family and its focal point is $(\pm L,0)$.

$\rho_0$ is infinitely growing from 1 and so as the value of $a$ and $b$. Then the external part of unit circle on plane $z$ is changed to be the entire $w$ plane. There is a secant along the real axis from $-L$ to $L$.

The radians $\arg z = \varphi_0$ on plane $z$ are changed to be the following on plane $w$:

$$\xi=\frac{L}{2}\left(\rho+\frac{1}{\rho}\right)\cos\varphi_0$$
$$\eta=\frac{L}{2}\left(\rho-\frac{1}{\rho}\right)\sin\varphi_0$$

The following expression can be obtained after canceling the parameter $\rho$:

$$\frac{\xi^2}{a^2}-\frac{\eta^2}{b^2}=1$$

In the formula above, $a=|L\cos\varphi_0|$, $b=|L\sin\varphi_0|$, this is a hyperbolic curve family, the real and imaginary semi-axis respectively are $|L\cos\varphi_0|$ and $|L\sin\varphi_0|$, and its focal points are $(\pm L,0)$.

Zhukovsky transformation changes circle on plane $z$ to be ellipse on plane $w$, changes radial slit to be hyperbolic curve, changes concentric circles family to be confocal ellipse family and changes radial slit to be confocal hyperbolic curve family.

Assuming that the long and short axis of actual supply boundary (plane $w$) respectively are $a$ and $b$, the vertical fracture located on the connection of two focal points, the focal distance is $L/2$. Assuming that it is a constant pressure line that the flow resistance is zero. Then the elliptical supply boundary on plane $w$ is changed to be the circular supply boundary on plane $z$ after Zhukovsky transformation, and its radius is:

$$\rho_e = \frac{a+b}{L/2}$$

The connection of focal points is the vertical fracture, and it is changed to be the unit circle $\rho_w = 1$ on plane $z$, and then the production of vertical fracture with infinite flow conductivity is:

$$Q = \frac{2\pi Kh(p_e - p_w)}{\mu \ln \frac{a+b}{L/2}}$$

If the supply boundary is very close to circle, which is $a \approx b \approx r_e$, then the production of vertical fracture can be expressed approximately as:

$$Q = \frac{2\pi Kh(p_e - p_w)}{\mu \ln \frac{r_e}{L/4}}$$

The vertical fractures obtained is amount to enlarge the radius of well to be 1/4 of the fracture length. If the length of fracture is close to that of supply boundary, or the permeability of fracture is not too big, then the formula above will be of great error.

6. The production formula of horizontal well

Assuming that there is a horizontal well with radius $r_w$ and length $L$ in the middle of formation with height $h$. There is a constant pressure supply boundary parallel with the horizontal well with the distance $L_e$ as shown in Figure 3-35. Please solve the production of the horizontal well.

Using mirror image to the top and bottom of formation, the infinitely well array with well spacing $h$ will be obtained. If the production of unit length of horizontal well is the same, the length of horizontal well is correspond to the height of formation. It is easy to obtain the inner and outer resistance with the method of equivalent flow resistance, which respectively are:

$$R_1 = \frac{\mu}{KhL}L_e; \quad R_2 = \frac{\mu}{2\pi KL}\ln \frac{h}{2\pi r_w}$$

So the production $Q$ of horizontal well with bottom pressure $p_w$ and length $L$ is:

$$Q = \frac{p_e - p_w}{\frac{\mu}{KhL}L_e + \frac{\mu}{2\pi KL}\ln\frac{h}{2\pi r_w}}$$

According to this, if the supply boundary is elliptical, the horizontal well lies on the connection of focal points, and the focal distance is half of the length of horizontal well as shown in Figure 3 - 36. The inner resistance should be the same as that of linear supply boundary, and only the outer resistance is changed to be:

按照这个思路,如果供给边缘是椭圆形的,水平井位于其焦点连线上,焦距为水平井长度的一半,如图 3 - 36 所示。内阻应当与直线供给边缘的一样,只是外阻改为:

$$R_1 = \frac{\mu}{2\pi Kh}\ln\frac{a+b}{L/2}; \quad R_2 = \frac{\mu}{2\pi KL}\ln\frac{h}{2\pi r_w}$$

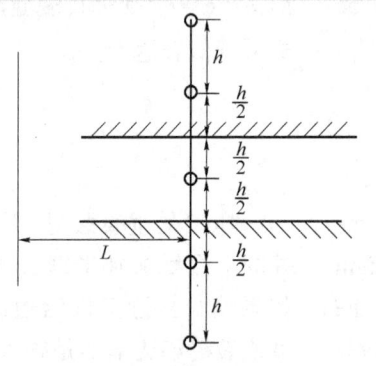

Figure 3 - 35　The sketch map of horizontal well with linear supply boundary
图 3 - 35　直线供给边界水平井示意图

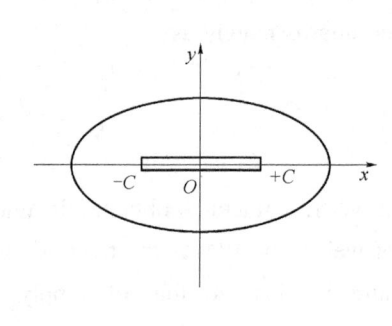

Figure 3 - 36　The sketch map of horizontal well with elliptical supply boundary
图 3 - 36　椭圆形供给边界水平井示意图

Because $a^2 - b^2 = (L/2)^2$, we can get:

因为 $a^2 - b^2 = (L/2)^2$,可得:

$$R_1 = \frac{\mu}{2\pi Kh}\ln\frac{a + \sqrt{a^2 - (L/2)^2}}{L/2}$$

Then the formula of production is:

于是产量公式为:

$$Q = \frac{p_e - p_w}{\frac{\mu}{2\pi Kh}\ln\frac{a + \sqrt{a^2 - (L/2)^2}}{L/2} + \frac{\mu}{2\pi Kh}\ln\frac{h}{2\pi r_w}}$$

Assuming that $a \approx r_e$ which means the supply boundary is circular and $r_e \gg L/2$, then the formula of production can be shortly written as:

设 $a \approx r_e$,即供给边界为圆形,且 $r_e \gg L/2$,则产量公式可简写为:

$$Q = \frac{2\pi Kh(p_e - p_w)}{\mu\left(\ln\frac{r_e}{L/4} + \frac{h}{L}\ln\frac{h}{2\pi r_w}\right)}$$

The physical significance to do this is very clear, and it is consistent with the complicated and approximate deducing result and also identical with the physical and mathematical simulating result. It is wrong to loss a $\pi$ in the second item of nominator in popular Joshi formula.

这样做物理意义很清楚,与复杂、近似的推导结果一致,与物理、数学模拟的结果一致。流行的 Joshi 公式分母第二项中缺一个 $\pi$ 是不对的。

## 3.4.5 Production formula of the multi-fracture well and multi-branched horizontal well

1. The production formula of multi-fracture well

Assuming that there are $n$ vertical fractures with length $l$ distributing evenly near the oil well and the height is the same as the height $h$ of formation. Then the formation is divided into $2n$ fans by the angle-bisector lines of adjacent fractures and the flow net of each sector is the same. Four vertical lines distributed symmetrically are given in Figure 3-37, adopting transformation:

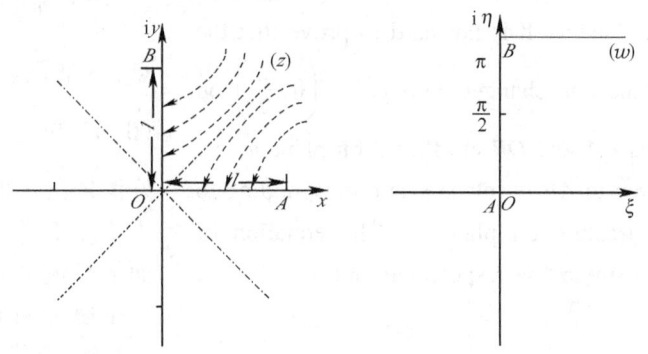

$$(z/l)^{n/2} = \text{ch}\, w \tag{3-66}$$

Figure 3-37 The distributing figure of 4 vertical fractures

The sector of $2\pi/n$ on plane $z$ is changed to be semi-infinite strip area with width $\pi$ on plane $w$. After parting the real part and imaginary part of formula (3-66), the following expression can be obtained:

$$\left(\frac{r}{l}\right)^{n/2}\left(\cos\frac{n\theta}{2} + i\sin\frac{n\theta}{2}\right) = \text{ch}\,\xi\cos\eta + i\,\text{sh}\,\xi\sin\eta$$

$$\left(\frac{r}{l}\right)^{n/2}\cos\frac{n\theta}{2} = \text{ch}\,\xi\cos\eta \tag{3-67}$$

$$\left(\frac{r}{l}\right)^{n/2}\sin\frac{n\theta}{2} = \text{sh}\,\xi\sin\eta \tag{3-68}$$

The real axis $OA$ on plane $z$ ($\theta = 0$ and $0 \leqslant r < \infty$) is changed by the following expression on plane $w$:

$$\text{ch}\,\xi\cos\eta = \left(\frac{r}{l}\right)^{n/2}; \quad \text{sh}\,\xi\sin\eta = 0$$

The coordinate of point $A$ on plane $z$ is $(l, 0)$, then transform it on plane $w$, the following expression can be obtained:

$$\text{ch}\,\xi\cos\eta = 1; \quad \text{sh}\,\xi\sin\eta = 0; \quad \xi = \eta = 0$$

That is the origin on plane $w$.

The line $OB$ on plane $z$: $\theta = \dfrac{2\pi}{n}$. The coordinate of point $B$ is changed from $\left(l, \dfrac{2\pi}{n}\right)$ to:

$$\text{ch}\,\xi\cos\eta = \left(\frac{r}{l}\right)^{n/2}\cos\pi = -\left(\frac{r}{l}\right)^{n/2}, \quad \text{sh}\,\xi\sin\eta = \left(\frac{r}{l}\right)^{n/2}\sin\pi = 0$$

Apparently, the formula above can be met only when $\eta = \pi$. $\text{ch}\,\xi \geq 1$ will never be negative, it means the axis $y$ on plane $z$ is changed to be the straight line $\eta = \pi$ on plane $w$.

The coordinate of point $B$ is changed from $\left(l, \dfrac{2\pi}{n}\right)$ on plane $z$ to $(0, \pi)$ on plane $w$. It is not hard to prove that the origin $(0,0)$ on plane $z$ is changed into point $\left(0, \dfrac{\pi}{2}\right)$ on plane $w$. It is like that $OA$ and $OB$ are flatted on plane $w$. So the planar parallel flow to $AB$ on plane $w$ is corresponding to the flow towards a fracture on plane $z$. The equation of equipotential line and streamline respectively are:

$$\xi = \xi_0; \quad \eta = \eta_0$$

After canceling $\xi$ and $\eta$ respectively in formula (3-67) and formula (3-68) the equation of equipotential line and streamline can be obtained:

$$\frac{\left(\dfrac{r}{l}\right)^{n/2}\cos^2\dfrac{n\theta}{2}}{\text{ch}^2\xi} + \frac{\left(\dfrac{r}{l}\right)^{n/2}\sin^2\dfrac{n\theta}{2}}{\text{sh}^2\xi} = 1 \qquad (3-69)$$

$$\frac{\left(\dfrac{r}{l}\right)^{n/2}\text{ch}^2\dfrac{n\theta}{2}}{\cos^2\eta} - \frac{\left(\dfrac{r}{l}\right)^{n/2}\text{sh}^2\dfrac{n\theta}{2}}{\sin^2\eta} = 1 \qquad (3-70)$$

It can be known from the formula (3-66) that:

$$\omega = \ln\left[\left(\frac{z}{l}\right)^{n/2} + \sqrt{\left(\frac{z}{l}\right)^{n/2} - 1}\right] \quad \left(\frac{z}{l} \geq 1\right)$$

Assuming that the modular of $z$ is very big and $\left|\dfrac{z}{l}\right| \gg 1$, the approximate expression can be obtained:

$$\omega = \frac{n}{2}\ln 2^{\frac{n}{2}}\left(\frac{z}{l}\right)$$

Assuming the supply boundary on plane $z$ is $r_e e^{i\theta}$ and $r_e$ is of great value, then:

$$\omega = \frac{n}{2}\ln 2^{\frac{n}{2}}\left(\frac{r_e}{l}\right) + \frac{n}{2}i\theta = \xi + i\eta$$

So

$$\xi = \frac{n}{2}\ln 2^{\frac{n}{2}}\frac{r_e}{l}$$

The potential distribution of planar parallel flow is:

$$\Phi = \frac{q}{\pi}\xi + C$$

It can be known from the formula above: $\xi = 0$, $\Phi = \Phi_w$, $r = r_e$, $\Phi = \Phi_e$. So the following expression can be obtained:

$$\Phi_e - \Phi_w = \frac{nq}{2\pi}\ln 2^{\frac{n}{2}}\frac{r_e}{l}$$

It has been said before that the strip area with width $\pi$ on plane $w$ represents for the production $q$ of a fracture on plane $z$, so $nq$ is the production of all fractures, it can be obtained at last that:

$$Q = \frac{2\pi Kh(p_e - p_w)}{\mu \ln \frac{2^{\frac{n}{2}} r_e}{l}} \qquad (3-71)$$

The formula above is the production formula of multi-fracture well. The situation of the two vertical fractures $n = 2$ with semi-length $l$ caused by common hydraulic fracturing is a special case of the multi-fracture well. The stimulation ratio $B$ of multi-fracture well is:

$$B = \frac{Q}{q} = \frac{\ln \frac{r_e}{r_w}}{\frac{1}{n}\ln 4 + \ln \frac{r_e}{r_w}} \qquad (3-72)$$

$B$ depends on the number of fractures $n$ and the length of fracture $l$. Assume $\frac{r_e}{r_w} = 10^3$, $\frac{r_e}{l} = 50, 20, 10$, $n = 2, 3, 4, 6$, the value of $B$ obtained is shown in Table 3–1. It can be seen from Table 3–1, if the number of fractures is increased one time, the production will be increased no more than 10%~30%, while if the length of fracture is increased one time, the production will be increased 50%~60%. The main effect on the production of oil well is the length of fracture not the number of it. Therefore, no more than four long fractures are better to make in actual treatment.

**Table 3-1　The stimulation ratio $B$ of multi-fracture well**

**表 3-1　多裂缝井的增产倍数 $B$**

| $\dfrac{r_e}{l}$ ＼ $n$ | 2 | 3 | 4 | 6 |
|---|---|---|---|---|
| 50 | 1.50 | 1.56 | 1.62 | 1.67 |
| 20 | 1.87 | 2.00 | 2.07 | 2.14 |
| 10 | 2.31 | 2.50 | 2.61 | 2.73 |

2. The production formula of multi-branched horizontal well

The geometric resistance $\ln \dfrac{4^{\frac{1}{n}} r_e}{l}$ in production formula of multi-fracture well is also the outer resistance of multi-branched horizontal well. The inner resistance of multi-branched horizontal well is the parallel connection of $n$ branches. so the inner resistance of a horizontal well when it is located in the formation center is:

$$\frac{1}{n}\ln \frac{h}{2\pi r_w}$$

The total production of multi-branched horizontal well is:

$$Q = \frac{2\pi Kh(p_e - p_w)}{\mu\ln \dfrac{4^{\frac{1}{n}} r_e}{l} + \dfrac{h}{nl}\ln\dfrac{h}{2\pi r_w}} \qquad (3-73)$$

When the anisotropy of permeability is taken into account, all the length of vertical direction should multiply $\beta = \sqrt{\dfrac{K_h}{K_v}}$ and the length of horizontal direction stay unchanged. And now the circular well with radius $r_w$ is changed to be an ellipse with major semi-axis $\beta r_w$ and minor semi-axis $r_w$, and the effective radius $r'_w$ is the average value of the two:

$$r'_w = \frac{1}{2}(1+\beta)r_w \qquad (3-74)$$

The production formula of multi-branched horizontal well in anisotropic formation is:

$$Q = \frac{2\pi K\beta h(p_e - p_w)}{\ln\left[\dfrac{2r_e}{l} + \sqrt{\dfrac{(2r_e)^2}{l^2} - 1}\right] + \dfrac{\beta h}{l}\ln\dfrac{\beta h}{\pi(1+\beta)r_w}} \qquad (3-75)$$

Under the same condition, the ratio of the production of horizontal well $Q_h$ and the production of vertical well $Q_v$ is:

2. 多分支水平井产量公式

多裂缝井产量公式中几何阻力 $\ln \dfrac{4^{\frac{1}{n}} r_e}{l}$ 也是多分支水平的外阻。多分支水平井的内阻为 $n$ 个分支的并联，所以水平井位于油层中部时的内阻为：

分支水平井的总产量为：

考虑渗透率各向异性时，所有垂直方向的长度均应乘以 $\beta = \sqrt{\dfrac{K_h}{K_v}}$，而水平方向长度不变，此时井由半径为 $r_w$ 的圆形变长半轴为 $\beta r_w$、短半轴为 $r_w$ 的椭圆，有效半径 $r'_w$ 为二者的平均值，即：

多分支水平井在各向异性地层中的产量公式为：

同样条件下，水平井产量 $Q_h$ 与直井产量 $Q_v$ 比为：

$$\frac{Q_h}{Q_v} = \frac{\ln \frac{r_e}{r_w}}{\ln\left[\frac{2r_e}{l} + \sqrt{\left(\frac{2r_e}{l}\right)^2 - 1}\right] + \frac{\beta h}{l}\ln \frac{\beta h}{\pi(1+\beta)r_w}} \qquad (3-76)$$

Generally, the change of radius after transformation is not taken into account; the reference value of production ratio is given in the Table 3-2.

一般不考虑变换后半径的变化，表 3-2 给出了产量比的参考值。

Table 3-2　The production ratio of horizontal well and vertical well
表 3-2　水平井与直井产量比

| Dimensionless length $\frac{l}{r_e}$ | The output ratio of horizontal well and vertical well ||||
|---|---|---|---|---|
| | branch number $n=1$ | branch number $n=2$ | branch number $n=3$ | branch number $n=4$ |
| 0.05 | 1.384 | 1.845 | 2.027 | 2.117 |
| 0.10 | 1.825 | 2.422 | 2.636 | 2.731 |
| 0.20 | 2.430 | 3.277 | 3.576 | 3.697 |
| 0.30 | 2.884 | 4.044 | 4.448 | 4.605 |
| 0.40 | 3.332 | 4.815 | 5.351 | 5.555 |
| 0.50 | 3.787 | 5.635 | 6.334 | 6.599 |

**Exercises**

3-1　There are one source and one sink in planar infinite formation with distance $2\sigma$ each other and intensity $q$. Please prove by the analytical method that the absolute value of the seepage velocity of an arbitrary point in formation is $v = q\sigma/(\pi r_1 r_2)$, $r_1$ and $r_2$ is the distance of an arbitrary point in formation to the two wells.

3-2　Please solve the moving law of fluid mass point along the connection of source and sink in the question above that is the relationship of time and distance.

3-3　Before the well $A_2$ is put into production, well $A_1$ has been put into production, the distance between the two wells is $2\sigma = 100\text{m}$, for well $A_1$, $p_{w1} = 4\text{MPa}$, the radius of two wells is $r_{w1} = r_{w2} = 7.5\text{m}$, $r_e = 15\text{km}$, $p_e = 6\text{MPa}$. Please solve the problem: when well $A_1$ stops producing what is $p_{w2}$ in well $A_2$ then?

3-4　There is a circular supply boundary with $r_e = 10\text{km}$ in a given oil producing formation, a production well is drilled far from the center of formation with distance $d = 2\text{km}$, $r_w = 10\text{cm}$, $h = 5\text{m}$, $K = 0.5\mu\text{m}^2$, $p_e = 25\text{MPa}$, $p_w = 23\text{MPa}$, $\mu = 2\text{mPa}\cdot\text{s}$. Please solve the following problems: (1) The production of oil well. (2) Assuming that the oil well lies in

**习　题**

3-1　平面无穷大地层上有一源一汇，相距 $2\sigma$，强度为 $q$，试用分析法证明地层任一点处的渗流速度的绝对值为 $v = q\sigma/(\pi r_1 r_2)$，$r_1$、$r_2$ 为地层中任意一点到两口井的距离。

3-2　求液体质点沿上题的源汇连线运动的规律，即时间与距离的关系。

3-3　在 $A_2$ 井投产前，$A_1$ 井已经投产，两口井间距离 $2\sigma = 100\text{m}$，$A_1$ 井的 $p_{w1} = 4\text{MPa}$，两井半径 $r_{w1} = r_{w2} = 7.5\text{cm}$，$r_e = 15\text{km}$，$p_e = 6\text{MPa}$。求 $A_2$ 的 $p_{w2}$ 为多少时，$A_1$ 井停止生产。

3-4　某产油层有 $r_e = 10\text{km}$ 的圆形供给边线，距地层中心 $\sigma = 2\text{km}$ 处钻了一口生产井，$r_w = 10\text{cm}$，$h = 5\text{m}$，$K = 0.5\mu\text{m}^2$，$p_e = 25\text{MPa}$，$p_w = 23\text{MPa}$，$\mu = 2\text{mPa}\cdot\text{s}$。(1) 求油井产量。(2) 假设油井位于地

the centre of formation with other parameters stay unchanged, what is the production?

3-5 The distance from a given well to linear supply boundary is $a=1\text{km}$, $h=8\text{m}$, $K=0.3\mu\text{m}^2$, $\mu_0=4\text{mPa}\cdot\text{s}$, $r_w=0.1\text{m}$, the producing pressure drop is $\Delta p=2\text{MPa}$. Please solve the following problems: (1) The production of oil well. (2) If the oil well lies in the centre of circular supply boundary line with $r_e=1\text{km}$ and other parameters stay unchanged, what is the production?

3-6 There are two impermeable faults with intersection angle 30°, on their angle bisector there is a production well with the distance $R$ from the apex. Assuming that there is a circular supply boundary away the intersection point of faults with the distance $r_e$ and $r_e \gg r_w$, please solve the production formula of this well.

3-7 Assuming that the diameter of semicircular supply boundary is an impermeable boundary, and there is a production well with radius $r_w$ and bottom pressure $p_w$ on the vertical line with impermeable boundary and passing through the centre of supply boundary. The distance from the well to impermeable boundary is $d$, the supply radius is $r_e$, the supply pressure is $p_e$, the permeability of formation is $K$, the effective thickness is $h$, the viscosity of fluid is $\mu$. Please solve the production formula of oil well.

3-8 Assuming that there are two point sinks with uniform intensity $q$ and distance $2a$ each other in planar homogeneous isopachous infinite formation. Please solve the moving law of fluid mass point on the connection of two point sinks.

3-9 Please try to prove that the solutions to one source and one sink and two sinks in planar infinite formation meets the Laplace's equation.

3-10 There is an injection well in the centre of circular supply boundary and a production well away from the centre with distance $d=200\text{m}$. Assume the radius of supply boundary is $r_e=1\text{km}$, the supple pressure is $p_e=20\text{MPa}$, the fluid injection of injection well is the same as the production of production well, the radius of bottom is $r_w=0.1\text{m}$, $K=0.4\mu\text{m}^2$, the viscosity of fluid is $\mu=2\text{mPa}\cdot\text{s}$, the formation height is $h=4\text{m}$, if the production of production

3-5 某井距直线供给边线的距离 $a=1\text{km}$, $h=8\text{m}$, $K=0.3\mu\text{m}^2$, $\mu=4\text{mPa}\cdot\text{s}$, $r_w=0.1\text{m}$, 生产压差 $\Delta p=2\text{MPa}$。(1)求油井产量。(2)若井位于 $r_e=1\text{km}$ 的圆形供给边线中心, 其余参数不变, 油井产量为多少?

3-6 两不渗透断层, 交角为30°, 在它们的角平分线上有一口生产井距离顶点为 $R$, 假设离断层交点为 $r_e$ 处, 有一圆形供给边界, 且 $r_e \gg r$, 求这口井的产量。

3-7 设半圆形供给边线的直径为不渗透边界, 在通过供给边线的中心且与不渗透边线垂直的垂线上有一口井半径为 $r_w$、井底压力为 $p_w$ 的生产井, 该井到不渗透边界的距离为 $d$, 供给半径为 $r_e$, 供给压力为 $p_e$, 地层渗透率为 $K$, 有效厚度为 $h$, 流体黏度 $\mu$, 求油井的产量公式。

3-8 设平面均质、等厚、无穷大地层中有两个相距 $2a$ 强度均为 $q$ 的汇点, 试求两汇连线上液体质点的运动规律。

3-9 试证明平面无穷大地层上一源一汇和两汇的解满足拉普拉斯方程。

3-10 圆形供给边线中心有一口注水井, 距中心 $d=200\text{m}$ 处有一生产井, 设供给边线的半径为 $r_e=1\text{km}$, 供给压力为 $p_e=20\text{MPa}$, 注水井的注水量与生产井的产量相等, 井底半径为 $r_w=0.1\text{m}$, $K=0.4\mu\text{m}^2$, 流体黏度 $\mu=2\text{mPa}\cdot\text{s}$, 地层厚度 $h=4\text{m}$,

well is $Q = 30\text{m}^3/\text{d}$. Please solve the following problems: (1) What the bottom pressure of the two wells respectively are? (2) Write out the pressure distribution formula in the formation.

若生产井的产量$Q = 30\text{m}^3/\text{d}$。
(1) 两井的井底压力各为多少？
(2) 写出该地层内的压力分布公式。

Chapter 3 Study Guide
第三章学习指南

# 4 Unsteady seepage of slightly compressible fluid

微可压缩液体的不稳定渗流

# 第四章知识图谱

## 微可压缩液体的不稳定渗流

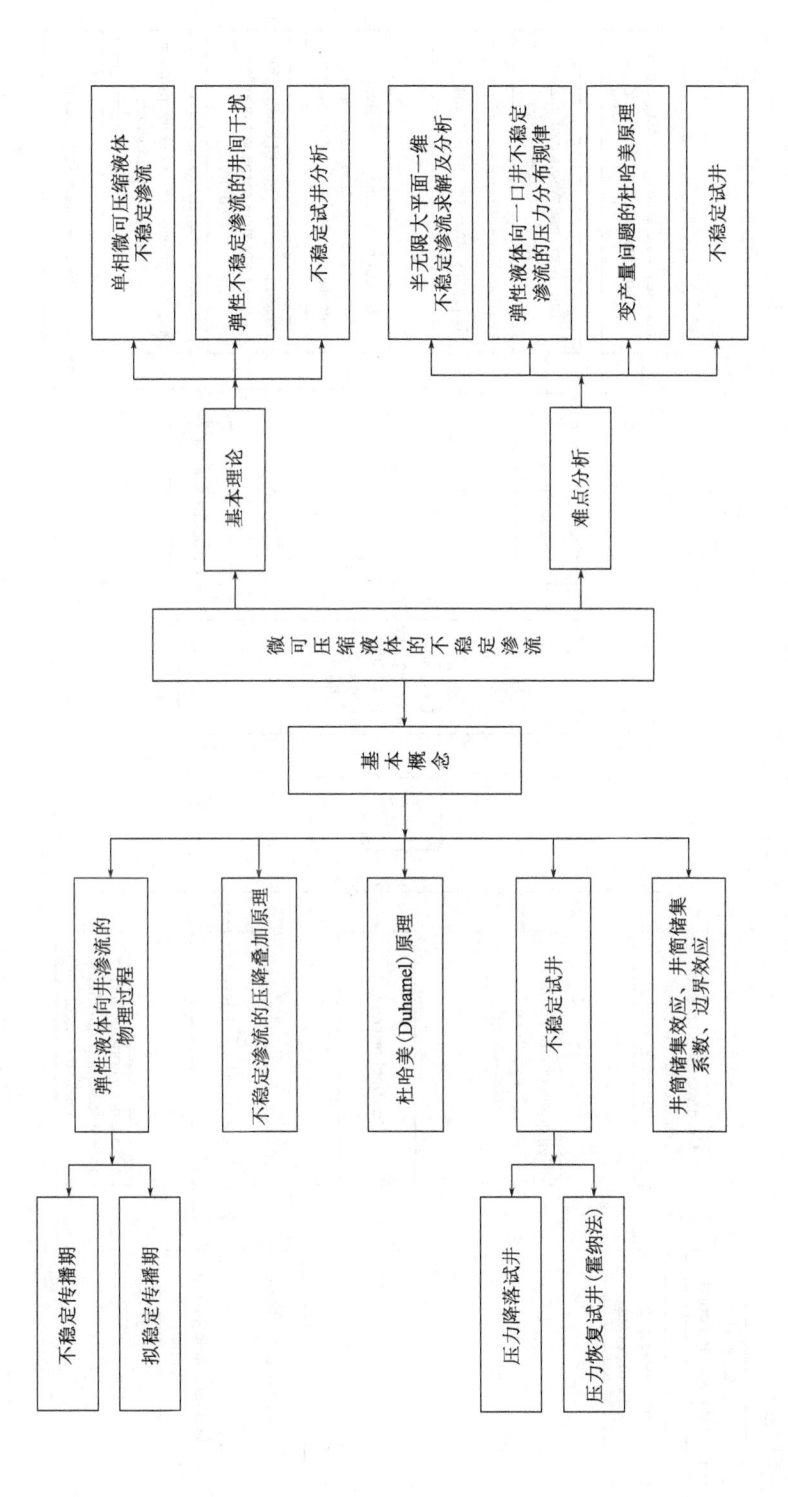

### 基本理论
- 单相微可压缩液体不稳定渗流
- 弹性不稳定渗流的井间干扰
- 不稳定试井分析

### 难点分析
- 半无限大平面一维不稳定渗流求解及分析
- 弹性液体向一口井不稳定渗流的压力分布规律
- 变产量问题的杜哈美原理
- 不稳定试井

### 基本概念
- 弹性液体向井渗流的物理过程
  - 不稳定传播期
  - 拟稳定传播期
- 不稳定渗流的压降叠加原理
- 杜哈美（Duhamel）原理
- 不稳定试井
  - 压力降落试井
  - 压力恢复试井（霍纳法）
- 井筒储集效应、井筒储集系数、边界效应

Chapter 4 Knowledge Graph

As to the reservoirs with closed boundary without other energy supply or the supply boundary is very far and without enough edge water, the fluid is driven to the wells mainly through the elastic of the formation and the fluid itself. Because reservoirs always lie in the deeper formation which bears higher pressure, although the compressibility of the formation and the fluid is not big, but they are compressible. It is determined that the elastic action is not to be neglected within the time after switching wells and changing the work. So the seepage law under elastic drive mode must be researched. The method obtaining the parameters of the reservoir with the theory of the elastic unsteady seepage as its foundation is called transient (unsteady state) well testing. By the method of unsteady well testing, the parameters of the formation can be determined, the formation pressure and the boundary of reservoirs can be calculated, and the behavior of reservoir can be analyzed.

## 4.1 The physical process of elastic fluid seepage towards wells

Because the formation and the fluid in it are all of definite compressibility, when a well is just put into production, its driving force mainly comes from the elastic expansion of the formation and the fluid nearby. Initially, only the fluid in a small area near a well flows towards the well bottom. With time passing by, the influencing area of the well becomes bigger and bigger, and the funnel of depression formed near the production well will be enlarged and deepened as shown in Figure 4 - 1 (Dynamic graph 4 - 1). This period is called the first period of pressure transient, that is, the unsteady propagation period. At this time, within the area of pressure drop, because different point is of different pressure, so the fluid flows towards the well bottom under the pressure difference. While beyond the area of pressure drop, the fluid does not move for lack of the pressure difference.

在边缘封闭没有外来能量补充或供给边缘很远、边水补充不济的油田,主要依靠地层和液体本身的弹性作用将液体驱向井中。由于油气藏一般深埋地下,承受压力较高,地层及液体虽然压缩性不大但均是可压缩的。这样就决定了在开关井和改变工作制度后的一段时间内,弹性作用不容忽视。因此,必须研究弹性驱动方式下的渗流规律。以弹性不稳定渗流理论为基础反求油气藏参数的方法称为不稳定试井方法。运用不稳定试井可以确定地层参数,推算地层压力、油气藏边界,分析油田的动态变化等。

## 4.1 弹性液体向井渗流的物理过程

由于地层及其中所含的流体都有一定的压缩性,当井以定产量刚投产时,其驱动力主要来自井附近的地层及其中所含液体的弹性膨胀。最初只有井附近小范围内的液体向井底流动,随着时间的推移,井所影响的范围越来越大,在生产井附近形成的压降漏斗将逐渐扩大并加深,如图4-1(动态图4-1)所示。这一阶段称为压力传导的第一阶段,即不稳定传播期。此时在压降范围内,由于不同点的压力不一样,因此,液体在压差作用下流向井底。而在压降范围外,由于没有压差,液体不流动。

After the pressure drop transmits to the supply boundary with constant pressure, as the bottom pressure decreases, the pressure gradient near boundary constantly increases tends to steady state finally as the increase of fluid supplied from the boundary into the formation.

If the production well near the supply boundary maintains constant bottom pressure after putting into production, initially, only the fluid near the well flows towards the well bottom which is of small resistance and high production. With the expansion of influencing area, the pressure difference is unchanged, the resistance increases, and the production decreases with time. The situation of pressure transient in formation is shown as Figure 4 – 2 (Dynamic graph 4 – 2). In the first phase, the pressure depression funnel increases gradually and tends to steady state after attaining the fixed constant pressure boundary.

当压力降传到压力为恒定的供给边界后，随着井底压力下降，边缘附近压力梯度不断增大，通过边界往地层中补充的液体逐渐增加最终趋于稳定。

若供给边界附近的生产井投产后保持恒定的井底压力，最初只是井附近的液体流向井底，阻力小、产量高。随着影响范围的扩大，压差不变，阻力增大，产量随时间延长而减小。地层中压力传导状态如图 4 – 2（动态图 4 – 2）所示。第一阶段压降漏斗逐渐扩大，传到定常压力边界后逐渐趋于稳定。

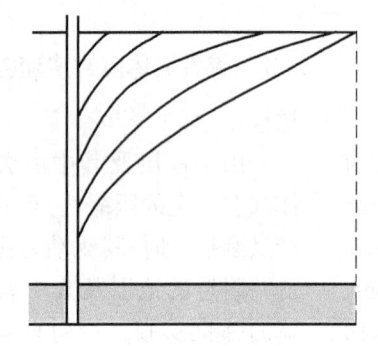

Figure 4 – 1  The curve of pressure distribution producing by single constant rate of well in constant pressure boundary

图 4 – 1  定压边界，井以定产量生产的压力分布曲线

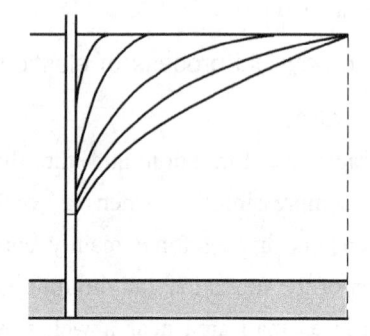

Figure 4 – 2  The curve of pressure distribution producing by constant pressure of well with constant pressure boundary

图 4 – 2  定压边界，井以恒定压力生产的压力分布曲线

Dynamic graph 4 – 1  The curve of pressure distribution producing by single constant rate of well in constant pressure boundary

动态图 4 – 1  定压边界，井以定产量生产的压力分布曲线

Dynamic graph 4 – 2  The curve of pressure distribution producing by constant pressure of well with constant pressure boundary

动态图 4 – 2  定压边界，井以恒定压力生产的压力分布曲线

If the production well lies near the closed boundary (fault), when maintaining constant production, the first phase of pressure transient (Figure 4-3、Dynamic graph 4-3) is the same as that shown in Figure 4-1, that is, the properties of boundary have no influence on the first phase of pressure transient. After the pressure is transmitted to the closed boundary, because there is no other energy supply, the bottom pressure and the pressure on outer boundary decrease gradually. After a while, when the pressure decreasing velocity at an arbitrary point on boundary and in formation are the same, it enters the second phase of pressure transient which is called the pseudo-steady state of seepage. Of course, there is a transitional period between the propagation period and pseudo-steady state period. If the well maintains constant bottom pressure, the two phases of pressure transient are shown in Figure 4-4(Dynamic graph 4-4). In practice, the compressibility is not taken into account under the condition of constant boundary pressure. The production is unchanged, the pressure is also unchanged and vice versa, so the flow is steady state. While if compressibility is taken into account, the production of the oil well and bottom pressure either or both vary with time, so seepage is unsteady. The unsteady seepage law of planar one-dimensional in semi-infinite formation and that of a point sink in infinite formation will be carefully introduced below.

## 4.2　The planar one-dimensional unsteady seepage in semi-infinite formation

It has been known that the change of formation pressure reflects the seepage process in reservoirs, so the dynamic analysis of reservoirs needs to know the distribution of formation pressure under definite condition and the changing law of bottom pressure.

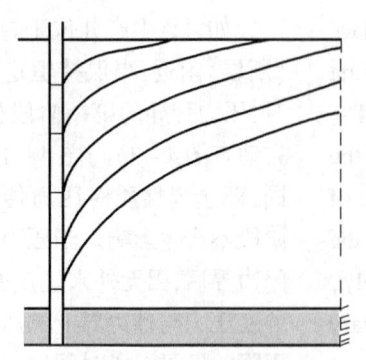

Figure 4-3  The curve of pressure distribution producing with constant rate in closed boundary

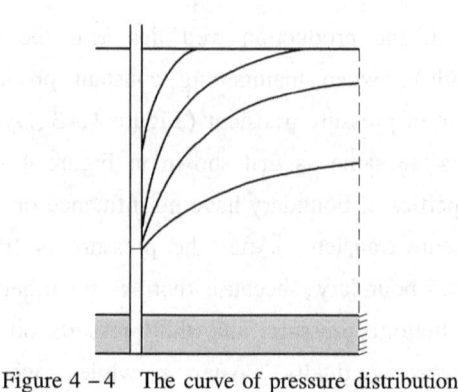

Figure 4-4  The curve of pressure distribution producing with constant pressure in closed boundary

Dynamic graph 4-3  The curve of pressure distribution producing with constant rate in closed boundary

Dynamic graph 4-4  The curve of pressure distribution producing with constant pressure in closed boundary

## 4.2.1 Unsteady seepage with constant bottom pressure

As to the semi-infinite strip reservoir with constant cross-section area $A$ as shown in Figure 4-5, assume that fluid flows out from one end, the formation is homogeneous, isopachous and isotropic, fluid and formation are slightly compressible, seepage obeys the Darcy's law, and the pressure gradient is small. So the flow of fluid in formation is planar one-dimensional unsteady state seepage which obeys the following equation:

$$\frac{\partial^2 p}{\partial x^2} = \frac{1}{\eta}\frac{\partial p}{\partial t} \tag{4-1}$$

If the researched well produces with constant bottom pressure, then the corresponding initial condition, inner and outer boundary condition are:

$$p(x,0) = p_i \quad (0 \leqslant x < +\infty) \tag{4-2}$$

$$p(0,t) = p_w \tag{4-3}$$

$$p(\infty,t) = p_i \tag{4-4}$$

The Boltzmann transformation is used to solve the equation (4-1) that is making:

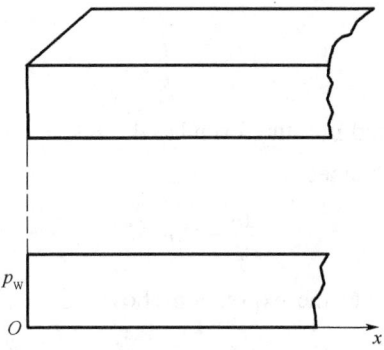

Figure 4-5  Planar one-dimensional elastic unsteady state seepage

图 4-5  平面一维弹性不稳定渗流

$$\xi = \frac{x}{2\sqrt{\eta t}}$$

Make pressure $p$ to be the only function of $\xi$, while $\xi$ is the function of $x$ and $t$.

According to the differential rule of compound function, the following expressions can be obtained:

使压力 $p$ 仅为 $\xi$ 的函数,而 $\xi$ 是 $x$ 和 $t$ 的函数。

按照复合函数的微分法则,有:

$$\frac{\partial p}{\partial x} = \frac{\mathrm{d}p}{\mathrm{d}\xi}\frac{\partial \xi}{\partial x} = \frac{1}{2\sqrt{\eta t}}\frac{\mathrm{d}p}{\mathrm{d}\xi}$$

$$\frac{\partial^2 p}{\partial x^2} = \frac{1}{4\eta t}\frac{\mathrm{d}^2 p}{\mathrm{d}\xi^2} \tag{4-5}$$

$$\frac{\partial p}{\partial t} = -\frac{x}{4t\sqrt{\eta t}}\frac{\mathrm{d}p}{\mathrm{d}\xi} \tag{4-6}$$

The formula (4-5) and formula (4-6) are substituted into the equation (4-1), the following expression can be obtained:

将式(4-5)、式(4-6)代入方程(4-1)中,得:

$$\frac{\mathrm{d}^2 p}{\mathrm{d}\xi^2} + 2\xi\frac{\mathrm{d}p}{\mathrm{d}\xi} = 0 \tag{4-7}$$

After transformation, the partial differential equation (4-1) describing seepage law is changed into the ordinary differential equation (4-7), and its corresponding initial and boundary conditions are transformed to be the following forms:

经过变换,描述渗流规律的偏微分方程(4-1)转化成了常微分方程(4-7),其相应的初始条件和边界条件也可转化成如下形式:

$$p\mid_{\xi \to \infty} = p_\mathrm{i}; p\mid_{\xi = 0} = p_\mathrm{w}$$

Make the depression of order of ordinary differential equation (4-7) that is making:

将二阶常微分方程(4-7)降阶,即令:

$$U = \frac{\mathrm{d}p}{\mathrm{d}\xi} \tag{4-8}$$

Then the formula (4-7) is changed to be:

则方程(4-7)变为:

$$\frac{\mathrm{d}U}{\mathrm{d}\xi} + 2\xi U = 0$$

Its solution is:

$$U = C_1 e^{-\xi^2} \qquad (4-9)$$

The expression above is substituted into the formula (4 – 8), the flowing expression can be obtained:

$$\frac{dp}{d\xi} = C_1 e^{-\xi^2} \qquad (4-10)$$

Separate the variables and integrate the expression above, $\xi$ is $0\to\infty$ and $p$ is $p_w \to p_i$, then:

$$p_i - p_w = C_1 \int_0^\infty e^{-\xi^2} d\xi \qquad (4-11)$$

Because $\int_0^\infty e^{-\xi^2} d\xi = \frac{\sqrt{\pi}}{2}$, so the following expression can be obtained:

$$C_1 = \frac{2}{\sqrt{\pi}}(p_i - p_w)$$

If the integrating interval of $\xi$ is changed from $0\to\infty$ to $\xi \to \infty$, then after separating the variables and integrating the equation (4 – 10), the following expression can be obtained:

$$p_i - p(x,t) = C_1 \int_\xi^\infty e^{-\xi^2} d\xi = \frac{2(p_i - p_w)}{\sqrt{\pi}} \left( \int_0^\infty e^{-\xi^2} d\xi - \int_0^\xi e^{-\xi^2} d\xi \right)$$

$$= (p_i - p_w)\left( 1 - \frac{2}{\sqrt{\pi}} \int_0^\xi e^{-\xi^2} d\xi \right)$$

In the expression above: $\frac{2}{\sqrt{\pi}} \int_0^\xi e^{-\xi^2} d\xi$ is called the error function or probability integral (with known function table), make:

$$\frac{2}{\sqrt{\pi}} \int_0^\xi e^{-\xi^2} d\xi = \mathrm{erf}\left( \frac{x}{2\sqrt{\eta t}} \right)$$

Because of $0 \le \mathrm{erf}\left(\frac{x}{2\sqrt{\eta t}}\right) \le 1$, the value of pressure at an arbitrary point on arbitrary time in formation meets (Figure 4 – 6、Dynamic graph 4 – 5):

$$p_i - p(x,t) = (p_i - p_w)\left[ 1 - \mathrm{erf}\left( \frac{x}{2\sqrt{\eta t}} \right) \right] \qquad (4-12)$$

The value of error function $\mathrm{erf}\left(\frac{x}{2\sqrt{\eta t}}\right)$ changes with $\frac{x}{2\sqrt{\eta t}}$. $\mathrm{erf}(z)$ can be expanded to be the following series:

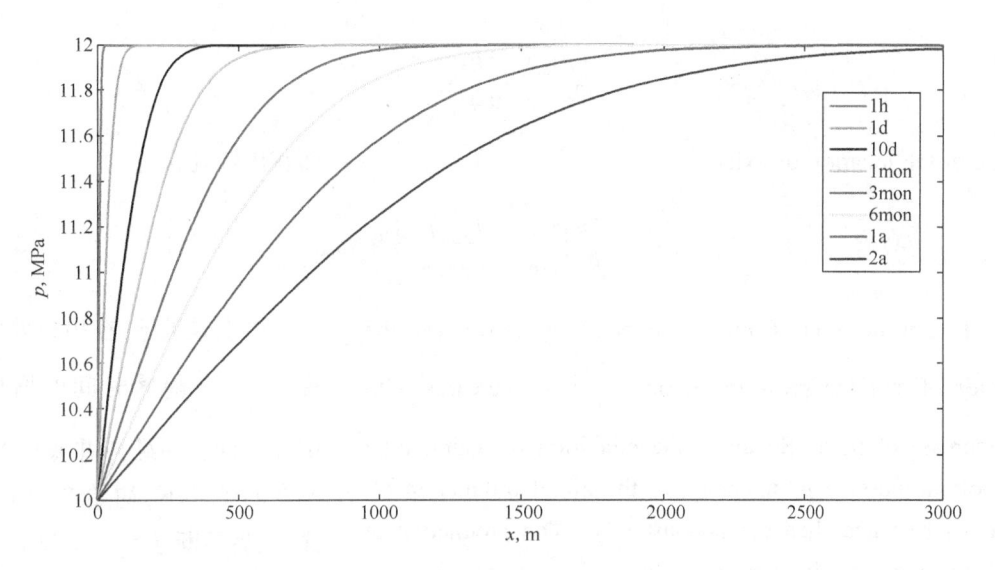

Figure 4 – 6  Pressure Vs xray Plot
图 4 – 6  地层任一点任一时刻的压力

Dynamic graph 4 – 5  The pressure distribution curve of unstable seepage flow with constant bottom hole pressure
动态图 4 – 5  井底压力为定值的不稳定渗流压力分布曲线

$$\mathrm{erf}(z) = \frac{1}{\sqrt{\pi}}\left(z - \frac{z^3}{1!\ 3} + \frac{z^5}{2!\ 5} - \frac{z^7}{3!\ 7} + \cdots\right) = \frac{2}{\sqrt{\pi}}\sum_{n=0}^{\infty}(-1)^n \frac{z^{2n+1}}{n!\ (2n+1)}$$

It is not hard to compile a calculating program according to the expression above and obtain the value of erf($z$) with computer. It can be proved that the value of error function erf($z$) increases with the increase of value of $z$. Therefore, as to the semi-infinite planar one-dimensional elastic unsteady seepage, when the pressure of delivery end is constant, it can be known from the formula (4 – 12), at the same time, the farther from the exit, the bigger the value of erf($z$) is and the smaller the value of pressure drop is at this point. While as to a given location, the longer the time of producing, the smaller the value of erf($z$) and so the bigger the value of pressure drop of this point which is apparent.

The magnitude of seepage velocity of fluid at any time and location in formation can also be obtained form the formula (4 – 12). After partially differentiate both sides of the formula (4 – 12) with respect to $x$, the following expression can be obtained:

根据上式不难编一计算程序，用计算机求得 erf($z$)值。可以证明误差函数值 erf($z$)随着 $z$ 值的增大而增大。所以，对于半无限大平面一维弹性不稳定渗流，当出口端压力恒定时，由式（4 – 12）可知，在同一时刻，离出口处越远，erf($z$)越大，该点的压力下降值越小。而对于某一确定的位置，生产时间越长，erf($z$)则越小，因而该点的压降值越大，这是显而易见的。

由式（4 – 12）还可以求出地层内任一时刻任一位置处流体渗流速度的大小。将式（4 – 12）两边对 $x$ 求偏导数可得：

$$\frac{\partial p}{\partial x} = \frac{p_i - p_w}{\sqrt{\pi \eta t}} e^{-\frac{x^2}{4\eta t}}$$

While at the location of exit:

而在出口处，有：

$$\left.\frac{\partial p}{\partial x}\right|_{x=0} = \frac{(p_i - p_w)}{\sqrt{\pi \eta t}}$$

It can be seen from the expression above: at the location of exit the pressure gradient $\left.\frac{\partial p}{\partial x}\right|_{x=0}$ decreases with the increase of time. So under the condition of maintaining the bottom pressure $p_w$ unchanged, the initial production $Q_0$ is the highest and then down constantly. The production at the location of exit is (Figure 4-7):

由上式可见，在出口处压力梯度 $\left.\frac{\partial p}{\partial x}\right|_{x=0}$ 随着时间的增加而下降，因此，在维持井底压力 $p_w$ 不变的条件下，初始产量 $Q_0$ 最高，然后不断递减。出口处的产量为（图4-7）：

$$Q = \frac{K}{\mu} A \left.\frac{\partial p}{\partial x}\right|_{x=0} = \frac{KA(p_i - p_w)}{\mu \sqrt{\pi \eta t}}$$

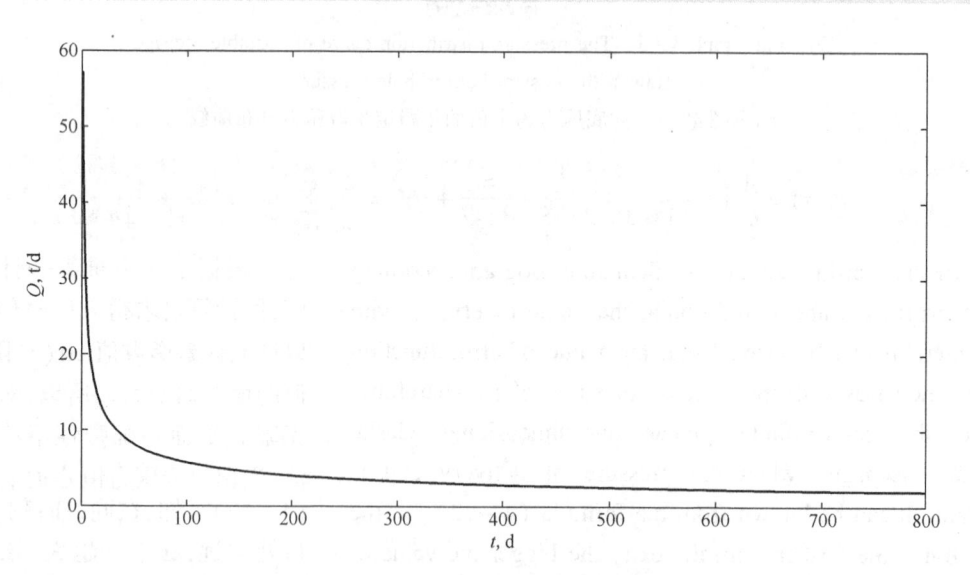

Figure 4-7  Production Vs Time Plot
图 4-7  出口处的产量

Make $l = \sqrt{\pi \eta t}$ which can be considered as the pressure swept scope. Apparently, the formula above is similar to the production formula of planar one-dimensional steady state seepage of fluid, merely the $l$ here is a variable.

令 $l = \sqrt{\pi \eta t}$，可以认为它是压力波及的范围。很明显，上式与液体作平面一维稳定渗流时的产量公式是相类似的，只不过这里的 $l$ 是个变量罢了。

## 4.2.2 Unsteady state seepage when production is constant

The planar one-dimensional unsteady state seepage law in semi-infinite formation when the pressure at the exit is constant is talked about before. When the production of exit is constant $Q_0$, the seepage of fluid in formation as shown in Figure 4-5 still obeys the equation (4-1) and the corresponding initial condition formula (4-2) and the outer boundary condition (4-4), the difference is that the inner boundary condition is changed to be:

$$Q_0 = \frac{K}{\mu} A \frac{\partial p}{\partial x}\bigg|_{x=0} = \text{constant}$$

Because the boundary condition is changed, it is very troublesome to find the solution of the new question by finding the solution of the partial differential equation. So it is wise to deform the formula (4-1) that is partially differentiating with respect to $x$ of two sides of the formula (4-1) and multiply $\frac{KA}{\mu}$, and then the following expression can be obtained:

$$\frac{KA}{\mu} \frac{\partial}{\partial x}\left(\frac{\partial^2 p}{\partial x^2}\right) = \frac{1}{\eta} \frac{KA}{\mu} \frac{\partial}{\partial x}\left(\frac{\partial p}{\partial t}\right)$$

According to the properties of partial differentiation of continuous function, the following expression can be obtained:

$$\frac{\partial^2}{\partial x^2}\left(\frac{KA}{\mu} \cdot \frac{\partial p}{\partial x}\right) = \frac{1}{\eta} \frac{\partial}{\partial t}\left(\frac{KA}{\mu} \frac{\partial p}{\partial x}\right)$$

That is:

$$\frac{\partial^2 Q}{\partial x^2} = \frac{1}{\eta} \frac{\partial Q}{\partial t} \tag{4-13}$$

The corresponding initial, inner and outer boundary conditions are:

$$Q(x,0) = 0 \tag{4-14}$$
$$Q(0,t) = Q_0 \tag{4-15}$$
$$Q(\infty,t) = 0 \tag{4-16}$$

By comparing the formula (4-13) to formula (4-16) with the formula (4-1) to formula (4-4), it can be known that the solution of unsteady seepage with constant production is:

## 4.2.2 产量恒定时的不稳定渗流

以上讨论了出口处压力为定值时,半无限大地层平面一维不稳定渗流规律。当出口处产量为定值$Q_0$时,图4-5所示的地层中的液体渗流仍服从方程(4-1)和相应的初始条件式(4-2)及外边界条件式(4-4),所不同的是内边界条件变成

由于边界条件发生了变化,因此要求得这一新问题的解如果重新从解偏微分方程入手,就显得十分麻烦。为此,不妨先将方程(4-1)作一些变形,即将方程(4-1)两边同时对$x$求偏导数,并乘以$\frac{K}{\mu}A$,得:

根据连续函数求偏导数的性质有:

即:

相应的初始及内、外边界条件可以写成:

比较式(4-13)~式(4-16)与式(4-1)~式(4-4)可知,恒定产量不稳定渗流的解为:

$$Q(x,t) = Q_0\left[1 - \mathrm{erf}\left(\frac{x}{2\sqrt{\eta t}}\right)\right] \qquad (4-17\mathrm{a})$$

Or (Figure 4-8、Dynamic graph 4-6)　　　　　　或(图4-8、动态图4-6)

$$\frac{Q(x,t)}{Q_0} = 1 - \mathrm{erf}\left(\frac{x}{2\sqrt{\eta t}}\right) \qquad (4-17\mathrm{b})$$

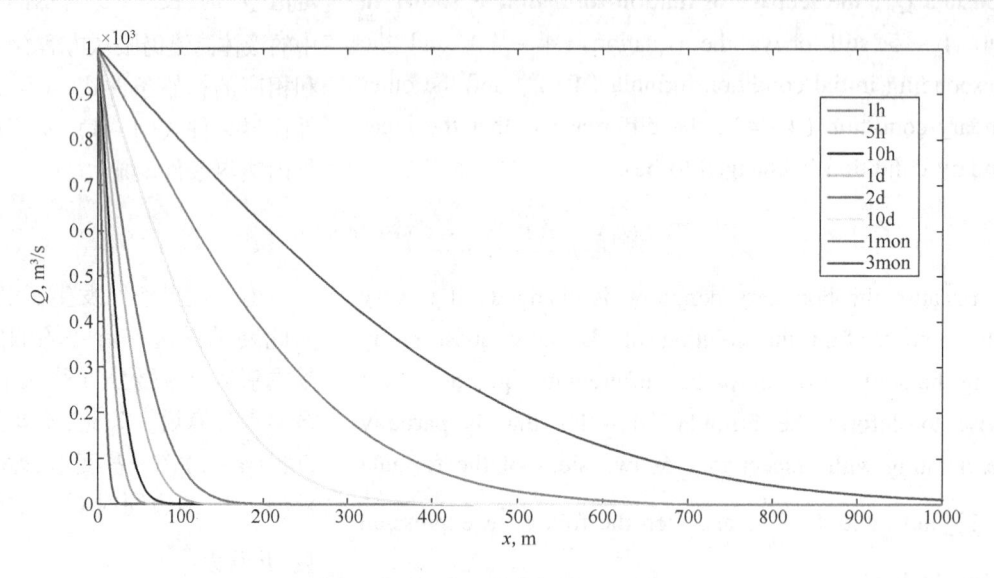

Figure 4-8　Production Vs xray Plot

图4-8　地层内任一点任一时刻的产量

Dynamic graph 4-6　The production curve of unsteady seepage

动态图4-6　不稳定渗流产量分布曲线

It can be known from the formula (4-17), at the same time, the farther from the exit is, the bigger the value of error function $\mathrm{erf}\left(\frac{x}{2\sqrt{\eta t}}\right)$ is, so the flow rate passing through the cross section is smaller. At a given definite location, with the increase of time, the value of error function $\mathrm{erf}\left(\frac{x}{2\sqrt{\eta t}}\right)$ decreases, so the flow rate passing through an arbitrary cross section increases with time. Besides, the formula (4-17) can also be written as:

由式(4-17)可知,在同一时刻离出口处越远,误差函数值$\mathrm{erf}\left(\frac{x}{2\sqrt{\eta t}}\right)$越大,因而通过该截面的流量越小。而对于某一确定的位置,随着时间的增大,误差函数值$\mathrm{erf}\left(\frac{x}{2\sqrt{\eta t}}\right)$减小,因此通过任一断面的流量是随时间增大的。此外,式(4-17)还可以写作:

$$Q = \frac{KA}{\mu} \cdot \frac{\partial p}{\partial x} = Q_0\left[1 - \mathrm{erf}\left(\frac{x}{2\sqrt{\eta t}}\right)\right]$$

As to an arbitrary time $t$, pressure $p$ only has a relationship with location $x$, so after separating the variables of the expression above and integrating it and using limit changing method of multiple integral, the formula of pressure distribution at arbitrary time can be obtained that is (Figure 4-9):

对于任一确定的时刻 $t$，压力 $p$ 仅与位置 $x$ 有关，所以对上式分离变量积分，使用重积分换限法，可得任一时刻压力分布公式，即（图4-9）：

$$p(x,t) - p_w(t) = \frac{Q_0 \mu x}{KA}\left[1 - \mathrm{erf}\left(\frac{x}{2\sqrt{\eta t}}\right) + \frac{2\sqrt{\eta t}}{x\sqrt{\pi}}(1 - e^{-\frac{x^2}{4\eta t}})\right]$$

$$p_i - p_w(t) = \frac{2Q_0 \mu}{AK}\sqrt{\frac{\eta t}{\pi}}$$

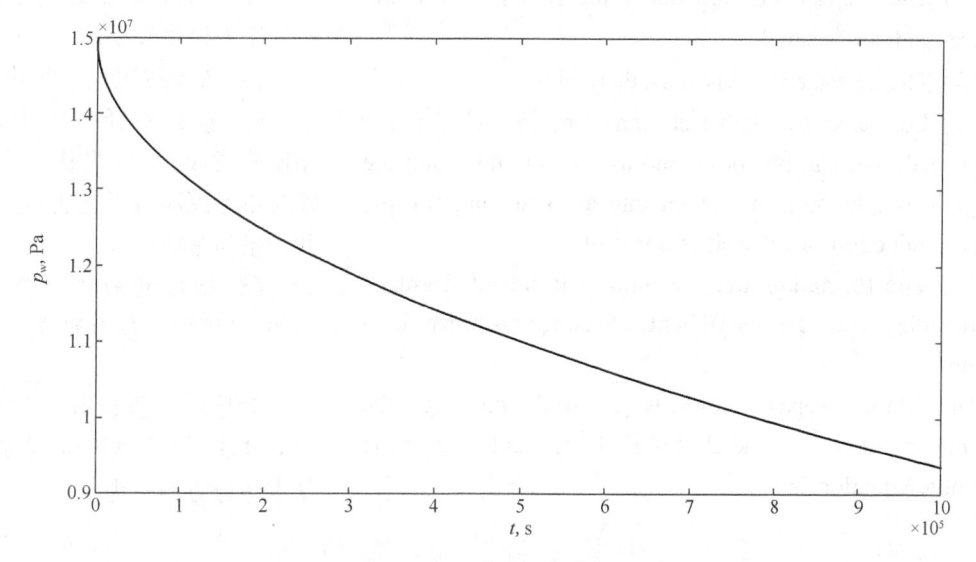

Figure 4-9　$p_w$ Vs Time Plot

图4-9　任一时刻的压力

As to steady flow, the seepage law is the same, no matter if the production is known or the bottom pressure is known or not. While as to unsteady seepage, there is apparent difference between the two.

The solution of planar one-dimensional unsteady seepage of semi-infinite formation is obtained by the Boltzmann transformation above, the same result can also be obtained if the Laplace transformation is adopted.

对于稳定流动，无论是产量已知还是井底压力已知，渗流规律是一样的，对于不稳定流动，二者有明显区别。

以上通过 Boltzmann 变换得到了半无限地层平面一维不稳定渗流的解，若采用拉普拉斯变换，可以获得相同的结果。

## 4.3　Pressure transient law of elastic fluid unsteady seepage towards wellbore

## 4.3　弹性液体向一口井不稳定渗流的压力传导规律

As to a relatively big reservoir, because there are a small number of wells at the beginning of the development, so the boundary of the reservoir and the interference of

如果在一个较大的油田，由于开发初期井数较少，因此，油田边界和井间干扰可以暂不考

wells will not be taken into account in the beginning. No matter how big the well spacing is, before the pressure drop transmit to the boundary, the influence of boundary is very small. So as to an arbitrary well, it can be similarly considered as the condition that there is a well in infinite formation in transient phase. In order to obtain the widely used solution in reservoir engineering, there is no harm to make the following assumptions:

(1) The formation is homogeneous isopachous, isotropic and infinite in which there is only one well producing.

(2) The seepage of fluid obeys the Darcy's law and pressure gradient is small.

(3) The seepage process is isothermal.

(4) Compared with the whole formation, the well diameter is very small and can be considered as a point sink, and the seepage incompletion is not taken into account temporarily, and the production of oil well is constant.

(5) The formation and the fluid in it are all slightly compressible, and the coefficient of compressibility is a constant.

The fluid seepage towards a well meeting the assumptions above is radial fluid flow, and the basic differential equation is:

$$\frac{\partial^2 p}{\partial x^2} + \frac{\partial^2 p}{\partial y^2} = \frac{1}{\eta}\frac{\partial p}{\partial t}$$

If the equation is expressed with polar coordinates, the following expression can be obtained:

$$\frac{\partial^2 p}{\partial r^2} + \frac{1}{r}\frac{\partial p}{\partial r} = \frac{1}{\eta}\frac{\partial p}{\partial t} \qquad (4-18)$$

Because the production of oil well is constant, the corresponding initial and boundary conditions are:

$$p(r,0) = p_i \quad (0 \leqslant r \leqslant +\infty)$$
$$p(\infty, t) = p_i$$
$$Q = \frac{2\pi Kh}{\mu}\left(r\frac{\partial p}{\partial r}\right)\bigg|_{r\to 0} = \text{constant} \quad (t>0)$$

In order to solve the equation (4-18), the Boltzmann transformation is used that is assuming:

$$\xi = \frac{r}{2\sqrt{\eta t}}$$

Make $p$ only the function of $\xi$, according to the rule of derivation of compound function:

使 $p$ 仅为 $\xi$ 的函数，按复合函数求导法则，有：

$$\frac{\partial p}{\partial t} = \frac{dp}{d\xi}\frac{\partial \xi}{\partial t} = -\frac{\xi}{2t}\frac{dp}{d\xi}$$

$$\frac{\partial p}{\partial r} = \frac{dp}{d\xi}\frac{\partial \xi}{\partial r} = \frac{1}{2\sqrt{\eta t}}\frac{dp}{d\xi}$$

$$\frac{\partial^2 p}{\partial r^2} = \frac{1}{4\eta t}\frac{d^2 p}{d\xi^2}$$

Substitute $\frac{\partial p}{\partial t}, \frac{\partial p}{\partial r}$ and $\frac{\partial^2 p}{\partial r^2}$ into the formula (4-18), the following expression can be obtained:

将 $\frac{\partial p}{\partial t}$、$\frac{\partial p}{\partial r}$ 和 $\frac{\partial^2 p}{\partial r^2}$ 代入式(4-18)得：

$$\frac{d^2 p}{d\xi^2} + \left(\frac{1}{\xi} + 2\xi\right)\frac{dp}{d\xi} = 0 \qquad (4-19)$$

So the partial differential formula (4-18) is changed into the ordinary differential formula (4-19). Make $U = \frac{dp}{d\xi}$ and substitute it into the formula (4-19), then the following expression can be obtained:

通过变换，将偏微分方程(4-18)变成常微方程(4-19)。进一步令 $U = \frac{dp}{d\xi}$ 并代入方程(4-19)得：

$$\frac{dU}{d\xi} + \left(\frac{1}{\xi} + 2\xi\right)U = 0$$

After separating the variables of the formula above and integrating it, the following expression can be obtained:

将上式分离变量积分得：

$$\ln U = -\ln \xi - \xi^2 + \ln C_1$$

In the formula above, $C_1$ is the integrating constant. After collating the formula above, the following expression can be obtained:

式中，$C_1$ 为积分常数。将上式合并整理得：

$$U = C_1 \frac{e^{-\xi^2}}{\xi} \qquad (4-20)$$

Substitute $U = \frac{dp}{d\xi}$ into the formula above, the following expression can be obtained:

将 $U = \frac{dp}{d\xi}$ 代入上式，得：

$$\frac{dp}{d\xi} = C_1 \frac{e^{-\xi^2}}{\xi} \qquad (4-21)$$

According to the inner boundary condition:

根据内边界条件：

$$Q = \frac{2\pi Kh}{\mu}\left(r\frac{\partial p}{\partial r}\right)\bigg|_{r\to 0} = \frac{2\pi Kh}{\mu}r\frac{dp}{d\xi}\frac{1}{2\sqrt{\eta t}}\bigg|_{r\to 0} = \frac{2\pi Kh}{\mu}\left(\xi\frac{dp}{d\xi}\right)\bigg|_{\xi\to 0}$$

That is:

即：

$$\xi \frac{\mathrm{d}p}{\mathrm{d}\xi}\bigg|_{\xi\to 0} = \frac{\mu Q}{2\pi Kh} \qquad (4-22)$$

After multiplying $\xi$ to both sides of formula (4-21) and taking the limit $\xi \to \infty$, and then using the formula (4-22), the following expression can be obtained:

$$C_1 = \frac{\mu Q}{2\pi Kh} \qquad (4-23)$$

Substitute the constant $C_1$ into the formula (4-21), separate the variables of the equation and integrate it, the interval of $\xi$ is $\xi \to \infty$, to $p$ is $p(r,t) \to p_\mathrm{i}$, so the following expression can be obtained:

$$p_\mathrm{i} - p(r,t) = \frac{\mu Q}{2\pi Kh}\int_\xi^\infty \frac{\mathrm{e}^{-\xi^2}}{\xi}\mathrm{d}\xi \qquad (4-24)$$

Make $x = \xi^2$, then $\xi = \sqrt{x}$ and $\mathrm{d}\xi = \frac{1}{2\sqrt{x}}\mathrm{d}x$; and when the changing interval of $\xi$ is $\frac{r}{2\sqrt{\eta t}} \to \infty$, of $x$ is $\frac{r^2}{4\eta t} \to \infty$, substitute the two changing intervals into the expression above, the following expression can be obtained:

$$p(r,t) = p_\mathrm{i} - \frac{\mu Q}{4\pi Kh}\int_{\frac{r^2}{4\eta t}}^\infty \frac{\mathrm{e}^{-x}}{x}\mathrm{d}x \qquad (4-25)$$

The formula (4-25) is the formula of pressure distribution of a point sink producing with constant production in infinite formation. In the formula above, $\int_{\frac{r^2}{4\eta t}}^\infty \frac{\mathrm{e}^{-x}}{x}\mathrm{d}x$ is an exponential integral function, and make:

$$\int_{\frac{r^2}{4\eta t}}^\infty \frac{\mathrm{e}^{-x}}{x}\mathrm{d}x = -\mathrm{Ei}\left(-\frac{r^2}{4\eta t}\right)$$

The functional value of exponential integral function $-\mathrm{Ei}\left(-\frac{r^2}{4\eta t}\right)$ changes with $\frac{r^2}{4\eta t}$ and the curve is shown as the Figure 4-10.

So the expression (4-25) can also be written as the following form (Figure 4-11、Dynamic graph 4-7):

$$p_\mathrm{i} - p(r,t) = -\frac{\mu Q}{4\pi Kh}\mathrm{Ei}\left(-\frac{r^2}{4\eta t}\right) \qquad (4-26)$$

It can be known from the formula (4-26) and Figure 4-10, when $x$ increases ($r$ increases or $t$ decreases), $-\mathrm{Ei}(-x)$ decreases and $\Delta p(r,t)$ decreases, that is to say, the farther from the well or the shorter the producing time is, the smaller the pressure drop is. It is apparent and it is vice versa.

间越短,压降越小,这是显而易见的,反之亦然。

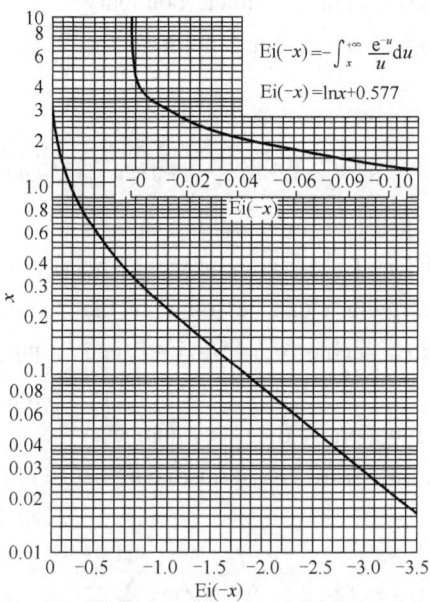

Figure 4-10　The curve of exponential integral function

图 4-10　指数积分函数曲线

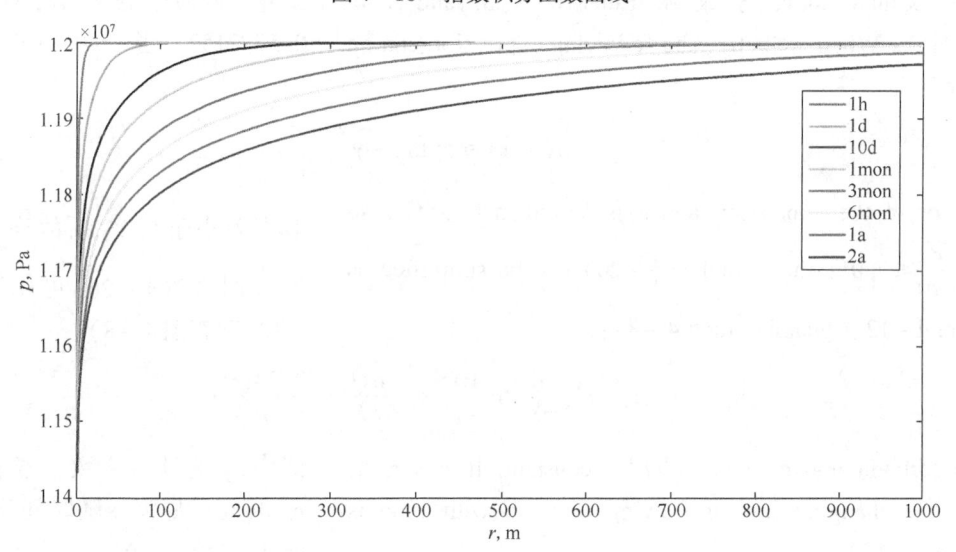

Figure 4-11　Pressure Vs Radius Plot

图 4-11　地层任一点任一时刻的压力

Dynamic graph 4-7　The pressure distribution curve of unstable seepage with constant production

动态图 4-7　产量恒定时的不稳定渗流压力分布曲线

If the researched subject is an injection well, because its only difference of the working condition with that of production well is the negative sign in inner boundary condition, so the pressure drop of an arbitrary point in the formation is:

$$p_i - p(r,t) = \frac{\mu Q}{4\pi Kh} \text{Ei}\left(-\frac{r^2}{4\eta t}\right) \quad (4-27)$$

It is apparent that the pressure drop caused by an injection well $\Delta p = p_i - p(r,t)$ is a negative. The relationship data of $-\text{Ei}(-x)$ with $x$ can be found in the widely used application software (such as the MATLAB and Maple).

The exponential integral function $-\text{Ei}(-x)$ can also be expressed with the following series, so it also can be calculated with programming:

$$-\text{Ei}(-x) = -\ln x - \gamma + \sum_{n=1}^{\infty} \frac{(-1)^{n-1} x^n}{n \cdot n!}$$

In the formula above, $\gamma$ is an Euler constant and $\gamma = 0.5772157$. When $x \leq 0.01$, the following expression can be obtained:

$$-\text{Ei}(-x) \approx -\ln x - \gamma$$

The error of the expression above is less than 0.25%, so when $\frac{r^2}{4\eta t} \leq 0.01$, the formula (4-26) can be simplified as (Figure 4-12、Dynamic graph 4-8):

$$p_i - p(r,t) = \frac{\mu Q}{4\pi Kh} \ln \frac{4\eta t}{e^\gamma r^2} \approx \frac{\mu Q}{4\pi Kh} \ln \frac{2.245\eta t}{r^2}$$

In the formula above, $\gamma$ is Euler's constant. If $r = r_w$ is taken, the changing law of flowing pressure with time is (Figure 4-13):

$$p_i - p_w(t) = \frac{\mu Q}{4\pi Kh} \ln \frac{2.245\eta t}{r_w^2} \quad (4-28)$$

If the oil well is an imperfect well, the radius $r_w$ should be replaced with the reduced radius of oil well. If the influences of formation damage or stimulation treatment are taken into account further, the skin factor $S$ should also be included. Because the relationship between reduced radius and actual radius is:

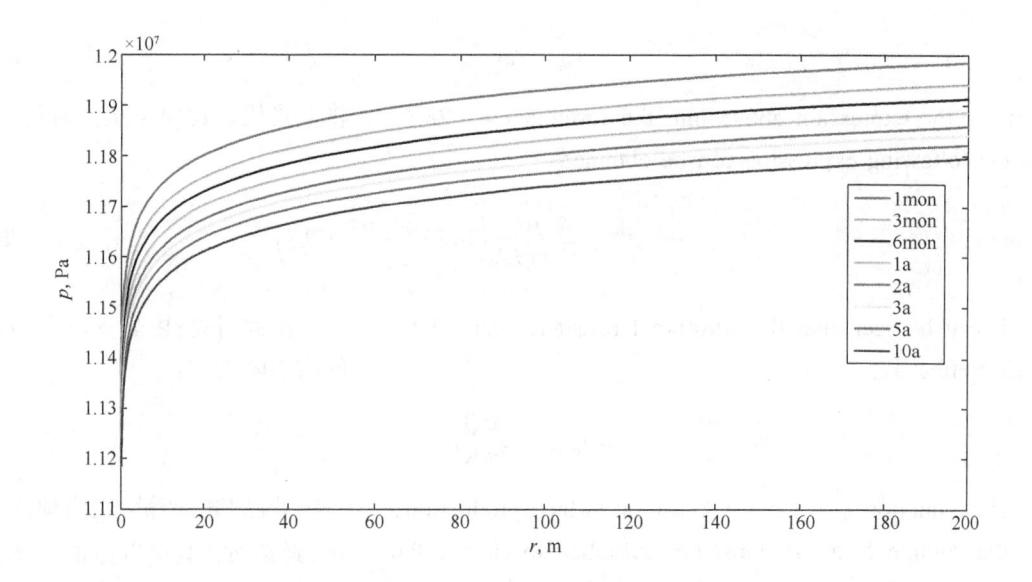

Figure 4-12　Pressure Vs Radius Plot
图 4-12　任一点任一时刻的压力

Dynamic graph 4-8　The pressure distribution
curve of injection well
动态图 4-8　注水井的压力分布曲线

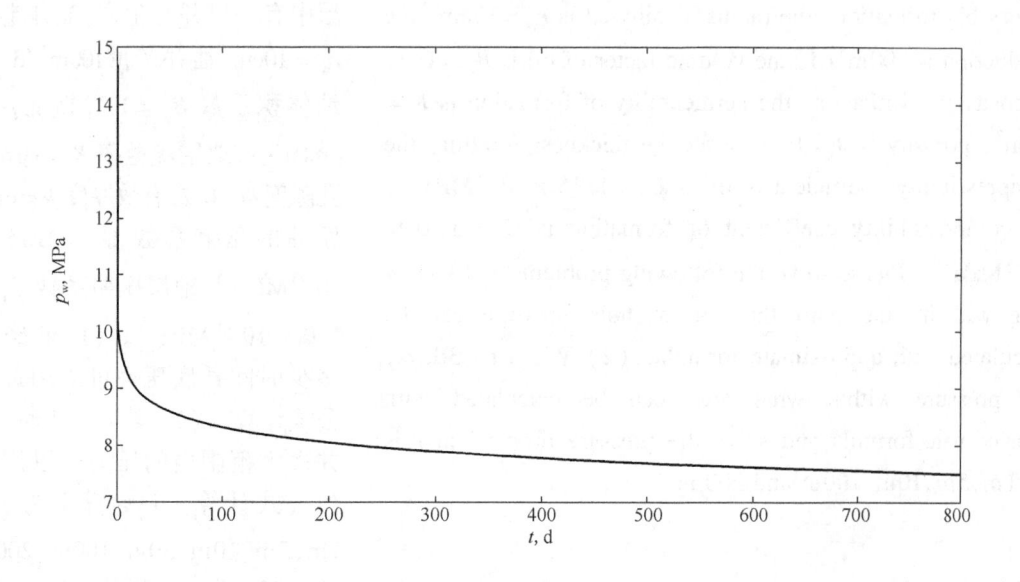

Figure 4-13　Flowing Pressure Vs Time Plot
图 4-13　任一时刻的井底流压

$$r_{we} = r_w e^{-S}$$

Substitute the expression above into the formula (4 – 28), then the following expression can be obtained:

$$p_i - p_w(t) = \frac{\mu Q}{4\pi Kh}\left(\ln\frac{2.245\eta t}{r_w^2} + 2S\right) \quad (4-29)$$

It can be seen that the additional resistance caused by the skin effect is:

$$\Delta p_{skin} = \frac{\mu Q}{2\pi Kh}S$$

If production $Q_{sc}$ that we know is surface production, then the volume factor $B_o$ must be multiplied to change the surface production to be the subsurface flow rate. So the formula (4 – 26) and the formula (4 – 29) are respectively changed to be:

$$p_i - p(r,t) = -\frac{\mu B_o Q_{sc}}{4\pi Kh}\mathrm{Ei}\left(-\frac{r^2}{4\eta t}\right)$$

$$p_i - p_w(t) = -\frac{\mu B_o Q_{sc}}{4\pi Kh}\left(\ln\frac{2.245\eta t}{r_w^2} + 2S\right)$$

**Example 4 – 1** Assume that there is a well producing in a very big formation, the radius of oil well is $r_w = 10\text{cm}$, the production is $100\text{m}^3/\text{d}$, the volume factor of oil is $B_o = 1.2$, viscosity $\mu = 4\text{mPa}\cdot\text{s}$, the permeability of formation is $K = 1\mu\text{m}^2$, porosity is $\phi = 0.2$, effective thickness $h = 6\text{m}$, the compressibility coefficient of oil is $C_o = 1.15\times10^{-3}\text{MPa}^{-1}$, the compressibility coefficient of formation is $C_f = 1.0\times10^{-4}\text{MPa}^{-1}$. Please solve the following problems: (1) How long will it take until the bottom hole pressure can be calculated with approximate formula. (2) When $t = 3\text{days}$, the pressure within what area can be calculated with approximate formula and solve the pressure drop when $r$ is $r_w$, 1m, 5m, 10m, 100m and 200m.

**Solution:** (1) The total compressibility coefficient is:

$$C_t = C_o + C_f = 1.25 \times 10^{-3} (\text{MPa}^{-1})$$

$$\eta = \frac{K}{\phi \mu C_t} = \frac{1 \times 10^{-12}}{0.2 \times 4 \times 10^{-3} \times 1.25 \times 10^{-9}} = 1 (\text{m}^2/\text{s})$$

When $\dfrac{r_w^2}{4\eta t} = \dfrac{0.1^2}{4 \times 1 \times t} < 0.01$ that is $t > 0.25$s, the bottom pressure can be calculated with approximate formula.

当 $\dfrac{r_w^2}{4\eta t} = \dfrac{0.1^2}{4 \times 1 \times t} < 0.01$ 时，即 $t > 0.25$s 时，井底压力即可用近似公式计算。

(2) When $t = 3$d, the pressure must meet the following condition if it can be calculated with approximate formula

(2) 当 $t = 3$d 时，若要用近似公式计算压力必须满足：

$$\frac{r^2}{4\eta t} = \frac{r^2}{4 \times 1 \times 3 \times 86400} \leq 0.01$$

It is precise enough to calculate pressure (pressure drop) with approximate formula in the area of $r \leq 101.82$m. Because

即 $r \leq 101.82$m 的范围内用近公式计算压力（压降）已足够准确。由于

$$\frac{\mu B_o Q_{sc}}{4\pi K h} = \frac{4 \times 10^{-9} \times 1.2 \times 100/86400}{4 \times \pi \times 1 \times 10^{-12} \times 6} = 0.07368 (\text{MPa})$$

So the following expression can be obtained in the area of $r < 101.82$m：

所以在 $r < 101.82$m 范围内有：

$$\Delta p = 0.07368 \ln \frac{2.245 \eta t}{r^2} = 0.07368 \ln \frac{2.245 \times 3 \times 86400}{r^2} = 0.07368 \ln \frac{5.819 \times 10^5}{r^2}$$

While when $r > 101.82$m：

而当 $r > 101.82$ 时：

$$\Delta p = -0.07368 \text{Ei}\left(-\frac{r^2}{4\eta t}\right) = -0.07368 \text{Ei}\left(-\frac{r^2}{1.0386 \times 10^6}\right)$$

The results are shown in Table 4-1.

计算结果列于表 4-1。

**Table 4-1 The results**

**表 4-1 计算结果**

| $r$, m | 0.1 | 1 | 5 | 10 | 50 | 100 | 200 |
|---|---|---|---|---|---|---|---|
| $\Delta p$, MPa | 1.3173 | 0.9780 | 0.7409 | 0.6387 | 0.4016 | 0.2994 | 0.1974 (approximant 0.1973) |

This example further illustrates that the bottom hole pressure can be calculated with the approximate formula when oil wells are put into production shortly. For the purpose of convenience, the formula (4-26) is always written as the dimensionless form that is:

$$r_D = \frac{r}{r_w}; \quad p_D = \frac{2\pi Kh(p_i - p)}{\mu Q}; \quad t_D = \frac{\eta t}{r_w^2} = \frac{Kt}{\phi\mu C_t r_w^2}$$

In the formulas above, $r_D$, $p_D$ and $t_D$ are respectively called the dimensionless distance, dimensionless pressure and dimensionless time. So the formula (4-26) and the formula (4-29) can be written as:

$$p_D(r_D, t_D) = -\frac{1}{2}\mathrm{Ei}\left(-\frac{r_D^2}{4t_D}\right)$$

$$p_{wD}(t_D) = \frac{1}{2}\left(\ln\frac{4t_D}{\gamma} + 2S\right)$$

This formula of pressure distribution expressed with dimensionless form does not relate to the concrete properties of the formation and the fluid. This formula only shows the relationship of the dimensionless variables, and the relationship of the dimensionless variables is unique, which has nothing to do with unit system. So it is especially convenient and widely used in unsteady well testing. It will not have a further discussion in this course.

## 4.4 Approximate solution of pressure change of fluid seepage towards well in finite closed elastic formation

The point sink solution in planar infinite formation is suitable for the first phase of pressure transient. While after the pressure drop transmit to the boundary, the point sink solution will not suit any more. As to the reservoir with finite boundary, it is very troublesome to solve the heat conduction equation. The similar solution of pressure distribution in the pseudo-steady period of reservoir with closed boundary will be introduced below.

Assume that the radius of circular closed reservoir is $r_e$ and there is a production well working with constant production in the center of the reservoir. After a long time production, the pressure drop has been transmitted to the boundary of the reservoir, the oil flow afterwards mainly depends on the elastic energy of the formation and the fluid within the boundary. If the production is unchanged, the decline rates of average formation pressure and the pressure on every point are definitely unchanged, that is, at the given time after pressure drop is transmitted to the boundary, the funnels of depression are parallel, the curve of pressure distribution at different time is parallel with each other, this is called the pseudo-steady seepage as shown in Figure 4-14. Assume that the initial formation pressure is $p_i$, current average formation pressure is $\bar{p}$, and then the total liquid drove out by elasticity is $V$, apparently the following expression can be obtained:

$$V = \phi \pi h (r_e^2 - r_w^2) C_t (p_i - \bar{p})$$

(a) The figure of perssure distribution when well with constant production in closed boundary
(a)封闭边界井以定产量生产时压力分布图

(b) The curve of pressure changes with time in the pseudo-steady period
(b)拟稳态阶段压力随时间变化曲线

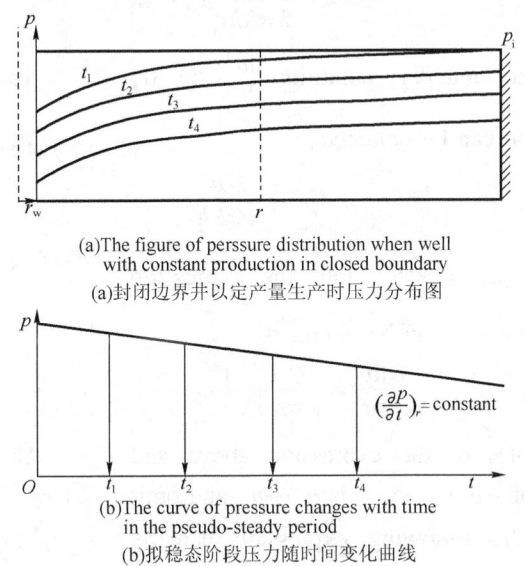

Figure 4-14  The figure of pressure change in pseudo-steady state
图 4-14  拟稳态压力变化图

Differentiate both sides of the formula with respect to $t$, the following expression can be obtained:

$$Q = \frac{dV}{dt} = -\phi \pi h C_t (r_e^2 - r_w^2) \frac{d\bar{p}}{dt}$$

Because $r_e^2 \gg r_w^2$, the following expression can be obtained from the expression above:

$$\frac{d\bar{p}}{dt} = -\frac{Q}{\phi \pi h C_t r_e^2}$$

When pressure drop is transmitted to the boundary and enters the pseudo-steady flow, the decline rate of pressure of every point in formation and that of the average formation pressure are equal, so that the following expression can be obtained:

$$\frac{\partial p}{\partial t} = -\frac{Q}{\phi \pi h C_t r_e^2} \qquad (4-30)$$

Because the seepage obeys the formula (4-19), the basic differential equation of pseudo-steady flow can be obtained by substituting the formula (4-30) into the formula (4-19), that is:

由于液体渗流服从方程(4-19),因此将方程(4-30)代入方程(4-18)可得拟稳态流动的基本微分方程:

$$\frac{\partial^2 p}{\partial r^2} + \frac{1}{r}\frac{\partial p}{\partial r} = -\frac{\mu Q}{\pi K h r_e^2} \qquad (4-31)$$

As to an arbitrary time $t$, the pressure in formation is only relevant to the distance $r$, so the partial derivative in the formula (4-31) can be written as total derivative. After separating the variables and integrating it, the following expression can be obtained:

对于任一确定的时刻 $t$, 地层中的压力只与位置 $r$ 有关, 因而方程(4-31)中的偏导数可以写成全导数, 并将其分离变量积分得:

$$r\frac{\mathrm{d}p}{\mathrm{d}r} = -\frac{Q\mu}{2\pi K h r_e^2}r^2 + C_1$$

According to the outer boundary condition $\left.\dfrac{\mathrm{d}p}{\mathrm{d}r}\right|_{r=r_e} = 0$, so the following expression can be obtained:

根据外边界条件 $\left.\dfrac{\mathrm{d}p}{\mathrm{d}r}\right|_{r=r_e} = 0$, 于是得:

$$C_1 = \frac{Q\mu}{2\pi K h}$$

Substitute the expression $C_1 = \dfrac{Q\mu}{2\pi K h}$ into the expression above, the following expression can be obtained:

并代入上式得:

$$\frac{\mathrm{d}p}{\mathrm{d}r} = \frac{Q\mu}{2\pi K h}\left(\frac{1}{r} - \frac{r}{r_e^2}\right)$$

Separate the variables of the expression above and integrate it, the interval of $r$ is $r_w \to r$, $p$ is $p_w \to p$, and omit the item $r_e^2/r_w^2$, then, the following expression can be obtained:

将上式分离变量积分, 即 $r$ 为 $r_w \to r$, $p$ 为 $p_{wf} \to p$, 且忽略 $r_w^2/r_e^2$ 项, 则:

$$p - p_{wf} = \frac{Q\mu}{2\pi K h}\left(\ln\frac{r}{r_w} - \frac{r^2}{2r_e^2}\right) \qquad (4-32)$$

If the integral interval is $(r, r_e)$, the interval of $p$ is $p \to p_e$, after integrating, the following expression can be obtained:

如果积分区间为 $[r, r_e]$, 则 $p$ 从 $p \to p_e$, 积分得:

$$p_e - p = \frac{Q\mu}{2\pi K h}\left(\ln\frac{r_e}{r} - \frac{r_e^2 - r^2}{2r_e^2}\right) \qquad (4-33)$$

The formula (4-32) and the formula (4-33) are the formulas of pressure distribution of pseudo-steady flow. Because $r_e^2 \gg r_w^2$, then the production formula of an oil well can also be further obtained. If the skin factor $S$ is taken into account, the formula of production is:

式(4-32)和式(4-33)就是拟稳定态流动的压力分布公式。由于 $r_e^2 \gg r_w^2$ 还可以进一步得到该油井的产量公式, 若考虑表皮系数 $S$, 则产量公式为:

$$Q = \frac{2\pi Kh(p_e - p_{wf})}{\mu\left(\ln\dfrac{r_e}{r_w} - \dfrac{1}{2} + S\right)} \qquad (4-34)$$

However, the boundary pressure in the expressions above is difficult to measure, so the boundary pressure is usually replaced with the average formation pressure (static pressure), and the average pressure is:

$$\bar{p} = \frac{1}{V_p}\iiint p\,dV_p \qquad (4-35)$$

Under the condition of radial flow, $V_p = \pi\phi h(r_e^2 - r_w^2)$, $dV_p = 2\pi\phi hr\,dr$ and $r_e^2 \gg r_w^2$, so the formula (4-35) can be written as:

$$\bar{p} = \frac{2}{r_e^2}\int_{r_w}^{r_e} pr\,dr$$

The following expression can be obtained after the formula (4-32) is substituted into the expression above:

$$\bar{p} = p_{wf} + \frac{Q\mu}{2\pi Kh}\frac{2}{r_e^2}\int_{r_w}^{r_e}\left(\ln\frac{r}{r_w} - \frac{r^2}{2r_e^2}\right)r\,dr$$

$$\int_{r_w}^{r_e}r\ln\frac{r}{r_w}dr = \left(\frac{r^2}{2}\ln\frac{r}{r_w}\right)\bigg|_{r_w}^{r_e} - \int_{r_w}^{r_e}\frac{r^2}{2}\frac{1}{r}dr \approx \frac{r_e^2}{2}\ln\frac{r_e}{r_w} - \frac{r_e^2}{4}$$

$$\int_{r_w}^{r_e}\frac{r^3}{2r_e^2}dr \approx \frac{r_e^2}{8}$$

Substitute the two integral values above into the formula (4-35), the following expression can be obtained:

$$\bar{p} = p_{wf} + \frac{\mu Q}{2\pi Kh}\left(\ln\frac{r_e}{r_w} - \frac{3}{4}\right) \qquad (4-36)$$

If the skin effect is taken into account, then the following expression can be obtained:

$$\bar{p} = p_{wf} + \frac{Q\mu}{2\pi Kh}\left(\ln\frac{r_e}{r_w} - \frac{3}{4} + S\right) \qquad (4-37)$$

The average formation pressure $\bar{p}$ can be ascertained by well testing. If the radius of reservoir $r_e$ is replaced with drainage area, the following expression can be obtained:

$$\bar{p} = p_{wf} + \frac{Q\mu}{4\pi Kh}\left(\ln\frac{4Ae^{-3/2}}{4\pi r_w^2} + 2S\right) \qquad (4-38)$$

In the formula above: $A = \pi r_e^2$, while $4\pi e^{3/2} = 32.6206e^\gamma$, $\gamma$ is an Euler's constant. Generally the shape of drainage area is not circular, so 31.6206 is replaced by shape factor $C_A$, so the formula (4-38) is changed to be:

$$\bar{p} = p_{wf} + \frac{Q\mu}{4\pi Kh}\left(\ln\frac{4A}{e^{\gamma}C_A r_w^2} + 2S\right) \qquad (4-39)$$

The shape factor $C_A$ in the formula above depends on the shape of reservoir boundary and the location of oil well. The corresponding value of $C_A$ can be ascertained from the Figure 4-15, the value of $C_A$ in the Figure is obtained by solving diffusion equation directly or using the image method.

式中，形状系数 $C_A$ 取决于不同的油气藏边界形状以及油井所处的位置。通过查图 4-15，就可确定相应的 $C_A$ 值，图中的 $C_A$ 值是直接解扩散方程或使用映射法得到的。

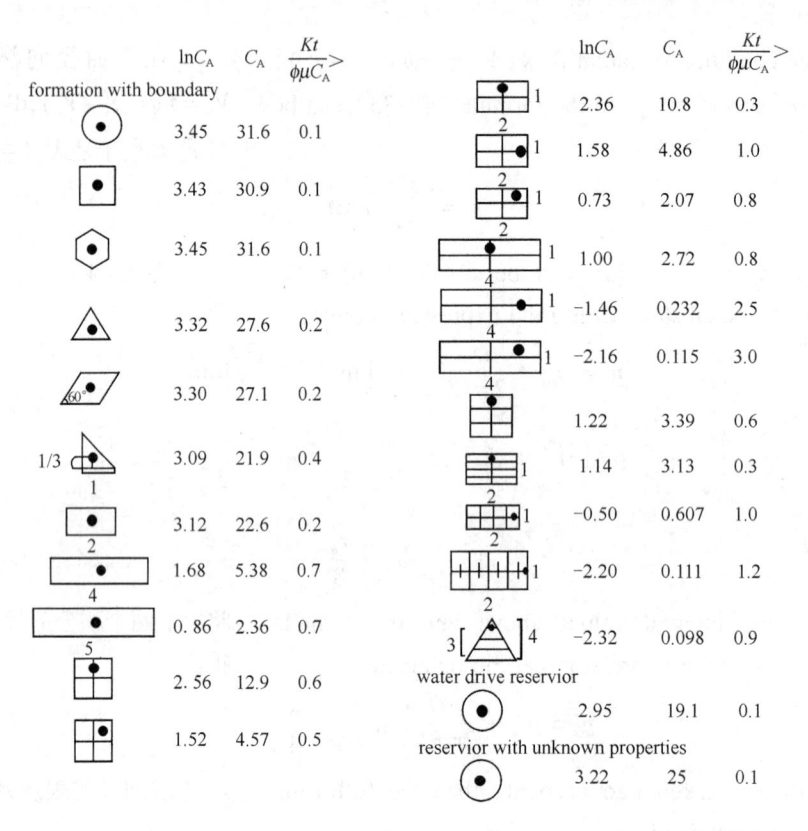

Figure 4-15　The value of $C_A$ of different well location and shape of boundary

图 4-15　不同边界形状及井径的 $C_A$ 值

According to the principle of material balance, the following expression can be obtained:

按照物质平衡原理有：

$$Qt = \pi r_e^2 \phi h C_t (p_i - \bar{p})$$

$$p_i - \bar{p} = \frac{Qt}{\pi r_e^2 \phi h C_t} \qquad (4-40)$$

The following expression can be obtained by combining the formula (4-39) and the formula (4-40):

联立式(4-39)和式(4-40)得：

$$\frac{2\pi Kh(p_i - p_{wf})}{\mu Q} = \frac{1}{2}\ln\frac{4A}{e^{\gamma}C_A r_w^2} + 2\pi\frac{\eta t}{A} + S \qquad (4-41)$$

The formula (4-41) reflects the changing law of pressure with time in the pseudo-steady phase. As to the reservoirs with closed boundary, the result shows that the transitional period is very short, so it can be approximately considered that the unsteady transient period is connected with the pseudo-steady state period directly. So after combining the formula (4-41) and the formula (4-29), the following expression can be obtained:

$$\frac{1}{2}\ln\frac{4A}{e^{\gamma}C_A r_w^2} + 2\pi\frac{\eta t}{A} = \frac{1}{2}\ln\frac{4\eta t}{e^{\gamma}r_w^2}$$

$$\frac{\eta t}{A}C_A = e^{\frac{4\pi\eta t}{A}}$$

Make $t_{DA} = \frac{\eta t}{A}$, then the following expression can be obtained:

$$t_{DA}C_A = e^{4\pi t_{DA}} \tag{4-42}$$

If the boundary of a reservoir is circular and the well lies in the geometric center of the reservoir, then it can be known that $C_A = 31.6206$ from the Figure 4-15. Substitute it into the formula (4-42), the following expression can be obtained:

$$t_{DA} \approx 0.1$$

According to this $t_{DA}$, the ending time of transient period or the beginning of pseudo-steady period can be ascertained. Apparently, the bigger the drainage area is, the smaller the diffusivity coefficient is, and the longer the period of transient period is. According to the data supplied in the Figure 4-15, the dimensionless transient time $t_{DA}$ of different shape of reservoir and well location can be calculated.

## 4.5 Multiple-well interference of elastic unsteady seepage

In an actual reservoir, there are always many wells working simultaneously, and the boundary of the reservoir is complicated, so the solutions of point (line) sink of a well working in infinite formation need to be spread into the multi-well system with complicated boundary.

## 4.5.1 Superposition principle of pressure drop

Under the assumption of the former section, two wells are put into production simultaneously at a given time, the distance between the two wells is $2a$, and the productions are $Q_1$ and $Q_2$ respectively. Now the pressure distribution is taken into consideration after wells are put into production. As shown in Figure 4-16 an appropriate coordinate system is chosen, in order to make sure the positions of two wells are $(a,0)$ and $(-a,0)$.

## 4.5.1 压降叠加原理

在上一节的假设下,在某时刻起同时有两口井投入生产,它们相距 $2a$,产量分别为 $Q_1$ 和 $Q_2$。现考虑投产后的压力分布。

如图 4-16 所示,选取适当的坐标系,使两井位置分别为 $(a,0)$ 和 $(-a,0)$。

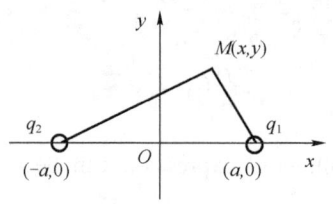

Figure 4-16 Interference of two wells in infinite formation
图 4-16 无穷大地层两口井干扰

If there is only a well with production $Q_1$ at location $(a,0)$ producing independently, then the pressure drop on an arbitrary point $M(x,y)$ in formation is:

若无穷大的地层中只有一口产量为 $Q_1$ 的井,位于 $(a,0)$,且单独生产,则地层中任一点 $M(x,y)$ 处的压力降为:

$$\Delta p_1 = -\frac{Q_1\mu}{4\pi Kh}\text{Ei}\left[-\frac{(x-a)^2+y^2}{4\eta t}\right]$$

If there is only a well with production $Q_2$ at location $(-a,0)$ producing independently, then the pressure drop of point $M$ is:

若无穷大地层中只有一口产量为 $Q_2$、井位 $(-a,0)$ 的井单独生产,则 $M$ 点的压降为:

$$\Delta p_2 = -\frac{Q_2\mu}{4\pi Kh}\text{Ei}\left[-\frac{(x+a)^2+y^2}{4\eta t}\right]$$

When the two wells work simultaneously with production $Q_1$ and $Q_2$, then the pressure drop on an arbitrary point $M(x,y)$ in the formation should be equal to the sum of pressure drop caused by the two wells on this point when they are working independently with own production, that is:

当以上两口井同时分别以产量 $Q_1$ 和 $Q_2$ 生产时,则地层中任一点 $M(x,y)$ 处的压力降应等于两井分别单独以各自的产量生产时在该点造成的压力降之和,即:

$$\Delta p = -\frac{Q_1\mu}{4\pi Kh}\text{Ei}\left[-\frac{(x-a)^2+y^2}{4\eta t}\right] - \frac{Q_2\mu}{4\pi Kh}\text{Ei}\left[-\frac{(x+a)^2+y^2}{4\eta t}\right] \quad (4-43)$$

Because the diffusion equation is linear, superposition is also suitable for a multi-well system. If there are $n$ wells working simultaneously in infinite elastic formation and the production of every well is $Q_i$, the production time is $\tau_i$, the

由于扩散方程是线性的,所以这种叠加方式同样也适合于多井系统。若无限大弹性地层中有 $n$ 口井同时生产,各井产量分别为 $Q_i$,投产时刻为 $\tau_i$,井的

well location is $(x_i, y_i)$, and $i = 1, 2, \ldots, n$, then the pressure drop on an arbitrary point $M(x, y)$ at any time in formation is:

$$\Delta p = -\frac{\mu}{4\pi Kh} \sum_{i=1}^{n} \pm Q_i \text{Ei}\left[-\frac{(x-x_i)^2 + (y-y_i)^2}{4\eta(t-\tau_i)}\right] \qquad (4-44)$$

As to oil well, positive sign should be chosen in front of $Q_i$ in formular (4-44), and negative sign is chosen in front of injection well.

### 4.5.2 Image method in the formation with linear boundary

As shown in Figure 4-17, if there are two wells working simultaneously in infinite formation, then the pressure drop on an arbitrary point $M(x, y)$ in the formation meets the formula (4-43). When $Q_1 = Q_2$ and the point $M$ is chosen on the axis $y$, and the seepage velocities caused by the two wells at the point $M$ are $v_1$ and $v_2$ respectively, while the seepage velocities in the direction $x$ are $v_{1x}$ and $v_{2x}$ respectively. It is easy to verify from the formula (4-43) that $\left.\frac{\partial p}{\partial x}\right|_{x=0} = 0$, that is:

$$v_{1x} = v_{2x}$$

That is to say that the axis $y$ is an impermeable boundary. So as to a well near linear impermeable boundary, it can still be transformed to be the problem of multi interference of wells in infinite formation with the method of sink point image introduced in the former section.

For the same reason, when $Q_1 = -Q_2$, and the axis $y$ is a constant pressure line. So the problem of a well working near linear supply boundary can be transformed to be the problem of multi interference of wells in infinite formation.

In short, the image method is also suitable for the problem of compressible fluid seepage with linear boundary which is not proved here.

**Example 4-2** A production well is set in the center of a closed rectangular area, the side lengths of which are $a$ and $b$

respectively. The production of well is $Q$, the initial formation pressure is $p_i$ and the volume factor of oil is $B_o$. Assume the other properties of formation and fluid are known, please solve the pressure distribution after the well is put into production.

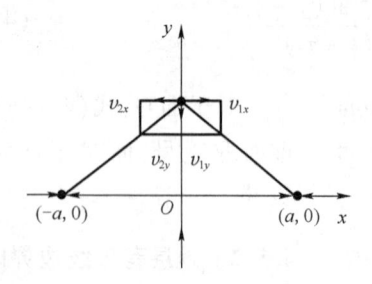

Figure 4 – 17   Two point-sink interference
in infinite formation
图 4 – 17   无穷大地层两汇干扰

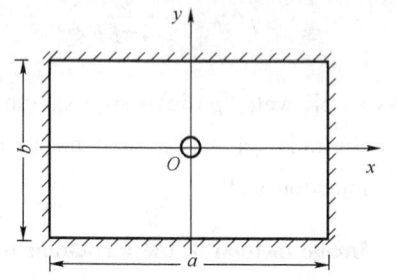

Figure 4 – 18   A well in the center of
closed rectangular reservoir
图 4 – 18   封闭矩形油藏中心一口井

**Solution**: Firstly, as to the reservoir, set up the coordinate shown in Figure 4 – 18 and transform it to be the production problem of infinite well-drain in infinite formation through the method of mirror. And the location of every well is ($\pm ma, \pm nb$) in which $m$ and $n$ respectively are integers of $-\infty \to +\infty$. It is easy to obtain the solution of this problem using superposition which is:

$$p(x,y,t) = p_i + \frac{Q_{sc}B_o\mu}{4\pi Kh}\sum_{n=-\infty}^{\infty}\sum_{m=-\infty}^{\infty}\text{Ei}\left[-\frac{(ma-x)^2+(nb-y)^2}{4\eta t}\right]$$

Because convergence of this series is rapid, when calculating the pressure distribution in the rectangular area $|x| \leq a/2$ and $|y| \leq b/2$, it only needs to consider the production well close to the rectangular area. For example:

$$p(x,y,t) = p_i + \frac{Q_{sc}B_o\mu}{4\pi Kh}\sum_{|n|+|m|\leq 2}\text{Ei}\left[-\frac{(ma-x)^2+(nb-y)^2}{4\eta t}\right]$$

It is generally enough.

## 4.5.3   Duhamel principle solving the problem of changing production

Let's think about a simple case like this: there is a production well in the infinite formation, the production is $Q_1$ from time $t=0$ to $t=t_0$, the production changes to be $Q_2$ from time $t=t_0$ (Figure 4 – 19), please solve the pressure distribution in the formation.

解:首先对该油气藏建立如图 4 – 18 所示的坐标,然后通过镜像反映法将其转化为无限大地层中无穷井排的生产问题,且各井井位为($\pm ma, \pm nb$),其中$m$、$n$是$-\infty \to +\infty$的整数。应用叠加原理易得该问题的解为:

由于这个级数收敛速度较快,因此在计算矩形区域$|x| \leq a/2$、$|y| \leq b/2$内的压力分布时,只需考虑与该矩形区域相邻的生产井就可以了。例如,取:

一般就足够了。

## 4.5.3   解变产量问题的杜哈美(Duhamel)原理

首先考虑这样一个简单情形:在无穷大地层中有一生产井,从$t=0$到$t=t_0$时刻产量为$Q_1$,从$t=t_0$时刻起产量变为$Q_2$(图 4 – 19),求地层内的压力分布。

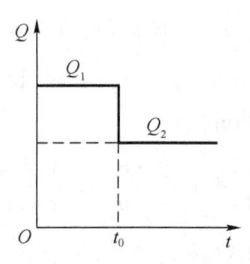

Figure 4-19  The relationship between step yield and time

图 4-19  台阶型产量随时间变化关系

Set up a coordinate system considering the well point as the origin, $p_1(r,t)$ is used to express the pressure distribution in formation in time $0 \leq t \leq t_0$, and $p_2(r,t)$ is used to express the pressure distribution in formation in time $t \geq t_0$. It is easy to know that the $p_1(r,t)$ of $0 \leq t \leq t_0$ suits the definite question:

$$\begin{cases} \dfrac{\partial^2 p_1}{\partial r^2} + \dfrac{1}{r}\dfrac{\partial p_1}{\partial r} = \dfrac{1}{\eta}\dfrac{\partial p_1}{\partial t} \\ p_1(r,0) = p_i \\ p_1(\infty,t) = p_i \\ \left. r\dfrac{\partial p_1}{\partial r}\right|_{r \to 0} = \dfrac{Q_1 \mu}{2\pi Kh} \end{cases}$$

While the solution of this definite question is the formula (4-26) that is:

$$p_1(r,t) = p_i + \dfrac{\mu Q_1}{4\pi Kh}\text{Ei}\left(-\dfrac{r^2}{4\eta t}\right) \qquad (4-45)$$

When $t \geq t_0$, $p_2(x,y,t)$ suits the following definite question:

$$\begin{cases} \dfrac{\partial^2 p_2}{\partial r^2} + \dfrac{1}{r}\dfrac{\partial p_2}{\partial r} = \dfrac{1}{\eta}\dfrac{\partial p_2}{\partial t} \\ p_2(r,t_0) = p_1(r,t_0) \\ p_2(\infty,t) = p_i \\ \left. r\dfrac{\partial p_2}{\partial r}\right|_{r \to 0} = \dfrac{Q_2 \mu}{2\pi Kh} \end{cases}$$

Starting from this definite question above, it is tough to obtain its solution. It may be assumed that the oil well still produces with production $Q_1$ since $t = t_0$, and at this moment the pressure distribution can be expressed with the formula (4-45), but there will be a pressure difference $p_3(r,t)$ compared with the actual pressure distribution that is:

— 181 —

$$p_3(r,t) = p_2(r,t) - p_1(r,t) \tag{4-46}$$

It is easy to prove that when $t > t_0$, $p_3(r,t)$ is the solution of the following definite question:

$$\begin{cases} \dfrac{\partial^2 p_3}{\partial r^2} + \dfrac{1}{r}\dfrac{\partial p_3}{\partial r} = \dfrac{1}{\eta}\dfrac{\partial p_3}{\partial t} \\ p_3(r,t_0) = 0 \\ p_3(\infty,t) = 0 \\ r\dfrac{\partial p_3}{\partial r}\bigg|_{r\to 0} = \dfrac{(Q_2 - Q_1)\mu}{2\pi Kh} \end{cases}$$

And its solution is:

$$p_3(r,t) = \frac{(Q_2 - Q_1)\mu}{4\pi Kh}\mathrm{Ei}\left[-\frac{r^2}{4\eta(t-t_0)}\right] \tag{4-47}$$

So when $t > t_0$, the following expression can be obtained according to the formula (4-45) ~ formula (4-47):

$$p_2(r,t) = p_i + \frac{Q_1\mu}{4\pi Kh}\mathrm{Ei}\left(-\frac{r^2}{4\eta t}\right) + \frac{(Q_2 - Q_1)\mu}{4\pi Kh}\mathrm{Ei}\left[-\frac{r^2}{4\eta(t-t_0)}\right] \tag{4-48}$$

The expression above shows, from time $t = t_0$ the pressure field of the formation can be considered as the super-position of two pressure fields. One pressure field caused by production well remains the production and the other is caused by the increased production from time $t = t_0$. Informally speaking, from time $t = t_0$, it is equivalent that a production well with production $(Q_2 - Q_1)$ is added on the origin oil well with the initial production $Q_1$ at the same location.

This kind of method is also used in the situation dealing with variable production. Assume that there is a production well in infinite formation, the variable production is shown as Figure 4-20, it is easy to obtain the formula of pressure distribution in formation when $t \geq t_{n-1}$ that is:

$$p(r,t) = p_i + \frac{\mu}{4\pi Kh}\sum_{i=1}^{n}(Q_i - Q_{i-1})\mathrm{Ei}\left[-\frac{r^2}{4\eta(t-t_{i-1})}\right] \quad (Q_0 = 0, t_0 = 0) \tag{4-49}$$

If there is a single well producing with variable rate $Q(t)$ in infinite formation (Figure 4-21), the situation of pressure distribution in formation should be taken into account first. This variable-rate problem still can be considered as the limitation of multi-production problem. Divide the time range $[0,t]$ into $n$ equal portions, $\Delta\tau = \dfrac{t}{n}, \tau_0 = 0, \tau_1 = \Delta\tau, \ldots, \tau_{n-1} = (n-1)\Delta\tau, t = n\Delta\tau$.

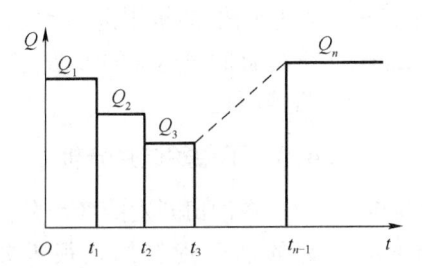

Figure 4-20  The figure of multi-flowrate production well
图 4-20  多产量生产井的产量变化图

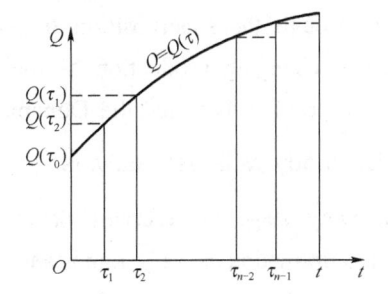

Figure 4-21  The figure of changing production well
图 4-21  变产量生产井的产量变化图

In every little time range, the oil well is considered to produce with constant production. Concretely speaking, in time $[\tau_0, \tau_1]$ the oil well producing with $Q(\tau_0)$, in time $[\tau_1, \tau_2]$ the oil well producing with $Q(\tau_1), \ldots$, under this assumption, in time $[0, t]$ the oil well produces with $n$ different production, thus, from the above formula of pressure distribution (4-49) producing with changing production and using the differential mean value theorem, the following expression can be obtained:

$$\begin{aligned}p(r,t) &= p_i + \frac{\mu}{4\pi Kh}\Big\{Q(\tau_0)\mathrm{Ei}\Big(-\frac{r^2}{4\eta t}\Big) + [Q(\tau_1) - Q(\tau_0)]\mathrm{Ei}\Big[-\frac{r^2}{4\eta(t-\tau_1)}\Big] + \cdots \\ &\quad + [Q(\tau_{n-1}) - Q(\tau_{n-2})]\mathrm{Ei}\Big[-\frac{r^2}{4\eta(t-\tau_{n-1})}\Big]\Big\} \\ &= p_i + \frac{\mu}{4\pi Kh}\Big(Q(\tau_0)\Big\{\mathrm{Ei}\Big(-\frac{r^2}{4\eta t}\Big) - \mathrm{Ei}\Big[-\frac{r^2}{4\eta(t-\tau_1)}\Big]\Big\} + \cdots \\ &\quad + Q(\tau_{n-2})\Big\{\mathrm{Ei}\Big[-\frac{r^2}{4\eta(t-\tau_{n-2})}\Big] - \mathrm{Ei}\Big[-\frac{r^2}{4\eta(t-\tau_{n-1})}\Big]\Big\} \\ &\quad + Q(\tau_{n-1})\mathrm{Ei}\Big\{-\frac{r^2}{4\eta(t-\tau_{n-1})}\Big\}\Big) \\ &= p_i + \frac{\mu}{4\pi Kh}\sum_{k=0}^{n-2}Q(\tau_i)\Big\{-\frac{\mathrm{d}}{\mathrm{d}\tau}\mathrm{Ei}\Big[-\frac{r^2}{4\eta(t-\tau)}\Big]\Big\}_{\tau=\tau_k+\theta_k\Delta\tau}\Delta\tau \\ &\quad + \frac{\mu Q(\tau_{n-1})}{4\pi Kh}\mathrm{Ei}\Big[-\frac{r^2}{4\eta(t-\tau_{n-1})}\Big]\quad (0<\theta_k<1)\end{aligned}$$

When $n\to\infty$, $\Delta\tau\to 0$ and $\tau_{n-1}\to t$, the following expression can be obtained:

$$p(r,t) = p_i + \frac{\mu}{4\pi Kh}\int_0^t Q(\tau)\frac{\mathrm{d}}{\mathrm{d}\tau}\left\{-\mathrm{Ei}\left[-\frac{r^2}{4\eta(t-\tau)}\right]\right\}\mathrm{d}\tau$$
$$= p_i - \frac{\mu}{4\pi Kh}\int_0^t \frac{Q(\tau)}{t-\tau}e^{-\frac{r^2}{4\eta(t-\tau)}}\mathrm{d}\tau \tag{4-50}$$

This kind of method that obtains the expression of solution through the superposition principle considering the problem of changing production as the limitation of multi-production problem is called the Duhamel principle.

## 4.6 Unsteady well-test analysis

In every stage of reservoir development, the reliable reservoir information is very important. In order to analyze the behavior of reservoir precisely and predict the production tendency under all ways of well producing methods, the reservoirs engineer must master lots of precise reservoir information. And the petroleum engineer also needs to know the working condition of the production well and the injection well, so the reservoir can be developed under the optimal situation. And the required information can mostly be obtained through the methods of unsteady state well testing.

The methods of unsteady state well testing such as the pressure build-up test, pressure fall-off test and the interference test and they are all critical parts of reservoir engineering. The unsteady well testing is to estimate the parameters of rock, fluid and well according to the relationship of measured bottom pressure with time. The actual information obtained from the unsteady state well testing (including the relevant data of formation damage and improvement, formation pressure, permeability, reserve and so on) helps to analyze, improve and predict the behavior of the reservoirs. Two simplest methods of well testing-pressure fall-off test and pressure build-up test will be introduced in this section.

### 4.6.1 The pressure fall-off test

The character of pressure decline when production well produces with constant production reflects the properties of the reservoir and the fluid in it. So the character of pressure decline should be researched to know the characters of reservoir and wells. The problem of pressure well testing (or pressure fall-off test) analysis in the first phase of elasticity and in the phase of pseudo-steady state period under the producing condition of constant production will be discussed here.

As shown in the formula (4-29), in infinite formation, if the initial formation pressure is $p_i$, then the bottom hole pressure of a production well with constant production is:

$$p_{wf}(t) = p_i - \frac{\mu B Q_{sc}}{4\pi Kh}\left(\ln\frac{2.245\eta t}{r_2^2} + 2S\right)$$

If the unit of pressure $p$ is MPa, viscosity $\mu$ is mPa·s, production $Q$ is m³/d, permeability $K$ is $10^{-3} \mu m^2$, effective thickness of oil formation $h$ is m, time $t$ is h, the radius of oil well is cm, the $C_t$ total compressibility is MPa$^{-1}$, and change the natural logarithm to be common logarithm, then the expression above can be further written as:

$$p_{wf}(t) = p_i - 2.1208\frac{\mu B Q_{sc}}{Kh}\left(\lg t + \lg\frac{K}{\phi\mu C_t r_w^2} + 0.8686S + 1.9077\right) \quad (4-51)$$

If the unit of parameters above changes, the constants 2.1208, 0.8686 and 1.9077 will be replaced by other constants. The formula (4-51) shows that $p_{wf}$ and $\lg t$ have a linear relationship. Make the bottom hole flowing pressure be $p_{1h}$ when $t=1h$, then the following expression can be obtained from the formula (4-51):

$$p_{1h} = p_i - 2.1208\frac{\mu B Q_{sc}}{Kh}\left(\lg\frac{K}{\phi\mu C_t r_w^2} + 0.8686S + 1.9077\right) \quad (4-52)$$

Make $m = -2.1208\frac{\mu B Q_{sc}}{Kh}$, and the formula (4-51) reduces formula (4-52), the following expression can be obtained:

$$p_{wf} = m\lg t + p_{1h} \quad (4-53)$$

Theoretically, the relation curve between bottom hole pressure and the log of producing time (usually be called the semi-log plot) is a straight line, its slope is $m$ and intercept is $p_{1h}$. The Figure 4-22 shows that after the disappearance of influence of bottom hole contamination and after flow, the linear segment appears indeed. If substituting the measured bottom hole pressure $p_{wf}(t)$ and its corresponding time on the linear segment as shown in Figure 4-15 into the formula (4-53) and solve the slope $m$ and intercept $p_{1h}$ by linear regression, the formation permeability can be further obtained:

$$K = -2.1208\frac{\mu B Q_{sc}}{mh} \qquad (4-54)$$

And the skin factor of well bore can also be obtained from the formula (4-52):

$$S = 1.1512\left(\frac{p_i - p_{1h}}{-m} - \lg\frac{K}{\phi\mu C_t r_w^2} - 1.9077\right) \qquad (4-55)$$

It will transmit to the phase of pseudo-steady seepage after the pressure drop of oil well is transmitted to the boundary. In the phase of pseudo-steady seepage, because the decline rate of pressure on every point in formation is equal, so under the condition of producing with constant production, the bottom hole pressure and time $t$ have a linear relationship as shown in Figure 4-23. So the linear segment equation can be obtained by linear regression to the straight line:

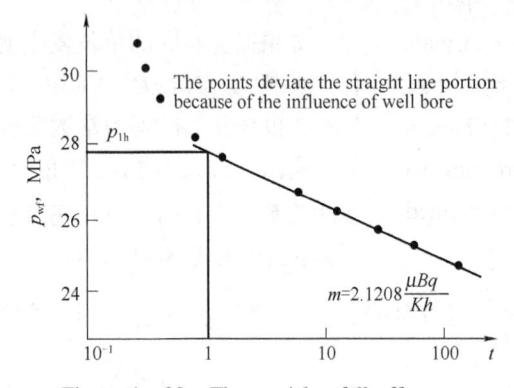

$$p_{wf} = p_0 + bt \qquad (4-56)$$

Figure 4-22　The semi-log fall-off curve
图 4-22　半对数压降曲线

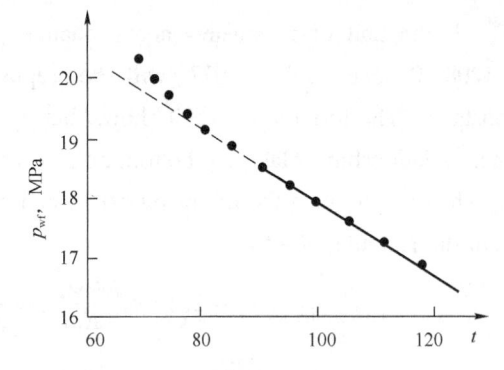

Figure 4-23　The fall-off curve in pseudo-steady state phase
图 4-23　拟稳态阶段的压降曲线

In the phase of pseudo-steady flow, the following expression can be obtained according to the formula (4-30):

$$\frac{dp_{wf}}{dt} = -\frac{QB}{\phi h C_t A}$$

The $\frac{dp_{wf}}{dt}$ in formula above is the slope $b$ of the straight line portion in Figure 4-23 and adopts the unit above, hence:

$$A = -\frac{0.04167 QB}{\phi h C_t b} \qquad (4-57)$$

It can be known from the formula (4-56), when $t = 0$, the shape coefficient $C_A$ can be further obtained, and so the shape of the reservoir boundary and well locations can also be ascertained.

**Example 4-3** According to the geological and seismic information, this reservoir may be a rectangular fault block with length of side 2∶1. In order to further verify this recognition, an exploration well is made to work 100h with a constant production $Q = 238.5 \text{m}^3/\text{d}$, and the different bottom pressure at different time is measured with manometer (Table 4-2), the data and pressure-measuring results of reservoir are as follows: $h = 6.1\text{m}$, $r_w = 10\text{cm}$, $\phi = 0.18$, $C_t = 2.13 \times 10^{-3}\text{MPa}^{-1}$, $\mu = 1.00\text{mPa} \cdot \text{s}$, $B = 1.20$. Please solve the effective permeability and the skin factor $S$, and estimate the area of the reservoir.

**Table 4-2  The measured results**

| $t$, h | $p_{wf}$, MPa | $t$, h | $p_{wf}$, MPa | $t$, h | $p_{wf}$, MPa |
|---|---|---|---|---|---|
| 0 | 24.61 | 7.5 | 20.02 | 50 | 18.26 |
| 1 | 20.51 | 10 | 19.90 | 60 | 17.89 |
| 2 | 20.38 | 15 | 19.64 | 70 | 17.54 |
| 3 | 20.30 | 20 | 19.42 | 80 | 17.18 |
| 4 | 20.24 | 30 | 19.00 | 90 | 16.82 |
| 5 | 20.17 | 40 | 18.63 | 100 | 16.46 |

**Solution**: Firstly, draw the measured data points on the semi-log paper as shown in Figure 4-24(a). It can be found that the former 4 points make a straight line, after linear regression, $m = -0.447\text{MPa}$, $p_{1h} = 20.51\text{MPa}$ can be obtained. And the following expression can be obtained after substituting them into the formula (4-54) and formula (4-55):

$$K = \frac{2.1208 \times 238.5 \times 1.2 \times 1}{0.447 \times 6.1} = 222.6 \times 10^{-3} (\mu\text{m}^2)$$

$$S = 1.15129 \times \left(\frac{24.61 - 20.51}{0.447} - \lg \frac{223}{1.0 \times 0.18 \times 2.13 \times 10^{-3} \times 10^2} - 1.90768\right) = 4.03$$

Draw the actually measured points on the rectangular coordinate plot as shown in Figure 4-24(b), it can be found that it transits into the pseudo-steady state period after 50h, and after linear regression, the following expression can be obtained:

(a)The semi-log coordinate system
(a)半对数坐标系

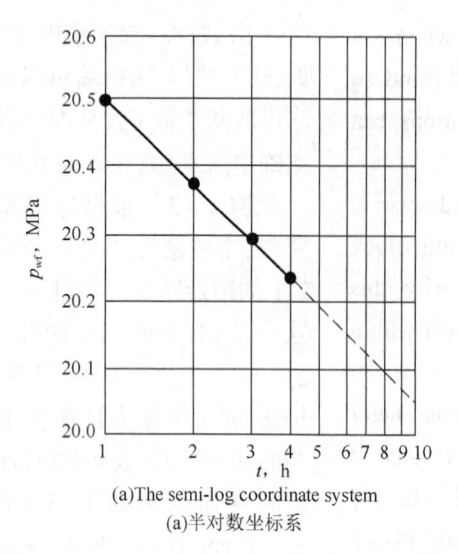

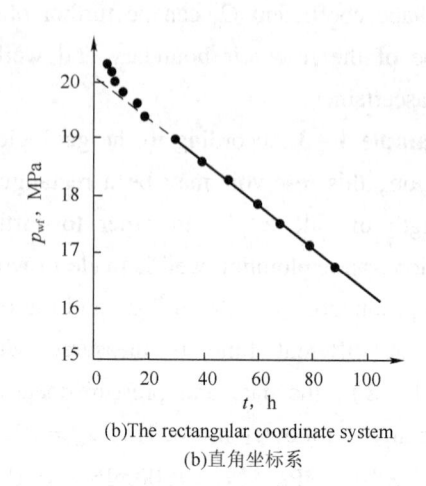

(b)The rectangular coordinate system
(b)直角坐标系

Figure 4 – 24  The drawdown curve of example 4 – 3
图 4 – 24  例 4 – 3 的压降曲线

$$p_{wf} = 20.51 - 0.0359t$$

From the formula (4 – 57), the following expression can be obtained:

由式(4 – 57)得:

$$A = -\frac{0.04167 \times 238.5 \times 1.2}{6.1 \times 0.18 \times 2.13 \times 10^{-3} \times (-0.0359)} = 141600\,(\text{m}^2)$$

Make $t = 0$, then $p_{wf} = p_0 = 20.5\text{MPa}$, and $C_A = 5.33$ can be obtained by substituting the data into the formula (4 – 41), and it also can be known from the Figure 4 – 15 that as to the rectangular reservoir with length of side 2:1, if the well lies in the center of the reservoir, the value of $C_A$ on the midpoint of a line between the short side and the center is 4.86 which is close to the calculated value. So the well-testing result verifies the geologic evaluation.

### 4.6.2  Pressure build-up test

The pressure build-up test is that before well testing the oil well produces with constant production $Q_{sc}$ for a time $t_p$ (Figure 4 – 25), and then abruptly shut down the well, and determine the parameters of the reservoir and the oil well according to the bottom pressure at different time after well shutdown.

The bottom pressure of oil well after shutting down as to the variable production as shown in Figure 4 – 25 can be expressed with superposition principle:

令 $t = 0$, $p_{wf} = p_0 = 20.051\text{MPa}$,代入式(4 – 41)得到 $C_A = 5.33$,查图 4 – 15 知边长为 2:1 的矩形油藏,井位于油藏中心到短边的连线中点处的 $C_A = 4.86$,与所求值比较接近,试井结果证实了地质推断。

### 4.6.2  压力恢复试井

压力恢复试井就是在试井前油井以稳定的产量 $Q$ 生产一段时间 $t_p$(图 4 – 25),然后突然关井,并根据关井后不同时刻的井底压力来求得油藏和油井的特征参数。

对于产量变化如图 4 – 25 所示的油井,关井后的井底压力可以用叠加原理来表示:

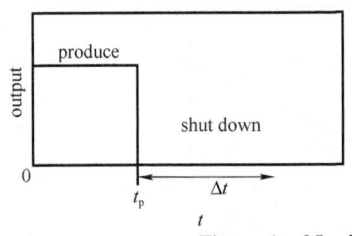

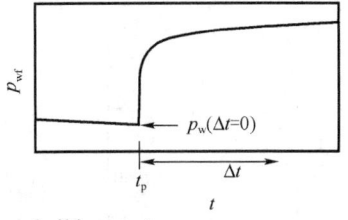

Figure 4 – 25  The pressure build-up test
图 4 – 25  压力恢复试井产量变化图

$$p_{ws}(\Delta t) = p_i + \frac{\mu B}{4\pi Kh}\left\{Q_{sc}\text{Ei}\left[-\frac{r_w^2}{4\eta(t_p+\Delta t)}\right] + (0-Q_{sc})\text{Ei}\left(-\frac{r_w^2}{4\eta\Delta t}\right)\right\}$$

As for most formation without fractures, its exponential integral function in short time can be expressed with logarithm, hence:

对于大多数无裂缝地层,其指数积分函数在短时间内就可以用对数形式表示,于是:

$$p_{ws}(\Delta t) = p_i - \frac{\mu B Q_{sc}}{4\pi Kh}\left[\ln\frac{2.245\eta(t_p+\Delta t)}{r_w^2} - \ln\frac{2.245\eta\Delta t}{r_w^2}\right] = p_i - \frac{\mu B Q_{sc}}{4\pi Kh}\ln\frac{\Delta t + t_p}{\Delta t}$$

If the unit of pressure is MPa, viscosity $\mu$ is mPa·s, production $Q_{sc}$ is m³/d, permeability $K$ is $10^{-3}$ μm², effective thickness $h$ is m, and the unit of bottom hole radius is cm, total compressibility $C_t$ is MPa$^{-1}$, and make:

如果压力的单位为MPa,黏度$\mu$的单位为mPa·s,产量$Q_{sc}$的单位为m³/d,渗透率$K$的单位为$10^{-3}$μm²,有效厚度$h$的单位为m,井底半径$r_w$的单位为cm,总压缩系数$C_t$的单位为MPa$^{-1}$,并令:

$$m = 2.1208\frac{\mu B Q_{sc}}{Kh}$$

Then the expression of $p_{ws}(\Delta t)$ can be expressed as:

则$p_{ws}(\Delta t)$可表示为:

$$p_{ws}(\Delta t) = p_i + m\lg\frac{\Delta t}{\Delta t + t_p} \tag{4-58}$$

The Figure 4 – 26 is the schematic figure of Horner build-up curve. $p_{ws}$ is considered as the y-axis and $\frac{\Delta t}{\Delta t + t_p}$ as x-axis, a straight line portion will appear after passing through the period of well bore storage influence. As shown in the equation (4 – 58), in order to measure $p_i$, the Horner curve can be extrapolated to $\frac{\Delta t}{\Delta t + t_p} = 1$ where $\lg\left(\frac{\Delta t}{\Delta t + t_p}\right) = 0$ that is amount to infinite shut down time. The calculation above when producing time is short is correct. However, in fact, after long-time development of reservoir, the extrapolated pressure $p_i^*$ can be used to solve the average formation pressure.

图 4 – 26 就是压力恢复资料的霍纳(Horner)曲线示意图,以$p_{ws}$为纵坐标,$\frac{\Delta t}{\Delta t + t_p}$为横坐标,过了井筒储存影响期后会呈现一直线段。如同式(4 – 58)所表明的,为了测定$p_i$,可以把霍纳曲线外推至$\frac{\Delta t}{\Delta t + t_p} = 1$,即$\lg\left(\frac{\Delta t}{\Delta t + t_p}\right) = 0$处,相当于无限长关井时间。上述计算对生产时间较短的情况是正确的。事实上,在油田经长时间开发后,这一外推的压力值$p_i^*$可用来求平均地层压力。

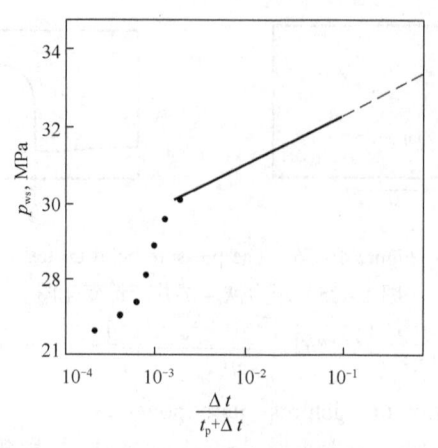

Figure 4 – 26  The pressure build-up curve
图 4 – 26  压力恢复曲线

The slope $m$ and $p_i^*$ can be solved through linear regression with the formula (4 – 58) and the bottom hole pressure $p_{ws}$ and the corresponding $\dfrac{\Delta t}{\Delta t + t_p}$ of the straight line portion in Figure 4 – 26, and the formation permeability $K$ can be further obtained:

可以利用式 (4 – 58) 由图 4 – 26 中直线部分的井底压力 $p_{ws}$ 及相应的 $\dfrac{\Delta t}{\Delta t + t_p}$，通过线性回归求出斜率 $m$ 和 $p_i^*$，然后进一步求出地层渗透率 $K$：

$$K = 2.1208\dfrac{\mu B Q_{sc}}{mh}$$

If the oil well produces for a long time with constant production $Q_{sc}$ before shutting down, and the shut-in time is short that is $t_p \gg \Delta t$, then the formula (4 – 58) can be approximately written as:

如果关井前油井以定产量 $Q_{sc}$ 生产了很长时间，而关井时间又较短，即 $t_p \gg \Delta t$，则式 (4 – 58) 可近似写成：

$$p_{ws} \approx p_i + m\lg\dfrac{\Delta t}{t_p} \tag{4 – 59}$$

As to the instant of shutting down that is $\Delta t = 0$, the following expression can be obtained:

对于油井在关井的瞬间，即 $\Delta t = 0$，有：

$$p_{wf} = p_i - m\left(\lg t_p + \lg\dfrac{K}{\phi\mu C_t r_w^2} + 1.9077 + 0.8686S\right) \tag{4 – 60}$$

Prolong the straight line portion of Horner curve and read out the bottom pressure $p_{ws}(1h)$ when $\Delta t = 1h$, and substitute it into the expression above, then the skin factor of well bore $S$ can be obtained:

将霍纳曲线上的直线段延长并读出 $\Delta t = 1h$ 的井底压力值 $p_{ws}(1h)$，代入上式即可求出井筒表皮系数 $S$ 值：

$$S = 1.152\left[\dfrac{p_{ws}(1h) - p_{wf}}{m} - \lg\dfrac{K}{\phi\mu C_t r_w^2} - 1.9077\right] \tag{4 – 61}$$

As to the concrete solving way please refer to the following example.

**Example 4-4** The first exploration well of a given reservoir produces nearly 100h, and then shut down and measure its build-up pressure. The production information and the properties of fluid are as follows: $Q_{sc} = 19.56 \text{m}^3/\text{d}$, $\phi = 0.2$, the cumulative production $N_p = 79.5 \text{m}^3$, $\mu = 1.0 \text{mPa} \cdot \text{s}$, $h = 6.1 \text{m}$, $B_{oi} = 1.22$, $r_w = 10 \text{cm}$, $C_t = 2.845 \times 10^{-3} \text{MPa}^{-1}$. The pressure-measuring data is shown in Table 4-3. Please solve the permeability, initial formation pressure, skin effect and the additional resistance caused by skin.

Table 4-3  The pressure-measuring data
表 4-3  测压数据

| $\Delta t$, h | $p_{ws}$, MPa | $\Delta t$, h | $p_{ws}$, MPa | $\Delta t$, h | $p_{ws}$, MPa |
|---|---|---|---|---|---|
| 0.00 | 31.68 | 2.00 | 33.45 | 8.00 | 33.56 |
| 0.50 | 32.87 | 2.50 | 33.46 | 10.00 | 33.57 |
| 0.66 | 33.08 | 3.00 | 33.49 | 12.00 | 33.59 |
| 1.00 | 33.28 | 4.00 | 33.51 | | |
| 1.50 | 33.40 | 6.00 | 33.54 | | |

**Solution**: Firstly, solve the reduced production time:

$$t_p = \frac{N_p}{Q_{sc}} \times 24 = \frac{79.5}{19.56} \times 24 = 97.5(\text{h})$$

And then calculate the $\dfrac{\Delta t}{\Delta t + t_p}$ of each time after shutting off, and then list the computation chart of $\dfrac{\Delta t}{\Delta t + t_p}$ and $p_{ws}$ (Table 4-4).

Table 4-4  The reference calculating list of each time after shutting off
表 4-4  关井各时刻对照计算表

| $\Delta t$, h | $\dfrac{\Delta t}{\Delta t + t_p}$ | $p_{ws}$, MPa | $\Delta t$, h | $\dfrac{\Delta t}{\Delta t + t_p}$ | $p_{ws}$, MPa |
|---|---|---|---|---|---|
| 0.00 | | 31.68 | 3.00 | 0.02985 | 33.49 |
| 0.50 | 0.00510 | 32.87 | 4.00 | 0.03941 | 33.51 |
| 0.66 | 0.00672 | 33.08 | 6.00 | 0.05797 | 33.54 |
| 1.00 | 0.01015 | 33.28 | 8.00 | 0.07583 | 33.56 |
| 2.00 | 0.02010 | 33.45 | 12.00 | 0.10960 | 33.59 |
| 2.50 | 0.2500 | 33.46 | | | |

— 191 —

Draw the data points $p_{ws} - \dfrac{\Delta t}{\Delta t + t_p}$ in Table 4 – 3 on semi-log paper (Figure 4 – 27) and make the linear regression of semi-log straight line portion, then the following expressions can be obtained:

将表 4 – 3 中 $p_{ws} - \dfrac{\Delta t}{\Delta t + t_p}$ 的数据点到半对数坐标纸上(图 4 – 27)并对半对数直线段进行线性回归得:

$$m = 0.1778 \text{MPa}; \quad p^* = 33.758 \text{MPa}$$

Then the permeability of formation can be further obtained: 进一步求得地层渗透率为:

$$K = \frac{2.1208 Q_{sc} B \mu}{hm} = \frac{2.1208 \times 19.56 \times 1.22 \times 1.0}{6.1 \times 0.1778} \approx 46.7 \times 10^{-3} (\mu m^2)$$

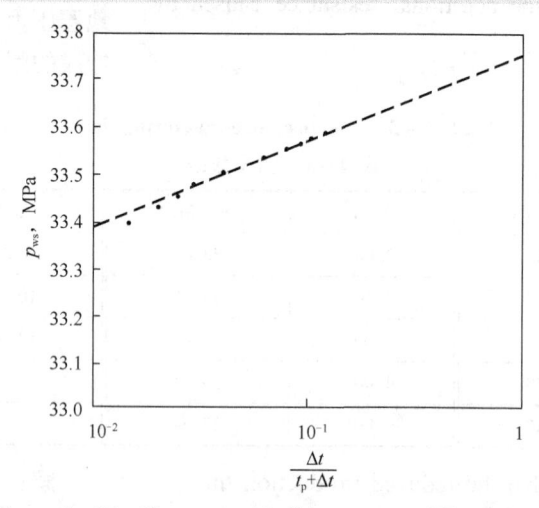

Figure 4 – 27  The pressure build up curve of Example 4 – 4

图 4 – 27  例 4 – 4 的压力恢复曲线

It can be found from Figure 4 – 27 that when $\Delta t = 1\text{h}$ the value of pressure is $p_{ws}(1\text{h}) = 33.40\text{MPa}$ on the extension line of semi-log straight line portion, so after substituting the value into the formula (4 – 61) the skin factor can be obtained as:

从图 4 – 27 中读得半对数直线段的延长线上 $\Delta t = 1\text{h}$ 的压力为 $p_{ws}(1\text{h}) = 33.40\text{MPa}$,所以代入式(4 – 61)得表皮系数为:

$$S = 1.1512 \times \left( \frac{33.40 - 31.68}{0.1778} - \lg \frac{46.7}{0.2 \times 1 \times 2.845 \times 10^{-3} \times 10^2} - 1.9077 \right) \approx 4.85$$

And the pressure drop caused by skin factor is: 由表皮系数造成的压力降为:

$$\Delta p_{skin} = \frac{\mu B Q_{sc}}{2\pi K h} S = 0.76 (\text{MPa})$$

### Exercises

### 习 题

4 – 1  Assume that there is only an injection well working in the homogeneous isopachous and infinite compressible formation, please try to derive the formula of pressure distribution in formation.

4 – 1  设均质、等厚、无限大地层中只有一口注入井生产,试推导地层中的压力分布公式。

4 – 2  Please try to prove that: the solution of $n$ wells working simultaneously in infinite elastic formation obtained utilizing superposition principle （4 – 44） meets the heat conduction equation.

4 – 3  Calculate with computer: （1） The error function is erf $(x)$, $x$ is from 0.01 to 3 and step is 0.01. （2） The exponential integral function is $-\mathrm{Ei}(-x)$, $x$ is from 0.01 to 5 and step is 0.01.

4 – 4  Assume formation is linear and semi-infinite; the exit production is a constant $Q_0$, the diffusivity coefficient is $\eta = 1.5 \mathrm{m}^2/\mathrm{s}$. （1） Please solve the ratio of production $Q$ and $Q_0$ when $t = 1\mathrm{d}$, 1mon and 1a from the exit $x = 2000\mathrm{m}$. （2） Please solve the ratio of production $Q$ and $Q_0$ when $t = 1\mathrm{a}$ and $x = 100\mathrm{m}, 500\mathrm{m}, 1000\mathrm{m}, 2000\mathrm{m}$ and $5000\mathrm{m}$. （3） Please try to interpret the results.

4 – 5  Assume there is a continuous point sink, and its production is $Q$, there is an observational well from the point sink at distance $r_0$. （1） Please solve the changing law of pressure in the observational well with time and analysis the law. （2） The relationship of flow velocity of fluid at the location of observational well with time.

4 – 6  An oil formation: $K = 0.8 \mu\mathrm{m}^2$, $\mu_0 = 3.0 \mathrm{mPa} \cdot \mathrm{s}$, $C_t = 8.3 \times 10^{-4} \mathrm{MPa}^{-1}$, $\phi = 0.2$, $h = 15\mathrm{m}$, after a well with $r_w = 10\mathrm{cm}$, $Q = 80\mathrm{m}^3/\mathrm{d}$ is put into production 10min, 1h, 1d, 10d, 1mon and 1a, please solve the pressure drop of bottom $r_w$, $r_1 = 400\mathrm{m}$ and $r_2 = 1\mathrm{km}$ from well.

4 – 7  There is an impermeable fault in planar infinite elastic formation and there is a well $A$ from the fault with distance $a = 2\mathrm{km}$, and $r_w = 10\mathrm{cm}$, $Q = 300\mathrm{m}^3/\mathrm{d}$. Please solve the pressure drop at point $B$ from the well $A$ to distance $b = 3\mathrm{km}$ after well $A$ is put into production in time 2a, and the distance $B$ to fault is also 2km, $\eta = 2.5 \mathrm{m}^2/\mathrm{s}$, $K = 0.4 \mu\mathrm{m}^2$, $h = 5\mathrm{m}$, $\mu = 1.2 \mathrm{mPa} \cdot \mathrm{s}$.

4 – 2  试证明运用叠加原理得到的无穷大弹性地层中 $n$ 口井同时生产的解式（4 – 44）满足热传导方程。

4 – 3  上机计算：（1）误差函数 erf$(x)$，其中 $x$ 从 0.01 到 3，步长为 0.01。（2）指数积分函数 $-\mathrm{Ei}(-x)$，$x$ 从 0.01 到 5，步长为 0.01。

4 – 4  设地层是线性半无限大的，出口产量为常数 $Q_0$，导压系数 $\eta = 1.5 \mathrm{m}^2/\mathrm{s}$。（1）求 $t = 1\mathrm{d}$、1mon 和 1a 时距出口 $x = 2000\mathrm{m}$ 处的流量 $Q$ 与 $Q_0$ 之比。（2）求 $t = 1\mathrm{a}$ 时 $x = 100\mathrm{m}$、500m、1000m、2000m、5000m 处的流量 $Q$ 与 $Q_0$ 之比。（3）试解释计算结果。

4 – 5  设平面无穷大地层上有一连续点汇，其产量为 $Q$，距离点汇 $r_0$ 处有一观测井。（1）求该观测井内的压力随时间的变化规律，并加以分析。（2）求观测井处液体的渗流速度与时间的关系。

4 – 6  油层 $K = 0.8 \mu\mathrm{m}^2$，$\mu = 3.0 \mathrm{mPa} \cdot \mathrm{s}$，$C_t = 8.3 \times 10^{-4} \mathrm{MPa}^{-1}$，$\phi = 0.2$，$h = 15\mathrm{m}$，井底半径 $r_w = 10\mathrm{cm}$，$Q = 80\mathrm{m}^3/\mathrm{d}$ 的井投产后 10min、1h、1d、10d、1mon、1a 时，求井底 $r_w$，以及距井 $r_1 = 400\mathrm{m}$、$r_2 = 1\mathrm{km}$ 各点的压力降。

4 – 7  平面无穷大弹性地层中有一不渗透断层，距断层 $a = 2\mathrm{km}$ 处有一口井 $A$，$r_w = 10\mathrm{cm}$，$Q = 300\mathrm{m}^3/\mathrm{d}$，试求 $A$ 井投入生产 2a 后距 $A$ 井 $b = 3\mathrm{km}$ 的 $B$ 点的压降，$B$ 点距断层也是 2km，$\eta = 2.5 \mathrm{m}^2/\mathrm{s}$，$K = 0.4 \mu\mathrm{m}^2$，$h = 5\mathrm{m}$，$\mu = 1.2 \mathrm{mPa} \cdot \mathrm{s}$。

4 – 8  In planar infinite formation two wells with same production are put into production simultaneously and the distance between them is $d = 300\text{m}$, and the production is $Q = 100\text{m}^3/\text{d}$, it is known that $h = 12\text{m}$, $K = 0.5\mu\text{m}^2$, $\phi = 0.2$, $C_t = 2.5 \times 10^{-3}\text{MPa}^{-1}$, $B_0 = 1.25$, $\mu = 1.5\text{mPa}\cdot\text{s}$, please solve the pressure drop on the midpoint of the connection of two wells after they are put into production 29d.

4 – 9  Assume the distance of a oil well from the linear impermeable boundary is $a = 120\text{m}$, the oil well initially produces with $Q_1 = 60\text{m}^3/\text{d}$ for 10d, and then produces with $Q_2 = 50\text{m}^3/\text{d}$. Assume that the height of oil layer is $h = 6.5\text{m}$, the permeability of formation is $K = 0.8\mu\text{m}^2$, the viscosity of fluid is $\mu = 2\text{mPa}\cdot\text{s}$, $B_0 = 1.2$, the radius of oil well is $r_w = 10\text{cm}$, porosity $\phi = 0.22$, total compressibility $C_t = 2 \times 10^{-3}\text{MPa}^{-1}$, please try to solve the bottom pressure drop after the production of well is changed 25d.

4 – 10  If the unit of pressure $p$ is MPa, viscosity $\mu$ is $\text{mPa}\cdot\text{s}$, production $Q$ is $\text{m}^3/\text{d}$, permeability $K$ is $10^{-3}\mu\text{m}^2$, effective thickness of oil layer $h$ is m, the radius of oil well is cm, total compressibility $C_t$ is $\text{MPa}^{-1}$, and the common logarithm is changed to be natural logarithm, please try to prove:

$$p_{wf}(t) = p_i - 2.1208\frac{\mu B Q}{Kh}\left(\lg t + \lg\frac{K}{\phi\mu C_t r_w^2} + 0.8686S + 1.9077\right)$$

4 – 11  The testing data of pressure drop of a given well producing with constant production is shown in Table 4 – 5, and other data are: $Q = 39.75\text{m}^3/\text{d}$, $B = 1.136$, $r_w = 6\text{cm}$, $\phi = 0.039$, $\mu = 0.8\text{ mPa}\cdot\text{s}$, $C_t = 2.4673 \times 10^{-3}\text{MPa}^{-1}$, please solve the permeability of formation and skin factor.

4 – 12  A pressure build-up test data run in the early developing stage of a given reservoir is shown in Table 4 – 6. The properties of rock and fluid are shown as follows:

$\mu=0.8\text{mPa}\cdot\text{s}$, $h=4.572\text{m}$, $r_w=10.2\text{cm}$, $C_t=2.177\times 10^{-3}\text{MPa}^{-1}$, $\phi=0.25$, $B_0=1.25$, the cumulative oil production is $N_p=198.75\text{m}^3/\text{d}$, the production of oil well is $Q=19.88\text{m}^3/\text{d}$. Please solve the permeability of formation $K$, skin factor $S$ and the initial formation pressure $p_i$.

**Table 4-5 The testing table of pressure-drop data**
表 4-5 压降数据测试表

| $t$, h | $p_{wf}$, MPa | $t$, h | $p_{wf}$, MPa | $t$, h | $p_{wf}$, MPa |
|---|---|---|---|---|---|
| 0.00 | 30.400 | 6.94 | 24.762 | 29.8 | 24.453 |
| 0.12 | 25.610 | 8.32 | 24.721 | 35.8 | 24.418 |
| 1.94 | 25.031 | 9.99 | 24.687 | 43.0 | 23.370 |
| 2.79 | 24.956 | 14.4 | 24.618 | 51.5 | 24.335 |
| 4.01 | 24.879 | 17.3 | 24.577 | 61.8 | 24.294 |
| 4.82 | 24.838 | 20.7 | 24.535 | 74.2 | 24.260 |
| 5.78 | 24.804 | 24.9 | 24.494 | | |

**Table 4-6 The pressure build-up test data**
表 4-6 压力恢复测试数据

| $\Delta t$, h | $p_{ws}$, MPa | $\Delta t$, h | $p_{ws}$, MPa | $\Delta t$, h | $p_{ws}$, MPa |
|---|---|---|---|---|---|
| 0 | 9.536 | 5 | 10.611 | 19 | 10.721 |
| 2 | 10.542 | 8 | 10.652 | 24 | 10.742 |
| 3 | 10.576 | 10 | 10.672 | 36 | 10.769 |
| 4 | 10.597 | 12 | 10.686 | | |

Chapter 4 Study Guide
第四章学习指南

# 5 Seepage law of natural gas
## 天然气的渗流规律

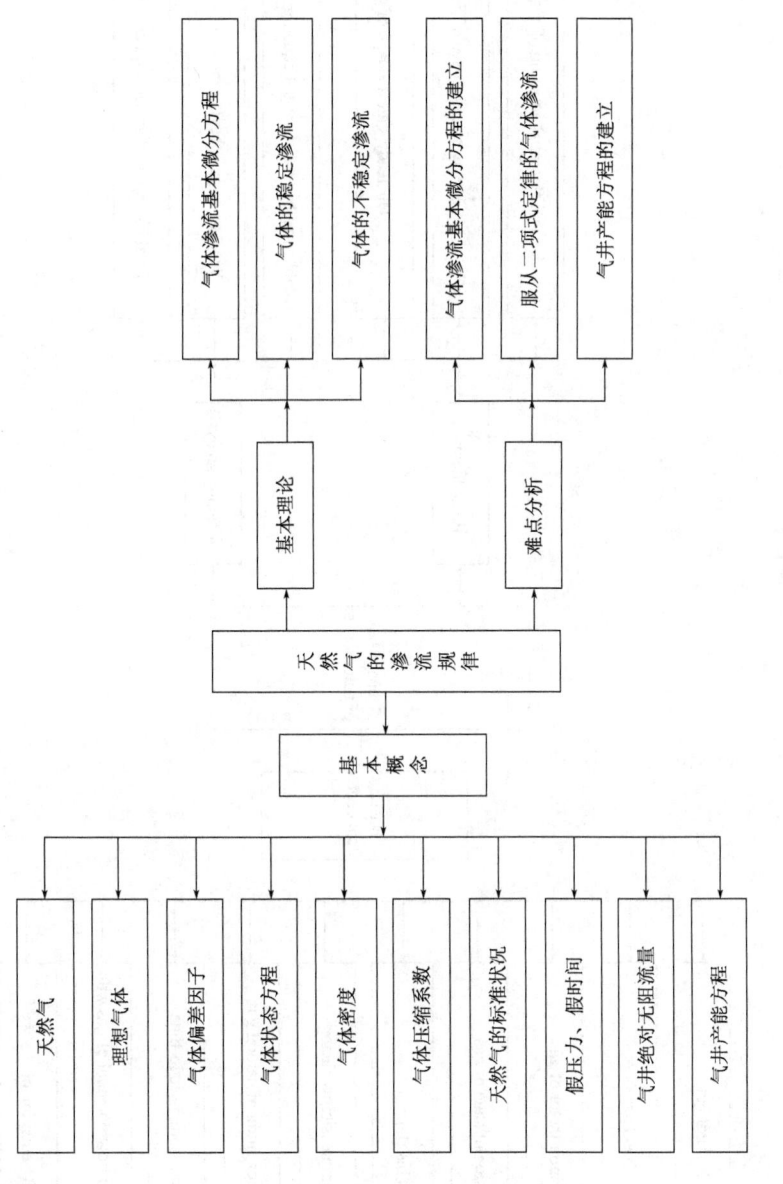

# 第五章知识图谱

```
天然气的渗流规律
├── 基本理论
│   ├── 气体渗流基本微分方程
│   ├── 气体的稳定渗流
│   └── 气体的不稳定渗流
├── 难点分析
│   ├── 气体渗流基本微分方程的建立
│   ├── 服从二项式定律的气体渗流
│   └── 气井产能方程的建立
└── 基本概念
    ├── 天然气
    ├── 理想气体
    ├── 气体偏差因子
    ├── 气体状态方程
    ├── 气体密度
    ├── 气体压缩系数
    ├── 天然气的标准状况
    ├── 拟压力、拟时间
    ├── 气井绝对无阻流量
    └── 气井产能方程
```

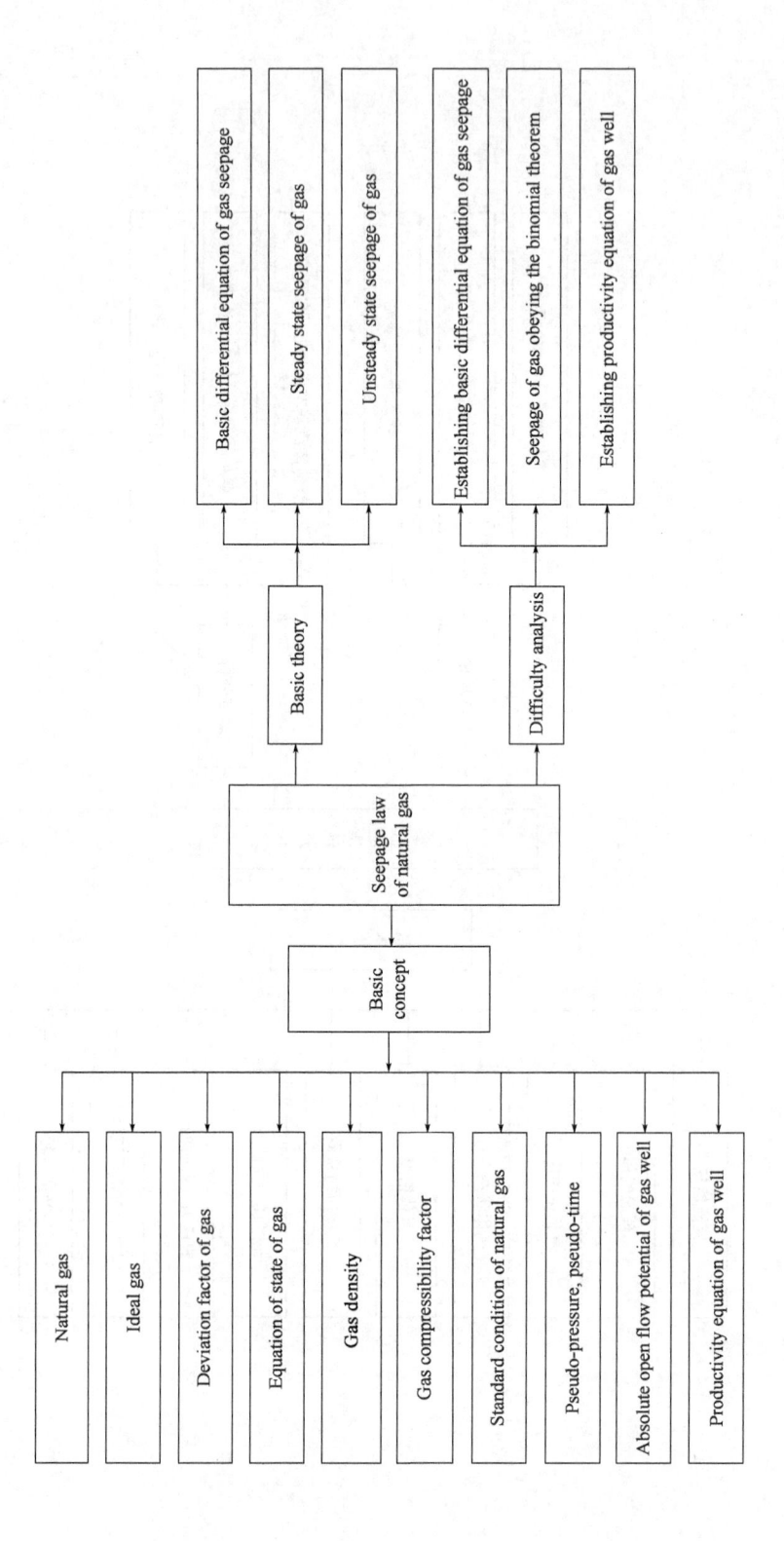

Chapter 5 Knowledge Graph

Natural gas is as same as oil, it is also a kind of important energy resource which will occupy more and more proportion in the structure of world fuel. Compared with oil, natural gas is clearer and will cause less pollution. With the development of economy and the improvement in the living conditions of people in China, the need for natural gas also increases continuously. The petroliferous basins in china have the regional geological conditions for the formation of large gas fields. Especially after 1980s, with the strategic westward transference of the oil industry in China and the further exploration and development in western region, the exploration of natural gas has obtained an amazing breakthrough. In the Shaan-Gan-Ning region, the world-class big gas fields have been found and big gas fields are also found in Talimu and Sichuan. The knowledge about the seepage of natural gas will be introduced in this chapter.

## 5.1 Properties of natural gas and its basic differential equation of seepage

Natural gas is the mixture of hydrocarbon gas (mainly methane) and some other contaminants. Hydrocarbon components commonly seen in natural gas are ethane ($C_2H_6$), propane ($C_3H_8$), butane ($C_4H_{10}$), pentane ($C_5H_{12}$) and some very slight hexane ($C_6H_{14}$), heptane ($C_7H_{16}$), octane ($C_8H_{18}$) and some heavier gases. The contaminants found in natural gas are carbon dioxide ($CO_2$), hydrogen sulfide ($H_2S$), nitrogen ($N_2$) and water vapor ($H_2O$) and so on. So natural gas is a mixture of many components. Because the volume of gas ($V$) bears a noticeable relationship with temperature ($T$) and absolute pressure ($p$), so some basic properties of gas will be reviewed first.

## 5.1.1 Equation of state of gas

Before researching the properties of true gas, let's think about a kind of ideal gas: its volume of molecules can be omitted compared with the volume that the whole gas occupies; there is no attractive and repulsive forces between molecules or molecules and walls of the container, all the collision between molecules is completely elastic, that is to say there is no loss of internal energy, this kind of gas is called ideal gas.

Under the condition of normal temperature and low pressure, most gases obey the law of ideal gas. But when the pressure increases, very big difference will appear between actual gas and ideal gas. Therefore, it is necessary to review the equation of state of ideal gas.

The equation of state of ideal gas is derived from the Boyle's law, the Charles-Gay-Lussac's law and the Avogadro' law:

$$pV = nRT \qquad (5-1)$$

Where: $p$— The absolute pressure, Pa;
$V$—Gas volume, m$^3$;
$T$—Absolute temperature, K;
$n$— Gas molar fraction, mol;
$R$—The universal gas constant.

Because the amount of substance of gas is equal to gas mass divided by the relative molecular mass of gas, the equation of state of ideal gas can also be expressed as:

$$pV = \frac{m}{M}RT \qquad (5-2)$$

Where: $m$—The mass of gas, g;
$M$—The molecular weight of gas, as to natural gas, it means the average molecular weight of gases.

The density of gas can be obtained after arranging the formula(5-2) again:

## 5.1.1 气体的状态方程

在开始研究真实气体的性质前,考虑这样一种假想的气体,其分子的体积相对于整个气体所占体积而言可忽略不计,分子之间或分子与容器壁之间没有吸引力和排斥力,所有分子间的碰撞都是完全弹性的,就是说碰撞中没有内能损失,这种气体称为理想气体。

常温低压下,多数气体服从理想气体定律,然而在压力增加时,气体的真实状态与理想气体的状态之间就会出现很大的差别。因此,有必要先重温一下理想气体状态方程。

理想气体状态方程是从波义耳、查理—盖吕萨克、阿伏加德罗定律结合而推导出的:

式中:$p$ = 绝对压力,Pa;
$V$ = 气体体积,m$^3$;
$T$ = 绝对温度,K;
$n$ = 气体的物质的量,mol;
$R$ = 通用气体常数。

由于气体的物质的量等于气体质量除以气体分子量,所以理想气体状态方程又可以表示为:

式中:$m$ = 气体质量,g;
$M$ = 气体分子量,对于天然气,它表示气体平均分子量。

将方程(5-2)重新整理后可给出气体的密度为:

$$\rho = \frac{m}{V} = \frac{Mp}{RT} \qquad (5-3)$$

Strictly speaking, no gas is ideal gas. There is always deviation to describe real gas by the law of ideal gas, lots of people have tried to obtain the calculation procedure of deviation between the real gas and ideal gas, and the most famous method in those is the Van der Waals equation. In recent years, researchers also have derived some more accurate equations, such as the Weattie-Brideman equation and the Benddict-Webb-Rubin equation, but at present the most commonly used equation in industry is:

严格讲没有一种气体是理想的,用理想气体定律来描述实际气体都有偏差,很多人试图得出真实气体与理想气体偏差的计算方法,其中最著名的是范德华(Van der Waals)方程。近年来人们也推导了一些更为精确的方程、如Weattie-Brideman方程、Benddict-Webb-Rubin方程,但目前工业上最常用的方程是:

$$pV = ZnRT \qquad (5-4)$$

In the formula above, Z is the deviation factor of gas, it means under the condition of a given temperature and pressure, the ratio is the volume of real gas to that of ideal gas, that is:

式中,Z是气体偏差系数,它表示在某一温度和压力条件下,真实气体的体积与理想气体的体积之比,即:

$$Z = \frac{V_t}{V_{sc}} = \frac{V_t}{nRT_{sc}/p_{sc}} \qquad (5-5)$$

In the formula above, $V_t$ is the volume of real gas. So Z is also called compressibility factor which is relevant to temperature and pressure. As to the concrete way to solve it, please refer to relevant books. If the equation of state is expressed with density, it can be written as:

式中,$V_t$为真实气体体积。因此Z也称为压缩因子,它与温度和压力有关,具体求法可参考有关书籍。若用密度来表示状态方程则为:

$$\rho = \frac{Mp}{ZRT} \qquad (5-6)$$

### 5.1.2 The compressibility of natural gas

The same as fluid, the definition of isothermal compressibility of natural gas is:

### 5.1.2 天然气的压缩性

和液体一样,天然气等温压缩系数的定义为:

$$C_g = -\frac{1}{V}\left(\frac{\partial V}{\partial p}\right)_T \qquad (5-7)$$

$$C_g = \frac{1}{\rho}\left(\frac{\partial \rho}{\partial p}\right)_T \qquad (5-8)$$

As to ideal gas, the following expression can be obtained:

$$V = \frac{nRT}{p}$$

And

$$\left(\frac{\partial V}{\partial p}\right)_T = -\frac{nRT}{p^2}$$

So the following expression can be obtained:

$$C_g = -\left(\frac{p}{nRT}\right)\left(-\frac{nRT}{p^2}\right) = \frac{1}{p} \qquad (5-9)$$

As to real gas:

$$V = \frac{nZRT}{p}$$

$$\left(\frac{\partial V}{\partial p}\right)_T = nRT\left(\frac{1}{p}\frac{\partial Z}{\partial p} - \frac{Z}{p^2}\right)$$

$$C_g = -\left(\frac{p}{nZRT}\right)nRT\left(\frac{1}{p}\frac{\partial Z}{\partial p} - \frac{Z}{p^2}\right)$$

That is:

$$C_g = \frac{1}{p} - \frac{1}{Z}\frac{\partial Z}{\partial p} \qquad (5-10)$$

### 5.1.3 The standard condition of natural gas

As to the natural gas, because its volume changes with temperature and pressure, so for the convenience of comparison, a standard condition needs to be regulated. In the SI system of reservoir engineering, the standard condition is regulated as: $p_{sc} = 0.101 \text{MPa}, t_{sc} = 20\text{°C}$. $p_{sc}$ and $t_{sc}$ are respectively called the standard pressure and the standard temperature.

### 5.1.4 The basic differential equation of gas seepage

The way to study the basic differential equation of gas seepage is the same as studying single-phase fluid seepage which needs to start from the continuing equation, the Darcy's law and the equation of state. The continuing equation can be obtained from the second chapter which is:

$$\nabla(\rho V) = -\frac{\partial(\rho\phi)}{\partial t} \qquad (5-11)$$

The Darcy equation is:

$$V = -\frac{K}{\mu_g}\nabla p \qquad (5-12)$$

After substituting the Darcy formula (5 – 12) and the state formula (5 – 6) into the formula (5 – 11), the following expression can be obtained:

将达西方程(5-12)和状态方程(5-6)代入式(5-11)得:

$$\nabla\left(\frac{Mp}{ZRT}\frac{K}{\mu_g}\nabla p\right) = \frac{\partial(\rho\phi)}{\partial t}$$

If the formation and gas are homogeneous, and the seepage process of gas is isothermal, then the viscosity of gas is relevant to pressure, so the formula above can be simplified as:

若地层和气体是均质的,气体的渗流过程是等温的,此时气体的黏度与压力有关,所以上式可化简为:

$$\frac{MK}{RT}\nabla\left(\frac{p}{\mu_g Z}\nabla p\right) = \frac{\partial(\rho\phi)}{\partial t} \qquad (5-13)$$

After expanding the right side of the formula above, the following expression can be obtained:

将上式右边展开得:

$$\frac{\partial(\rho\phi)}{\partial t} = \phi\frac{d\rho}{dp}\frac{\partial p}{\partial t} + \rho\frac{d\phi}{dp}\frac{\partial p}{\partial t} = \phi\rho\left(\frac{1}{\rho}\frac{d\rho}{dp} + \frac{1}{\phi}\frac{d\phi}{dp}\right)\frac{\partial p}{\partial t} = \rho\phi(C_g + C_f)\frac{\partial p}{\partial t}$$

Let $C_t = C_g + C_f$ and substitute the formula (5 – 6) into the expression above, then the following formula can be obtained:

令 $C_t = C_g + C_f$,并将式(5-6)代入上式得:

$$\frac{\partial(\rho\phi)}{\partial t} = \frac{\phi C_t p}{ZRT}\frac{\partial p}{\partial t} \qquad (5-14)$$

Combine the formula (5 – 13) and formula (5 – 14), and the following formula can be obtained:

联立式(5-13)和式(5-14)得:

$$\nabla\left(\frac{p}{\mu_g Z}\nabla p\right) = \frac{\phi C_t}{K}\frac{p}{Z}\frac{\partial p}{\partial t} \qquad (5-15)$$

Let:

令:

$$m = \int\frac{2p}{\mu_g Z}dp$$

In the formula above, $m$ is called pseudo-pressure. According to the derivation rule, the following expressions can be obtained:

式中,$m$ 称为假压力。根据复合函数求导法则,有:

$$\nabla m = \frac{2p}{\mu_g Z}\nabla p \qquad (5-16)$$

$$\frac{\partial m}{\partial t} = \frac{2p}{\mu_g Z}\frac{\partial p}{\partial t} \qquad (5-17)$$

After combining the formula (5 – 15), formula (5 – 16) and the formula (5 – 17), the following expression can be obtained:

联立式(5-15)、式(5-16)和式(5-17)得:

$$\nabla^2 m = \frac{\phi \mu_g C_t}{K} \frac{\partial m}{\partial t} \qquad (5-18)$$

Because the viscosity of gas and the coefficient of compressibility $C_g$ change with the pressure, so the formula (5 – 18) is nonlinear. Generally speaking, $C_g \gg C_f$, so $C_t \approx C_g$. Many scholars have done lots of work in order to solve the formula (5 – 18), it has been found after researching that the viscosity of gas is in direct proportion to pressure, while the coefficient of compressibility is in inverse proportion to pressure, so it is prone to reduce the correlation of their product with pressure. Under high pressure condition this beneficial effect is very obvious, and then the product of the two is very constant. At this time, the formula (5 – 18) and the basic differential equation of fluid seepage are the same linear equation which is very easy to solve and can even use the previous solution directly.

由于气体的黏度 $\mu_g$ 及压缩系数 $C_g$ 随着压力而变化,所以方程(5 – 18)是非线性的。通常情况下 $C_g \gg C_f$,所以 $C_t \approx C_g$。许多学者为了求解方程(5 – 18)做了大量工作,经研究发现,气体黏度与压力成正比,而压缩系数与压力成反比,这就倾向于减少它们的乘积同压力的相关性。在高压范围内这种有利的效果特别明显,这时两者的乘积相当恒定,此时,方程(5 – 18)与液体渗流的基本微分方程一样是一个线性方程,求解很方便,甚至可以直接利用前面的解。

When pressure is very low, $\mu_g C_t$ is not a constant anymore and changes with pressure. In fact, even when the pressure is very high, $\mu_g C_t$ and pressure still have definite correlation, but this change is not too much. So strictly speaking, the formula (5 – 18) is still a nonlinear partial differential equation, for the convenience of solving, it should be further linearized by some means. Let:

当压力较低时,$\mu_g C_t$ 不再是常数,而是随着压力变化而变化。事实上,即使压力较高时,$\mu_g C_t$ 与压力仍有一定的相关性,只是变化不大。因此,严格地说方程(5 – 18)仍是一个非线性的偏微分方程,为了便于求解,应该设法进一步将它线性化。令:

$$t_a = \int_0^t \frac{dt}{\mu_g C_t} \qquad (5-19)$$

In the formula above, $t_a$ is called the pseudo-time, so:

式中,$t_a$ 称为假时间,则:

$$\frac{\partial m}{\partial t} = \frac{\partial m}{\partial t_a} \frac{1}{\mu_g C_t} \qquad (5-20)$$

After substituting the formula (5 – 20) into the formula (5 – 18), the following expression can be obtained:

将式(5 – 20)代入式(5 – 18)中得:

$$\frac{K}{\phi} \nabla^2 m = \frac{\partial m}{\partial t_a} \qquad (5-21)$$

Because the coefficient $\frac{K}{\phi}$ is a constant, the equation (5 – 21) is a completely linearized differential equation of gas seepage. Now the dimension of pseudo-pressure and pseudo-time will be analyzed again, according to the definition:

由于系数 $\frac{K}{\phi}$ 是一个常数,因此,方程(5 – 21)是一个完全线性化了的气体渗流的微分方程。现在再来分析一下假压力和假时间的量纲,根据定义有:

$$[m] = \frac{[p]^2}{[\mu_g][Z]} = \frac{[M/(T^2L)][p]}{M/TL} = \frac{[p]}{T} \tag{5-22}$$

$$[t_a] = \frac{[t]}{[\mu_g][C_t]} = \frac{T}{M/TL \ T^2L/M} = \text{Dimension}$$

The calculation procedure of pseudo-pressure and pseudo-time will be introduced next. Assume the initial pressure of a reservoir is $p_i$, divide the pressure interval $[0, p_i]$ into several equal portions, and according to the components of gas in the reservoir and temperature in the reservoir and also the knowledge instructed in the course of Physics of Reservoir, the viscosity of gas and deviation factor $Z$ can be ascertained. The first three columns in Table 5-1 are the data of viscosity of gas and deviation factor of a given gas reservoir under different pressure.

下面再介绍一下假压力和假时间的计算方法。设油气藏的原始压力为 $p_i$，将压力区间 $[0, p_i]$ 分成若干等份，再根据气藏中气体的组分、气藏温度等因素运用油气藏物理学中所讲授的知识，确定各压力下的气体黏度和偏差系数 $Z$。表 5-1 中的前 3 列就是某气藏不同压力下的气体黏度和偏差系数数据。

Table 5-1  The calculation procedure of psuedo-pressure
表 5-1  某气藏不同压力下的气体黏度和偏差系数数据

| \multicolumn{3}{c}{$pVT$ data} | \multicolumn{3}{c}{numerical integration} | \multicolumn{2}{c}{pseudo-pressure} |
|---|---|---|---|---|---|---|---|
| $p$ MPa | $\mu_g$ mPa·s | $Z$ | $\frac{2p}{\mu_g Z}\cdot 10^2$ MPa/s | $\overline{\frac{2p}{\mu_g Z}}\cdot 10^2$ MPa/s | $\Delta p$ MPa | $\overline{\frac{2p}{\mu_g Z}}\Delta p \cdot 10^2$ MPa/s | $m(p)=\sum\overline{\frac{2p}{\mu_g Z}}\Delta p$ MPa/s |
| 2.76 | 0.01286 | 0.937 | 4.581 | 2.290 | 2.76 | 6.322 | 6.322 |
| 5.52 | 0.01390 | 0.882 | 9.005 | 6.793 | 2.76 | 18.747 | 25.069 |
| 8.28 | 0.01530 | 0.832 | 13.009 | 11.007 | 2.76 | 30.379 | 55.448 |
| 11.04 | 0.01680 | 0.794 | 16.553 | 14.781 | 2.76 | 40.796 | 96.244 |
| 13.80 | 0.01840 | 0.770 | 19.481 | 18.017 | 2.76 | 49.726 | 145.970 |
| 16.56 | 0.02010 | 0.763 | 21.596 | 20.538 | 2.76 | 56.685 | 202.655 |
| 19.32 | 0.02170 | 0.775 | 22.958 | 22.277 | 2.76 | 61.484 | 264.139 |
| 22.08 | 0.02340 | 0.797 | 23.660 | 23.309 | 2.76 | 64.332 | 328.431 |
| 24.84 | 0.02500 | 0.827 | 24.010 | 23.835 | 2.76 | 65.785 | 394.256 |
| 27.60 | 0.02660 | 0.860 | 24.130 | 24.070 | 2.76 | 66.460 | 460.716 |
| 30.36 | 0.02831 | 0.896 | 23.938 | 24.034 | 2.76 | 66.334 | 527.05 |

Because only the difference of pseudo-pressure is required in practical application, so when calculating, the pseudo-pressure can be written as:

$$m = \int_0^p \frac{2p}{\mu_g Z} \mathrm{d}p$$

Draw the point $\frac{2p}{\mu_g Z}$ under different pressure on the coordinate paper on which the pressure $p$ as the $x$-axis and $\frac{2p}{\mu_g Z}$ as the $y$-axis, and then the curve shows in Figure 5-1 can be obtained. The corresponding pseudo-pressure $m(p)$

实用时，由于只需用不同压力下假压力的差值，所以在计算时可以将假压力写成：

将不同压力下的 $\frac{2p}{\mu_g Z}$ 点在以压力 $p$ 为横坐标、以 $\frac{2p}{\mu_g Z}$ 为纵坐标的图上得到一条如图 5-1 所示的曲线。采用梯形积分就可

corresponding to every pressure can be solved by trapezoidal integration, and the calculation procedure is shown in Table 5-1. If the values of pseudo-pressure corresponding to different pressures are drawn on the figure which the pressure $p$ is used as the $x$-axis and pseudo-pressure $m(p)$ as the $y$-axis, the curve shows in Figure 5-2 can be obtained. With this curve, the corresponding value of pseudo-pressure of an arbitrary pressure will be easily found. And the solving way of pseudo-time is also the same. Table 5-2 is the calculation procedure of pseudo-time.

以求出各个压力相对应的假压力 $m(p)$，其计算步骤见表 5-1。如果将不同压力下的假压力点在以压力 $p$ 为横坐标、以假压力 $m(p)$ 为纵坐标的图上，就可以得到一条如图 5-2 所示的曲线。有了这条曲线，就能很方便地找到与任何一个压力相对应的假压力的值。假时间的求法也是如此，表 5-2 就是假时间的求解过程。

Table 5-2  The calculation procedure of pseudo-time

表 5-2  假时间的计算

| $t$ h | $p$ MPa | $\mu_g$ mPa·s | $C_t$ MPa$^{-1}$ | $A = \dfrac{1}{\mu C_t}$ s$^{-1}$ | $B = \dfrac{A_{n-1}+A_n}{2}$ | $C = B(t_n - t_{n-1})$ s$^{-1}$ | $t_a = \sum C_i$ s$^{-1}$ |
|---|---|---|---|---|---|---|---|
| 0.00 | 8.932 | 0.0144 | 0.1244 | 558.235 | 279.118 | 251.205 | 251.205 |
| 0.25 | 11.126 | 0.01536 | 0.0991 | 656.954 | 607.595 | 546.835 | 798.040 |
| 0.50 | 12.505 | 0.01596 | 0.08657 | 723.769 | 690.357 | 621.321 | 1419.361 |
| 0.75 | 13.677 | 0.01647 | 0.07756 | 782.832 | 753.301 | 677.970 | 2097.332 |
| 1.00 | 14.483 | 0.01689 | 0.07200 | 822.310 | 802.573 | 722.316 | 2819.648 |
| 1.25 | 15.186 | 0.01726 | 0.0675 | 858.332 | 840.194 | 756.175 | 3575.823 |
| 1.50 | 15.723 | 0.01755 | 0.06431 | 886.022 | 872.05 | 0.784.845 | 4260.668 |
| 1.75 | 16.150 | 0.01778 | 0.06189 | 908.757 | 897.389 | 807.651 | 5168.319 |
| 2.00 | 16.605 | 0.01802 | 0.05943 | 933.769 | 921.263 | 829.137 | 5997.456 |
| 2.25 | 17.025 | 0.01825 | 0.05724 | 957.277 | 945.523 | 850.970 | 6848.427 |
| 2.50 | 17.363 | 0.01843 | 0.05557 | 976.415 | 966.846 | 870.161 | 7718.588 |
| 2.75 | 17.666 | 0.01859 | 0.05412 | 993.946 | 985.181 | 886.662 | 8605.251 |
| 3.00 | 17.900 | 0.01872 | 0.05302 | 1007.52 | 1000.734 | 900.600 | 9505.912 |

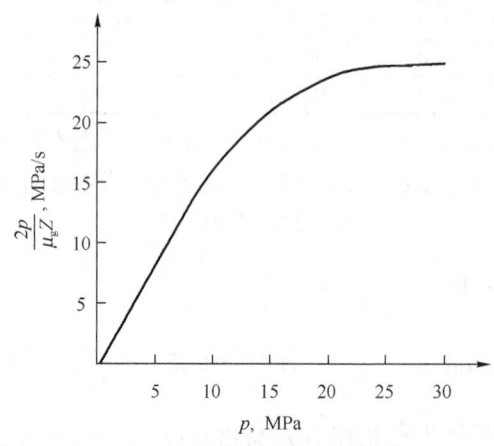

Figure 5-1  The integrated function curve of pseudo-pressure

图 5-1  假压力被积函数曲线

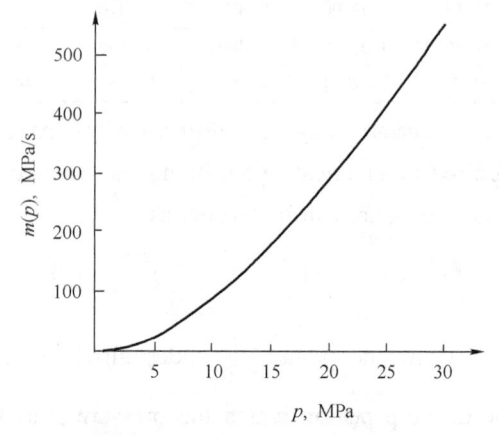

Figure 5-2  The relation curve of pseudo-pressure and pressure

图 5-2  假压力与压力的关系曲线

It will be found that in the following sections, the calculation of pseudo-pressure and pseudo-time is the foundation of solving problem.

## 5.2 Steady seepage of gas

If the produced gas is supplemented with other equivalent gas, then the gas well will attain a steady state after producing with constant production for some time. In fact, there is no gas resource outside and there is generally no steady flow in gas well production, it can only be considered as steady state flow in a short time. The planar one-dimensional steady seepage and the planar radial steady flow will be analyzed in the following.

### 5.2.1 The planar one-dimensional steady seepage

The gas reservoir with supply pressure $p_e$ and the outlet pressure $p_w$ as shown in Figure 5–3 will attain a steady state after producing for some time, and the Laplace's equation of the seepage of gas is:

$$\frac{d^2 m}{dx^2} = 0$$

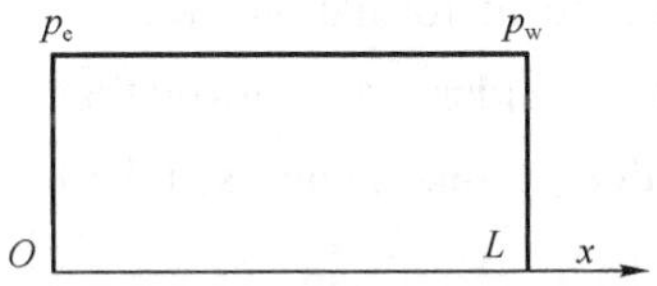

Figure 5–3  The planar one-dimensional steady state seepage

图 5–3  一维稳定渗流

The formula above is accordant with the Laplace's equation of the uncompressible steady seepage, and the only difference is that the real pressure is replaced by pseudo-pressure. The boundary condition of this problem expressed with the pseudo-pressure can be written as:

$$m\big|_{x=0} = m_e;\ m\big|_{x=L} = m_w$$

It is easy to know that the solution to this problem is:

$$m(x) = \begin{cases} m_e - \dfrac{m_e - m_w}{L} x \\ m_w + \dfrac{m_e - m_w}{L}(L - x) \end{cases} \quad (5-23)$$

The formula (5–23) is the distribution formula of pseudo-pressure, while the gas flow rate on an arbitrary cross section is:

## 5.2 气体的稳定渗流

如果气井采出多少气体外界就补充进等量的其他气体，则气井以恒产量生产一段时间后会达到稳定。事实上，外界不可能有气源，气井生产一般不存在稳定流，只是在一个短时期内可以把流动看作是稳定的。下面就平面一维稳定渗流和平面径向稳定渗流进行分析。

### 5.2.1 平面一维稳定渗流

如图 5–3 所示的供给压力 $p_e$、出口压力为 $p_w$ 的气藏，在生产一段时间后便会达到稳定，此时气体渗流的拉氏方程为：

它与液体作不可压缩稳定渗流的拉氏方程一致，只是用假压力代替了真实压力。这一问题的边界条件用假压力表示可写成：

易知，这一问题的解为：

式(5–23)是假压力的分布公式，而任一截面上的气体流量为：

$$Q = -\frac{KA}{\mu_g}\frac{dp}{dx} \qquad (5-24)$$

It can be known from the definition of pseudo-pressure that:

由假压力的定义知：

$$\frac{dp}{dx} = -\frac{\mu_g Z}{2p}\frac{m_e - m_w}{L} \qquad (5-25)$$

The following expression can be obtained after substituting the formula (5-25) into the formula (5-24):

将式(5-25)代入式(5-24)得：

$$Q = \frac{KA(m_e - m_w)Z}{2pL} \qquad (5-26)$$

The volume flow rate of gas $Q$ in the formula (5-26) is relevant with the pressure, the volume flow rate of gas on every cross section obtained by the formulas above is not constant which is usually conversed to be the volume flow rate under standard conditions. It can be obtained through the equation of state that:

式(5-26)中的气体体积流量 $Q$ 与压力的大小有关，由上式计算的各个截面上的体积流量不为常数，所以通常将它换算成标准条件下的体积流量。由状态方程得：

$$\frac{p_{sc}Q_{sc}}{T_{sc}} = \frac{pQ}{ZT} \qquad (5-27)$$

Where: $Q_{sc}$ —The volume flow rate under standard conditions, m³/s;

$T$— The absolute temperature of formation, K;

$Z$—The deviation factor under the formation condition and the value is one under standard conditions.

式中：$Q_{sc}$ = 标准条件下的体积流量, m³/s;

$T$ = 地层绝对温度, K;

$Z$ = 地层条件下的偏差系数, 标准条件下的偏差系数为1。

After combining the equation (5-26) and the formula (5-27) the following expression can be obtained:

联立方程(5-26)和式(5-27)得：

$$Q_{sc} = \frac{KA(m_e - m_w)}{2\mu_g L}\frac{T_{sc}}{p_{sc}T} \qquad (5-28)$$

The formula (5-28) is the production formula of gas when it is planar one-dimensional steady seepage. It can be seen that, under standard conditions, when gas flowing is isothermal, the standard volume flow rate of gas passing through an arbitrary cross section is a constant.

式(5-28)是气体作平面一维稳定渗流的产量公式。由此可见，气体作等温稳定渗流时，通过任一截面的标准条件下的气体体积流量为一常数。

When pressure is small, natural gas can be approximately considered as ideal gas, that is $Z=1$ and viscosity $\mu_g$ is constant. So the pseudo-pressure can be simplified as:

当压力较低时，则天然气可以近似当作理想气体来处理，即 $Z=1$, 黏度 $\mu_g$ 为常数，因而假压力可以简化为：

$$m = \int \frac{2p}{\mu_g}dp = \frac{p^2}{\mu_g} + C \qquad (5-29)$$

After substituting the formula above into the formula (5-28), the following expression can be obtained:

$$Q_{sc} = \frac{KA(p_e^2 - p_w^2)}{2\mu_g L} \frac{T_{sc}}{p_{sc} T} \tag{5-30}$$

The formula (5-30) is the production formula of the ideal gas when its action is planar one-dimensional isothermal steady seepage.

### 5.2.2 The planar radial fluid steady flow

As shown in Figure 5-4, assume that there is a gas well in the center of the circular formation, the supply pressure is $p_e$, the well bottom hole pressure is $p_w$, the supply radius is $r_e$ and the bottom hole radius is $r_w$, then the basic differential equation when gas attaining steady state seepage is:

$$\frac{1}{r}\frac{d}{dr}\left(r\frac{dm}{dr}\right) = 0$$

The boundary condition is:

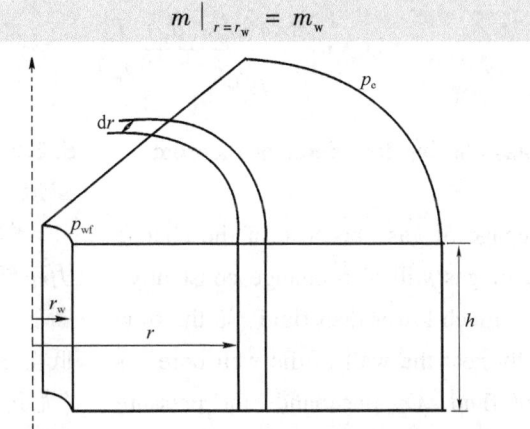

$$m\big|_{r=r_e} = m_e$$
$$m\big|_{r=r_w} = m_w$$

Figure 5-4 The planar radial steady state flow
图 5-4 平面径向稳定渗流

Obviously the solution of this problem is:

$$m(r) = \begin{cases} m_e - \dfrac{m_e - m_w}{\ln\dfrac{r_e}{r_w}}\ln\dfrac{r_e}{r} \\[2ex] m_w + \dfrac{m_e - m_w}{\ln\dfrac{r_e}{r_w}}\ln\dfrac{r}{r_w} \end{cases} \tag{5-31}$$

And the corresponding gradient of pseudo-pressure is:

$$\frac{dm}{dr} = \frac{m_e - m_w}{\ln\dfrac{r_e}{r_w}}\frac{1}{r} \tag{5-32}$$

The flow rate of gas passing through an arbitrary cross section is:

$$Q = \frac{2\pi r K h}{\mu_g}\frac{dp}{dr} = \frac{2\pi r K h}{\mu_g}\frac{\mu_g Z}{2p}\frac{dm}{dr} = \frac{\pi K h(m_e - m_w)Z}{p\ln\frac{r_e}{r_w}}$$

If the flow rate under the formation condition is converted to be the flow rate under the standard condition, the formula of gas flow rate under standard condition can be further obtained from the formula above and the formula (5-27) when gas is doing planar radial fluid isothermal steady flow is as follows:

$$Q_{sc} = \frac{\pi K h(m_e - m_w)}{\ln\frac{r_e}{r_w}}\frac{T_{sc}}{p_{sc}T} \qquad (5-33)$$

When the formation pressure is low, the natural gas can be considered as ideal gas, the corresponding production formula can be obtained by substituting the formula (5-29) into the formula (5-33):

$$Q_{sc} = \frac{\pi K h(p_e^2 - p_w^2)}{\mu_g \ln\frac{r_e}{r_w}}\frac{T_{sc}}{p_{sc}T} \qquad (5-34)$$

### 5.2.3 The seepage of gas obeying the binomial theorem

In the course of seepage of gas, because of the change of pressure, the volume of gas will also change constantly, and the viscosity of gas is much lower than fluid, so the flow velocity of gas (especially near the wall of the well bore) is much higher than that of fluid. On one hand, the pressure loss much more converges near the bore hole wall to protect the gas well from contaminating is more important; on the other hand, the inertia loss in the course of seepage of gas can not be neglected, and the Darcy's law can not be applied here any more, and now the radial flow of gas well needs to be described with the following binomial law:

$$\frac{dp}{dr} = \frac{\mu_g}{K}v - b\rho v^2 \qquad (5-35)$$

In the formula above, $\rho$ is the density of gas, and:

$$\rho = \frac{Mp}{ZRT} \qquad (5-36)$$

The flow velocity of gas on an arbitrary section in

formation can be expressed with the following expression:

$$v = \frac{Q}{2\pi rh} = \frac{p_{sc}Q_{sc}ZT}{2\pi p T_{sc} rh} \tag{5-37}$$

And the following expression can be obtained by substituting the formula (5-36) and formula (5-37) into the formula (5-35):

$$\frac{dp}{dr} = \frac{\mu_g p_{sc} Q_{sc} ZT}{2\pi Kh p T_{sc}} \frac{1}{r} - \frac{p_{sc}^2 Q_{sc}^2 bZTM}{4\pi^2 h^2 T_{sc}^2 pR} \frac{1}{r^2} \tag{5-38}$$

The result obtained from lots of calculation shows: when the ratio of bottom pressure $p_w$ and the formation pressure (or the boundary pressure $p_e$) is bigger than 0.8, it can be considered that $\mu_g Z$ will not change with pressure. In the process of gas production, the temperature drop of gas layer will be quickly made up by the surrounding formation, so the seepage can be considered as isothermal, so after separating the variables of formula (5-35) and integrating it the following expression can be obtained:

$$p_e^2 - p_w^2 = \frac{Q_{sc} p_{sc} \mu_g ZT}{\pi Kh T_{sc}} \ln\frac{r_e}{r_w} - \frac{Q_{sc}^2 p_{sc}^2 bZTM}{2\pi^2 h^2 T_{sc}^2 R}\left(\frac{1}{r_w} - \frac{1}{r_e}\right) \tag{5-39}$$

It can be known from the formula (5-36):

$$\rho_{sc} = \frac{p_{sc}M}{RT}$$

After substituting the expression above into the formula (5-39), the following expression can be obtained:

$$p_e^2 - p_w^2 = \left(\frac{\mu_g}{\pi Kh}\frac{ZTp_{sc}}{T_{sc}}\ln\frac{r_e}{r_w}\right)Q_{sc} - \frac{b\rho_{sc}p_{sc}ZT}{2\pi^2 h^2 T_{sc}}\left(\frac{1}{r_w} - \frac{1}{r_e}\right)Q_{sc}^2 \tag{5-40}$$

For $\mu_g$ and $Z$ in the formula (5-40) we choose the average viscosity and average gas deviation factor between $p_w \sim p_e$ generally. If the expression above is written as the form of binomial expression, it will be:

$$p_e^2 - p_w^2 = AQ_{sc} + BQ_{sc}^2 \tag{5-41}$$

$$A = \frac{\mu_g ZTp_{sc}}{\pi Kh T_{sc}}\ln\frac{r_e}{r_w} \tag{5-42}$$

$$B = -\frac{b\rho_{sc}p_{sc}ZT}{2\pi^2 h^2 T_{sc}}\left(\frac{1}{r_w} - \frac{1}{r_e}\right) \tag{5-43}$$

In the formulas above, $A$ and $B$ are respectively the Darcy flow coefficient and the non-Darcy flow coefficient.

In well testing, if $\dfrac{p_e^2 - p_w^2}{Q_{sc}}$ and $Q_{sc}$ are respectively considered as the Y-axis and X-axis, a straight line with intercept $A$ and slope $B$ will be obtained as shown in Figure 5-5. And the permeability of formation $K$ will be further obtained from the formula (5-42).

When the imperfect well is considered, the radius of gas well should be replaced by the reduced radius of gas well.

The absolute open flow potential of gas well is an important index reflecting the potential deliverability of gas well. Regardless of exploration well or brought-in gas producing well, the magnitude of absolute open capacity needs to be known on time. In order to realize the target of gas well working with high and steady production, one-fourth of the magnitude of absolute open flow potential is chosed as the reasonable producing standard of gas well. So, it is very important to ascertain precisely the magnitude of absolute open flow potential of gas well.

When 0.101MPa is chosen as the bottom hole flowing pressure, the biggest potential deliverability is the absolute open flow potential of gas well. So the following expression can be obtained from the formula (5-4):

$$p_e^2 - (0.101)^2 = AQ_{AOF} + BQ_{AOF}^2$$

In the formula above, $Q_{AOF}$ is the absolute open flow potential of gas well shown with $10^4 m^3/d$. Because $p_e^2 - (0.101)^2 \approx p_e^2$, so:

$$Q_{AOF} = \frac{\sqrt{A^2 + 4Bp_e^2} - A}{2B} \qquad (5-44)$$

**Example 5-1** In May 1971, multi-point well testing is run to a given gas well, the formation pressure $p_e = 28.15$ MPa, the value of bottom hole flowing pressure and gas production rate of four test points are shown in Table 5-3. Please solve the absolute open flow potential.

Table 5-3  The testing situation
表 5-3  测试情况表

| test point | $p_e$, MPa | $p_{wf}$, MPa | $Q, 10^4 m^3/d$ | $\dfrac{p_e^2 - p_{wf}^2}{Q}$, $(MPa^2 \cdot d)/m^3$ |
|---|---|---|---|---|
| 1 | 28.151 | 22.173 | 39.74 | 7.5699 |
| 2 | 28.151 | 23.307 | 35.46 | 7.0294 |
| 3 | 28.151 | 24.275 | 31.32 | 6.4878 |
| 4 | 28.151 | 25.269 | 26.19 | 5.8784 |

**Solution**: Firstly, work out the value of $\dfrac{p_e^2 - p_{wf}^2}{Q_{sc}}$ under different bottom pressure, and draw the relational data of $\dfrac{p_e^2 - p_w^2}{Q_{sc}} - Q_{sc}$ on the coordinate paper, and a very good straight line will be obtained (Figure 5-6). The intercept with coordinate of this straight line is $A = 2.5854$, slope $B = 0.1253$, the correlation factor $r = 0.9998$, the absolute open flow potential $Q_{AOF} = 69.88 \times 10^4 \text{m}^3/\text{d}$ will be obtained after substituting $A$ and $B$ into the formula (5-44).

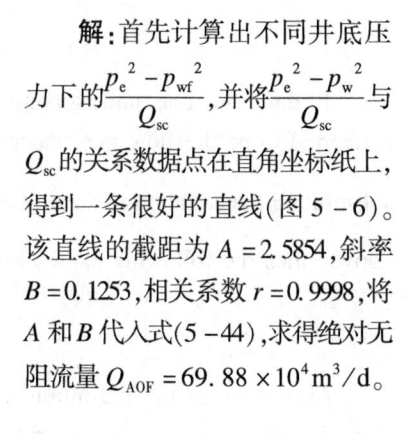

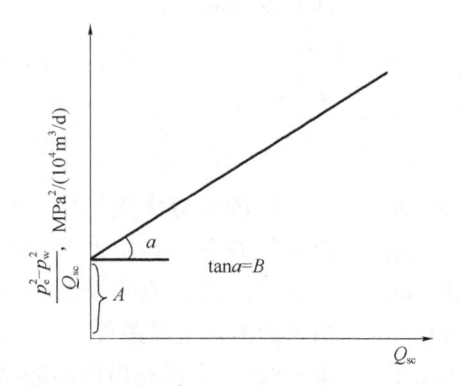

解：首先计算出不同井底压力下的 $\dfrac{p_e^2 - p_{wf}^2}{Q_{sc}}$，并将 $\dfrac{p_e^2 - p_w^2}{Q_{sc}}$ 与 $Q_{sc}$ 的关系数据点在直角坐标纸上，得到一条很好的直线(图5-6)。该直线的截距为 $A = 2.5854$，斜率 $B = 0.1253$，相关系数 $r = 0.9998$，将 $A$ 和 $B$ 代入式(5-44)，求得绝对无阻流量 $Q_{AOF} = 69.88 \times 10^4 \text{m}^3/\text{d}$。

Figure 5-5  The binomial well testing data
图 5-5  二项式试井数据

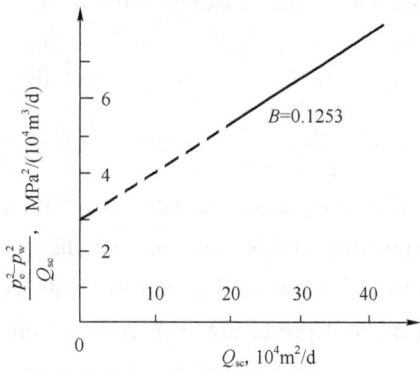

Figure 5-6  The binomial well testing curve
图 5-6  二项式试井曲线

## 5.3  Unsteady seepage of gas

In the first section of this chapter, the basic differential formula (5-21) of seepage of real gas has been derived, its corresponding radial diffusion equation is:

## 5.3  气体的不稳定渗流

在本章第一节中，已推导出了真实气体渗流的基本微分方程式(5-21)，其相应的径向扩散方程为：

$$\frac{1}{r}\frac{\partial}{\partial r}\left(r\frac{\partial m}{\partial r}\right) = \frac{\phi}{K}\frac{\partial m}{\partial t_a} \tag{5-45}$$

If the formation is homogeneous, isopachous and isotropic infinite formation, the gas is isothermal in the process of seepage, the initial pressure of formation is $p_i$, and there is only one production well with constant production (point sink), then the corresponding initial condition and the outer boundary condition are:

若地层是均质、等厚、各向同性的无限大地层，气体在渗流过程中是等温的，地层的原始压力为 $p_i$，地层中只有一口定产量的生产井(点汇)，则其相应的初始条件和外边界条件为：

$$m(r,0) = m_i \tag{5-46}$$
$$m(\infty,t) = m_i \tag{5-47}$$

In the expressions above, $m_i$ is the corresponding pseudo-pressure of initial formation pressure $p_i$. And the inner boundary condition of this problem is:

式中，$m_i$ 是与原始地层压力 $p_i$ 相对应的假压力。而这一问题的内边界条件为：

$$Q = \frac{2\pi Kh}{\mu_g}\left(r\frac{\partial p}{\partial r}\right)\bigg|_{r\to 0}$$

If the flow rate underground in the formula above is converted to be the flow rate under standard condition, that is substituting the equation of state $Q = \dfrac{p_{sc}Q_{sc}ZT}{pT_{sc}}$ into the formula above, then the following expression can be obtained:

$$\frac{2p}{\mu_g Z}r\frac{\partial p}{\partial r}\bigg|_{r\to 0} = \frac{p_{sc}Q_{sc}T}{\pi Kh T_{sc}}$$

According to the definition of pseudo-pressure, the formula above can be further written as:

$$r\frac{\partial m}{\partial r}\bigg|_{r\to 0} = \frac{G}{2\pi Kh} \qquad (5-48)$$

$$G = \frac{2p_{sc}Q_{sc}T}{T_{sc}} \qquad (5-49)$$

After comparing the basic differential equation and the corresponding initial condition and the outer-inner boundary condition of a gas well producing with constant production and a oil well producing with constant production in infinite formation, it can be found that the two are accordant with each other, the solution of formula (5-45) ~ formula (5-48) can be obtained by the results before:

$$m(r,t) = m_i + \frac{G}{4\pi Kh}\text{Ei}\left(-\frac{r^2}{4\eta t_a}\right) \qquad (5-50)$$

$$\eta = \frac{K}{\phi}$$

In the formula above, $\eta$ is the diffusivity coefficient, the formula (5-50) is the solution of line sink of a gas well producing with constant production in infinite formation. After substituting the formula (5-49) into the formula (5-50), the following expression can be obtained:

$$m(r,t) = m_i + \frac{p_{sc}Q_{sc}T}{2\pi Kh T_{sc}}\text{Ei}\left(-\frac{r^2}{4\eta t_a}\right) \qquad (5-51)$$

When $\dfrac{r^2}{4\eta t_a} \leq 0.01$, the formula above can be expressed with logarithm approximate formula:

$$m(r,t) = m_i - \frac{p_{sc}Q_{sc}T}{2\pi Kh T_{sc}}\ln\left(\frac{4\eta t_a}{e^\gamma r^2}\right) \qquad (5-52)$$

In the formula above, $\gamma$ is an Euler's constant. Because after a very short time the formula (5-52) can be used to calculate the bottom hole pseudo-pressure of gas well bottom, the formula can be further used to run the pressure fall-off test and the pressure build-up test, the way is the same as that of oil-well well test, here will not be introduced again.

Likewise, as to the infinite formation with multi gas wells producing simultaneously and the formation with linear boundary, the method of mirror and the superposition principle can also be used.

When gas pressure is very high, because $\mu_g C_t$ is approximate a constant, so as to the well bottom, the following expression can be obtained:

$$\Delta m = \int_{p_w}^{p_i} \frac{2p}{\mu_g Z} dp = \frac{p_i^2 - p_w^2}{\bar{\mu}_g \bar{Z}}$$

$$t_a = \frac{t}{\bar{\mu}_g \bar{C}_t}$$

用式(5-52)计算井底假压力，因此，可以进一步利用该公式进行压降试井和压力恢复试井，方法与油井试井完全相同，这里不再重复。

同样，对于无穷大地层多口气井同时生产以及含有直线边界的地层，仍可应用镜像反映法和叠加原理。

当气体压力较高时，由于$\mu_g C_t$近似为一常数，所以对于井底，有：

After substituting the expression above into the formula (5-52), the following expression can be obtained:

代入式(5-52)得：

$$p_w^2 = p_i^2 - \frac{\bar{\mu}_g p_{sc} Q_{sc} T \bar{Z}}{2\pi K h T_{sc}} \ln\left(\frac{4Kt}{e^\gamma r_w^2 \phi \bar{C}_t \bar{\mu}_g}\right) \qquad (5-53)$$

In the formula above, $\bar{\mu}_g$ and $\bar{Z}$ are respectively the average viscosity and the deviation factor of gas when the pressure of gas is between $p_w \sim p_e$. The pressure fall-off test and the pressure build-up test can be ran to gas well with the formula (5-52) or the formula (5-53). However, near the face of the well wall, the Darcy's law may be destroyed, generally in the area of 50~150 times of well diameter the Darcy's law will lose effectiveness. Assume that beyond the area that radius is $r$, the seepage of gas obeys the Darcy's law and the flow is unsteady state, then the following expression can be obtained from the formula (5-53):

式中，$\bar{\mu}_g$、$\bar{Z}$为气体在压力$p_w \sim p_e$的平均黏度和气体偏差系数。利用式(5-52)或式(5-53)可以对气井进行压降试井和压力恢复试井。然而，在井壁附近达西定律可能遭到破坏，一般认为在50~150倍井径范围内达西定律失效。设在半径为$r$的范围外气体渗流服从达西定律，且流动是不稳定的，则由式(5-53)得：

$$p_i^2 - p^2 = \frac{Q_{sc}\bar{\mu}_g p_{sc} \bar{Z} T}{2\pi K h T_{sc}} \ln\left(\frac{4Kt}{e^\gamma r^2 \phi \bar{\mu}_g \bar{C}_t}\right) \qquad (5-54)$$

While in the area that radius is $r$, gas obeys the binomial theorem and the flow is steady state. The following expression can be obtained from the formula (5-40):

而在半径为$r$的范围内，气体服从二项式定律，且流动是稳定的。由式(5-40)得：

$$p^2 - p_w^2 = \frac{\bar{\mu}_g p_{sc} Q_{sc} T \bar{Z}}{2\pi K h T_{sc}} \ln\left(\frac{r^2}{r_w^2}\right) + B Q_{sc}^2 \qquad (5-55)$$

After adding the formula (5-54) to the formula (5-55), the following expression can be obtained:

将式(5-54)和式(5-55)相加得：

$$p_i^2 - p_w^2 = \frac{\bar{\mu}_g p_{sc} Q_{sc} TZ}{2\pi KhT_{sc}} \ln\left(\frac{4Kt}{e^\gamma r_w^2 \phi \bar{\mu}_g \bar{C}_t}\right) + BQ_{sc}^2 \qquad (5-56)$$

If the skin factor of gas well is taken into account, and the natural logarithm is changed to be the common logarithm, then the expression above is changed to be:

$$p_i^2 - p_w^2 = A_1 Q_{sc} + BQ_{sc}^2 \qquad (5-57)$$

$$A_1 = \frac{2.303 \bar{\mu}_g p_{sc} \bar{Z} T}{2\pi KhT_{sc}}\left(\lg \frac{4Kt}{e^\gamma r_w^2 \phi \bar{\mu}_g \bar{C}_t} + 0.8686 S\right)$$

$$B = -\frac{b\rho_{sc} p_{sc} \bar{Z} T}{2\pi^2 h^2 T_{sc}}\left(\frac{1}{r_w} - \frac{1}{r_e}\right)$$

The content discussed above is the formula of pressure distribution of a gas well producing with constant production in infinite formation. While the actual formation is generally finite, after the pressure drop transmits to the boundary of gas field for a short while, it will go into the phase of pseudo-steady seepage. Assume the formation is circular and the accumulative gas production when formation pressure decreases from $p_i$ to $\bar{p}$ is:

$$V = C_t \pi r_e^2 h \phi (p_i - \bar{p})$$

Assume the gas production rate is $Q$, then:

$$Q = \frac{dV}{dt} = -\pi r_e^2 h \phi C_t \frac{\partial \bar{p}}{\partial t} = \frac{p_{sc} Q_{sc} 2T}{\bar{p} T_{sc}}$$

If the pseudo-pressure and the pseudo-time are adopted, the formula above can be written as:

$$\frac{\partial m(\bar{p})}{\partial t_a} = -\frac{2 p_{sc} Q_{sc} T}{\pi r_e^2 h \phi T_{sc}}$$

As to pseudo-steady state seepage, the decline rate of pseudo-pressure of every point in formation is same, so $\frac{\partial m}{\partial t_a} = \frac{\partial \bar{m}}{\partial t_a}$. In this phase, the basic differential equation of gas seepage is:

$$\frac{1}{r}\frac{\partial}{\partial r}\left(r\frac{\partial m}{\partial r}\right) = \frac{-1}{\pi Khr_e^2}\frac{2p_{sc}Q_{sc}T}{T_{sc}} = -\frac{G}{\pi Khr_e^2} \qquad (5-58)$$

By comparing the formula (5-58) and the basic differential equation (4-31) of pseudo-steady state liquid seepage, it can be known that they are accordant, and the boundary condition is also the same with each other, so the solution of equation (5-58) is:

$$m - m_w = \frac{1}{\pi Kh} \frac{p_{sc} Q_{sc} T}{T_{sc}} \left( \ln \frac{r}{r_w} - \frac{r^2}{2r_e^2} \right) \tag{5-59}$$

So if the boundary pressure $p_e$ is known, the expression above can be written as:

因此,若已知边界压力 $p_e$,则上式可写成:

$$p_e^2 - p_w^2 = \frac{\overline{\mu}_g \overline{Z}}{\pi Kh} \frac{p_{sc} Q_{sc} T}{T_{sc}} \left( \ln \frac{r_e}{r_w} - \frac{1}{2} + S \right) \tag{5-60}$$

If the average pressure of gas field $\bar{p}$ is known, a formula similar with the formula (4-37) can also be obtained:

若已知气藏的平均压力 $\bar{p}$,则同样可得与式(4-37)相类似的公式:

$$\bar{p}^2 - p_w^2 = \frac{\overline{\mu}_g \overline{Z}}{\pi Kh} \frac{p_{sc} Q_{sc} T}{T_{sc}} \left( \ln \frac{r_e}{r_w} - \frac{3}{4} + S \right) \tag{5-61}$$

If expressed with gas supply area, the expression above can be written as:

如果用供气面积表示,则上式可写成:

$$\bar{p}^2 - p_w^2 = \frac{\overline{\mu}_g \overline{Z}}{2\pi Kh} \frac{p_{sc} Q_{sc} T}{T_{sc}} \left( \ln \frac{4A}{e^{\gamma} C_A r_w^2} + 2S \right) \tag{5-62}$$

Assume that the Darcy's law is effective beyond the area that radius is $r$, in the area, the seepage obeys the binomial theorem. Assume the pressure is $p$ on the area contour, so it meets the formula (5-55), and according to the formula (5-62), it also meets the following expression:

设在半径为 $r$ 的范围以外,达西定律成立,而在 $r$ 范围内,渗流服从二项式定律。设半径为 $r$ 处的压力为 $p$,则它满足式(5-55),且根据式(5-62),它又满足下式:

$$\bar{p}^2 - p^2 = \frac{\overline{\mu}_g \overline{Z}}{2\pi Kh} \frac{p_{sc} Q_{sc} T}{T_{sc}} \left( \ln \frac{4A}{e^{\gamma} C_A r^2} + 2S \right) \tag{5-63}$$

It can be obtained from the formula (5-63) and the formula (5-55) that:

由式(5-63)和式(5-55)可得:

$$\bar{p}^2 - p_w^2 = \frac{\overline{\mu} \overline{Z}}{2\pi Kh} \frac{p_{sc} Q_{sc} T}{T_{sc}} \left( \ln \frac{4A}{e^{\gamma} C_A r_w^2} + 2S \right) + B Q_{sc}^2 \tag{5-64}$$

Rewrite the formula (5-64) to be the following form:

将式(5-64)改写成下列形式:

$$\bar{p}^2 - p_w^2 = A_2 Q_{sc} + B Q_{sc}^2 \tag{5-65}$$

$$A_2 = \frac{\overline{\mu} \overline{Z}}{2\pi Kh} \frac{p_{sc} T}{T_{sc}} \left( \ln \frac{4A}{e^{\gamma} C_A r_w^2} + 2S \right)$$

In the formula above, the constant $A_2$ and $B$ can be ascertained with the method of isochronal well test, and this method will be introduced in other relevant courses and will not be further discussed here. Once $A_2$ and $B$ are ascertained, the absolute open flow potential of gas well can be further ascertained from the formula (5-65).

式(5-65)中的常数 $A_2$ 和 $B$ 可以通过等时试井的方法来确定,这种方法在其他有关课程中将会介绍,这里不再深入讨论。一旦 $A_2$ 和 $B$ 确定以后,还可以进一步由式(5-65)确定气井的绝对无阻流量。

## Exercises

5-1  Please directly derive the production formula of ideal gas when it is doing planar one-dimensional isothermal steady state seepage.

5-2  Please directly derive the production formula of ideal gas when it is doing planar radial fluid isothermal steady state flow.

5-3  Assume that there is a given circular gas reservoir, its radius is $r_e = 1$km, the supply boundary pressure is $p_e = 10$MPa, $h = 10$m, $K = 0.1\mu m^2$, $\mu_g = 0.01$mPa·s, and the radius of well is $r_w = 0.1$m, $p_w = 7.9$MPa, the formation temperature is 85℃, the compressibility factor of gas is $Z = 0.84$, please solve the production of this gas well in the center of circular formation.

5-4  The following data in Table 5-4 are known, please try to calculate the pseudo-pressure and the pseudo-time (on the computer) and draw the relation curve of pseudo-pressure and pressure, and the pseudo-time and time respectively.

**Table 5-4  The testing data**
表 5-4  测试数据

| order number | $t$,h | $p$,MPa | $\mu$,mPa·s | $Z$ | $C_t$,MPa$^{-1}$ |
|---|---|---|---|---|---|
| 1 | 0 | 27.95 | 0.02428 | 0.9008 | 0.02679 |
| 2 | 0.25 | 26.91 | 0.02371 | 0.8904 | 0.02684 |
| 3 | 0.50 | 25.87 | 0.02314 | 0.8805 | 0.02687 |
| 4 | 0.75 | 24.83 | 0.02256 | 0.8712 | 0.02691 |
| 5 | 1.00 | 23.79 | 0.02198 | 0.8626 | 0.02702 |
| 6 | 1.25 | 22.75 | 0.02141 | 0.8548 | 0.02715 |
| 7 | 1.50 | 21.71 | 0.02083 | 0.8477 | 0.02727 |
| 8 | 1.75 | 20.67 | 0.02025 | 0.8416 | 0.02740 |
| 9 | 2.00 | 19.63 | 0.01968 | 0.8365 | 0.02753 |
| 10 | 2.25 | 18.59 | 0.01910 | 0.826 | 0.02773 |
| 11 | 2.50 | 17.55 | 0.01854 | 0.8299 | 0.02799 |
| 12 | 2.75 | 16.51 | 0.01799 | 0.8286 | 0.02832 |
| 13 | 3.00 | 15.47 | 0.01743 | 0.8288 | 0.02878 |

continue

| order number | $t$,h | $p$,MPa | $\mu$,mPa·s | $Z$ | $C_t$,MPa$^{-1}$ |
|---|---|---|---|---|---|
| 14 | 3.25 | 14.43 | 0.01688 | 0.8305 | 0.02904 |
| 15 | 3.50 | 13.39 | 0.01636 | 0.8339 | 0.02945 |
| 16 | 3.75 | 12.25 | 0.01592 | 0.8390 | 0.02987 |
| 17 | 4.00 | 11.21 | 0.01547 | 0.8457 | 0.03045 |

Chapter 5 Study Guide
第五章学习指南

# 6 Foundation of water/oil displacement theory

水驱油理论基础

# 第六章 知识图谱

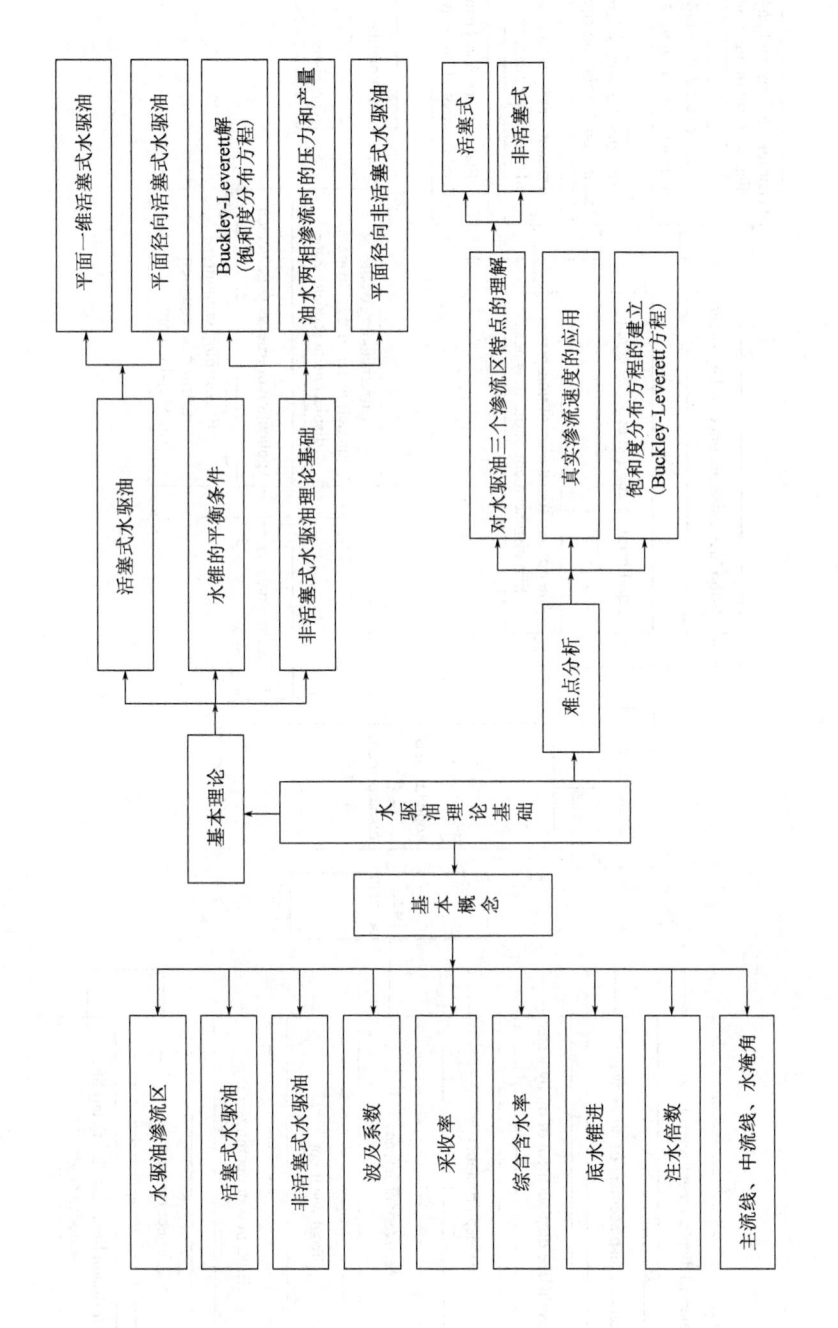

- 水驱油理论基础
  - 基本理论
    - 活塞式水驱油
      - 平面一维活塞式水驱油
      - 平面径向活塞式水驱油
    - 水锥的平衡条件
    - 非活塞式水驱油理论基础
      - Buckley-Leverett解（饱和度分布方程）
      - 油水两相渗流时的压力和产量
      - 平面径向非活塞式水驱油
  - 难点分析
    - 对水驱油三个渗流区特点的理解
      - 活塞式
      - 非活塞式
    - 真实渗流速度的应用
    - 饱和度分布方程的建立（Buckley-Leverett方程）
  - 基本概念
    - 水驱油渗流区
    - 活塞式水驱油
    - 非活塞式水驱油
    - 波及系数
    - 采收率
    - 综合含水率
    - 底水锥进
    - 注水倍数
    - 主流线、中流线、水淹角

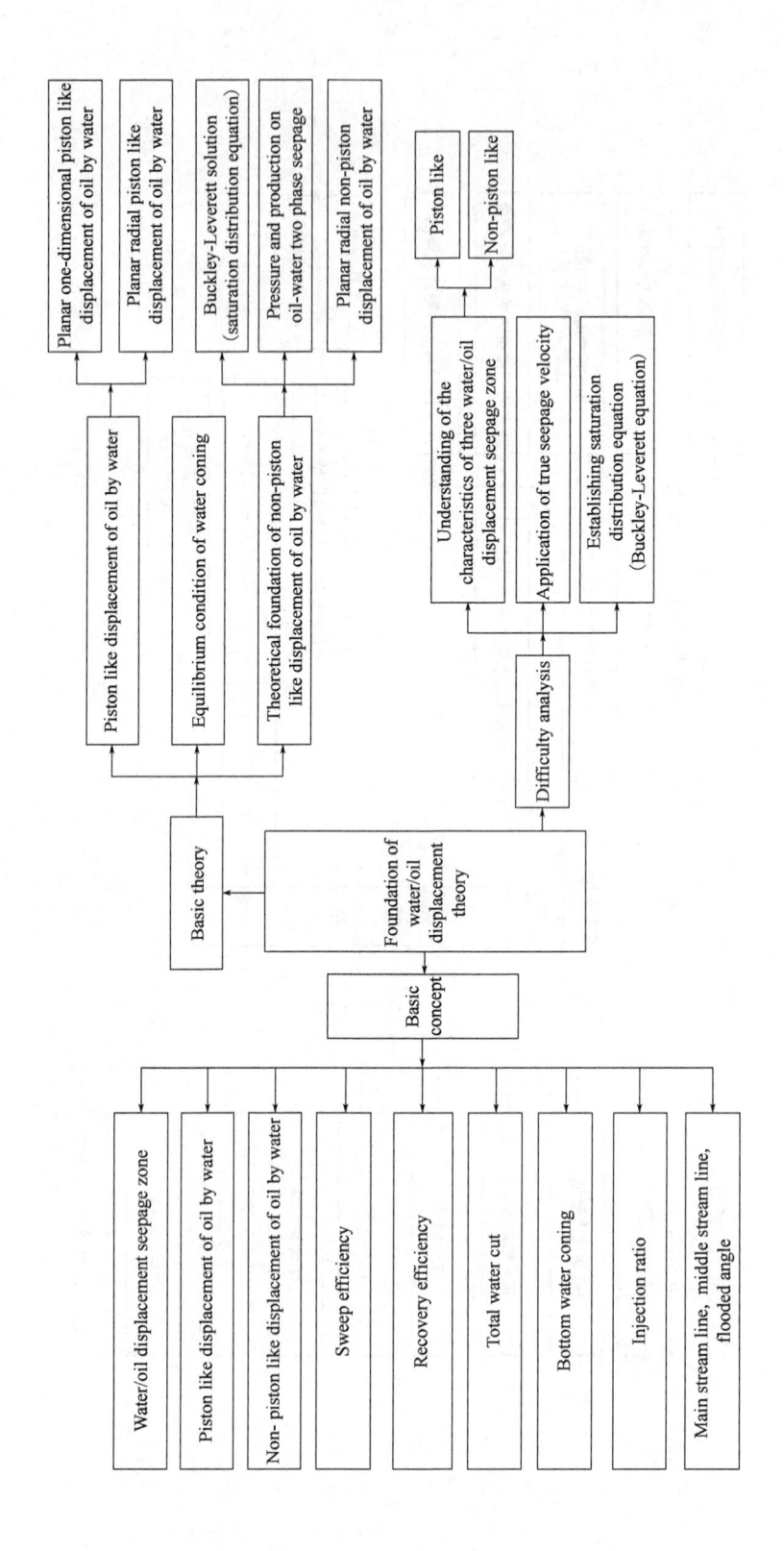

# Chapter 6 Knowledge Graph

Lots of reservoirs in the world are of the competency for natural water drive, while most of oil reservoirs are developed by the method of cheap and effective artificial water flooding. In China, to maintain pressure, all main oil reservoirs are developed by the method of artificial water flooding, and therefore the appearance of oil-water two-phase flow in the reservoir is inevitable. Therefore, studying the problem of oil-water two-phase seepage becomes a very necessary practical matter.

The process of water/oil displacement will appear in the reservoir under the developing method of natural water drive and artificial water flooding. At the beginning of oil reservoir development, water starts flowing into oil area and gradually approaches the bottom of production well. Because of the high heterogeneity of pore structure of oil reservoir and the difference of physical-chemical property of oil and water, water cannot displace all the oil in the area where it passed through, so there is residual oil. Between the initial oil/water boundary and the water front (present oil/water boundary), the oil-water two phases flow simultaneously and the water saturation increases gradually. Both the water/oil displacement laboratory experiment and the actual production verify that there is a water production period which is much longer than pure oil production period.

Under the condition of edge water drive, there are three seepage areas in oil reservoir. The first area is from the supply boundary to the initial oil/water boundary in which only the water moves. Of course, as to the situation of inner edge water flooding and pattern water flooding, the first area does not exist. The second area is from the initial oil/water boundary to the present oil/water boundary (generally is the water/oil front) in which oil - water two phases flow. The third area is from the water/oil front to the bottom hole of production well in which the pure oil flows as shown in Figure 6 - 1.

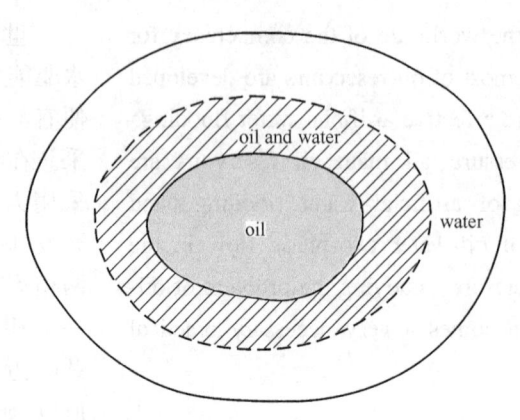

Figure 6 – 1　The zoning map of water/oil displacement seepage
图 6 – 1　水驱油渗流分区图

The motion law in the oil-water two-phase area is more complicated which is also very troublesome to process in mathematical. Although as early as 1942, the exact solution of two-phase flow of planar one-dimensional and radial fluid flow was found, it is widely known after 1950's. So at the beginning, it is assumed that oil-water two-phase area doesn't exist and the seepage area of water directly joins with that of oil. It means that the oil/water boundary is pushed like a piston, and all oil (at least all movable oil) is displaced once swept, which is generally called piston displacement. The assumption of piston like displacement does not fit the reality, but after making this assumption, the trouble of dealing with the two-phase area can be neglected, and the results obtained can also show the character of water/oil displacement in a certain sense, which can also be seen in some literatures.

## 6.1　Piston like displacement of oil by water

Under the assumption that the displacement of oil by water is piston-like, it is of great difficulty to discuss the general problem of water displacing oil. In order to illustrate the problem qualitatively, the problem of plane flooding oil is considered first. Now the oil-water interface is cylindrical surface, and its equation is the same as that of planar curve. After neglecting the compressibility of oil and water, the Laplace equation is all obeyed no matter whether seepage region of water or oil is involved or not.

油水两相区的运动规律比较复杂，数学处理也很麻烦。虽然早在1942年就已经获得平面一维和平面径向两相渗流的精确解，但广为人知的则是20世纪50年代以后的事了。所以开始先假设油水两相区不存在，水的渗流区与油的渗流区直接衔接。这就等于假设油水界面像活塞一样推进，一经扫过，全部油(至少是全部可动油)被驱替干净，习惯上称作活塞式驱油。活塞式驱油的假设是不符合实际的，但作了这个假设以后，省去了处理油水两相区的麻烦，所得结果在一定意义上也能揭露水驱油的特点，所以至今在文献上仍能见到。

## 6.1　活塞式水驱油

在水驱油是活塞式的假设下，要讨论一般水驱油问题，其难度也是很大的。为了定性地加以说明，先考虑平面水驱油问题。此时油水接触面是柱面，方程形式与平面曲线一样。忽略油和水的压缩性，无论水渗流区，还是油渗流区都服从Laplace方程：

$$\nabla^2 p_o = 0; \quad \nabla^2 p_w = 0$$

In the formula above, the subscript "o" and "w" respectively represent the parameters of region of water or oil.

The boundary condition is:

(1) Assume that the pressure is a constant $p_e$ on the supply boundary $c$.

$$p|_c = p_e$$

(2) All the bottom hole pressure of production wells is constant $p_{wf}$.

$$p_o|_{w_i} = p_{wf} \quad (i = 1, 2, \cdots, n)$$

(3) The pressure is equal on the oil-water interface.

$$p_o|_f = p_w|_f$$

At the same time according to the continuity principle, the normal component velocity of oil and water must be equal.

Generally, during water displacement, the position and shape of oil-water interface are constantly changing. The oil-water interface is not only the function of coordinate, but also has relationship with time which adds to the difficulty of the problem. So far, the general solution of this problem has not been obtained yet. Only the exact solution under two easiest conditions is obtained by now, and the common assumption is:

(1) The formation is homogeneous, iso-pachous and isotropic, and the rock and fluid (oil, water) are incompressible.

(2) Displacement of oil by water is piston-like; the gravity differentiation caused by the difference of the density in oil and water is not taken into consideration.

(3) The action of capillary force between oil and water is ignored.

(4) The oil-water interface is vertical to bed plane.

The situation of planar one-dimensional piston-like displacement of oil by water and planar radial piston like displacement of oil by water will be respectively discussed in the following.

式中,下角"o"和"w"分别表示油水区的参数。

边界条件为:

(1) 设供给边界 $c$ 上,压力为常数 $p_e$。

(2) 所有生产井的井底压力为常数 $p_{wf}$。

(3) 在油水接触面上,压力相等。

同时根据连续原理,油水的法向分速度也必须相等。

一般,在水驱油过程中,油水界面的位置、形状是不断变化的。油水界面不仅是坐标的函数,还与时间有关,这就增加了问题的难度。所以到目前为止,还没有关于这一问题的一般解。只有两个最简单条件下的精确解,共同的假设是:

(1) 地层是均质、等厚、各向同性的,岩石和流体(油、水)是不可压缩的。

(2) 水驱油是活塞式的,不考虑油水密度的不同而将引起的重力分异现象。

(3) 不计油水之间的毛管力作用。

(4) 油水界面与层面垂直。

下面将分别对平面一维活塞式水驱油和平面径向活塞式水驱油的情况加以讨论。

### 6.1.1 Planar one-dimensional piston-like displacement of oil by water

Firstly, assume that there is a rectangular reservoir as shown in Figure 6-2. The length, width and height of formation are respectively $L$, $B$ and $h$, the pressure on supply boundary is $p_e$, the pressure on the gallery is $p_w$ and in the process of water displacing oil the two pressures do not change. The position of initial oil-water interface in this oil layer is $x_0$, after time $t$, the oil-water interface is still vertical with bed plane and assumes its position is $x$, the pressure on the initial oil-water interface and the present oil-water interface respectively is $p_1$ and $p_2$. What we care about here is how the production and the oil-water interface change with time. It can be seen from the Figure 6-2 that at time $t$, the whole region can be divided into three intervals:

The first interval is $[0, x_0]$ which is pure water region, and the flow rate formula of water is:

$$Q_{w1} = \frac{KA(p_e - p_1)}{\mu_w x_0} \tag{6-1}$$

### 6.1.1 平面一维活塞式水驱油

首先,设有一长方形油气藏,如图6-2所示。地层的长、宽、厚度分别为 $L$、$B$、$h$,供给边缘的压力为 $p_e$,排液道的压力 $p_w$,并且在水驱油过程中这两个压力保持不变。该油层的原始油水界面的位置为 $x_0$,经过时间 $t$ 以后,油水界面仍与层面垂直,设其位置为 $x$,原始油水界面与目前油水界面上的压力分别为 $p_1$、$p_2$。这里所关心的问题是产量、油水界面是如何随时间而变化的。从图6-2中可见,在 $t$ 时刻,整个油层可以划分成三个区间:

第一区间为 $[0, x_0]$,是纯水区,水的流量公式为:

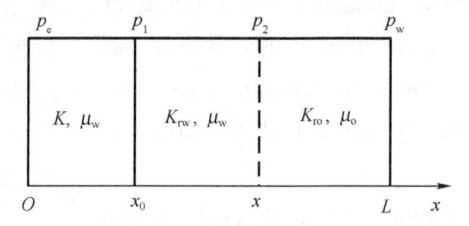

Figure 6-2 The planar one-dimensional piston displacement of oil by water
图6-2 平面一维活塞式水驱油

The second interval is $[x_0, x]$ in which oil saturation is residual oil saturation $S_{or}$ and oil cannot move. Assume that the relative permeability of water under residual oil saturation is $K_{rw}$, because in this interval flow is still single-phase incompressible fluid seepage, the flow rate formula of water is:

$$Q_{w2} = \frac{KK_{rw}A(p_1 - p_2)}{\mu_w(x - x_0)} \tag{6-2}$$

第二个区间为 $[x_0, x]$,含油饱和度为残余油饱和度 $S_{or}$,油不能流动。设在残余油饱和度下水的相对渗透率为 $K_{rw}$,由于这个区间仍是单相不可压缩液体渗流,所以,水的流量公式为:

The third interval is $[x, L]$, and formation is saturated with oil except water with irreducible saturation $S_{wc}$. Water is non-flowing phase and oil is flowing phase. The relative permeability to oil is $K_{ro}$ under the saturation of irreducible water, so the flow rate of oil is:

$$Q_o = \frac{KK_{ro}A(p_2 - p_w)}{\mu_o(L - x)} \qquad (6-3)$$

Because the formation and fluid are incompressible, $Q_{w1} = Q_{w2} = Q_o = Q$ must be obtained. After combining the formula (6-1), formula (6-2) and formula (6-3), the following expression can be obtained:

$$Q = \frac{KA(p_e - p_w)}{\mu_w x_0 + \frac{\mu_w}{K_{rw}}(x - x_0) + \frac{\mu_o}{K_{ro}}(L - x)} \qquad (6-4)$$

The formula (6-4) is the production formula of planar one-dimensional piston-like displacement of oil by water. $K_{rw}/\mu_w$ and $K_{ro}/\mu_o$ are respectively called the mobility of oil and water.

It can be seen from the formula (6-4) that, under the condition that the pressure difference does not change, because the position of oil-water interface $x$ moves forward with time changing in the process of water displacing oil, the production of oil also changes with time. When $K_{rw}/\mu_w > K_{ro}/\mu_o$, the production $Q$ increases with time; on the contrary, when $K_{rw}/\mu_w < K_{ro}/\mu_o$, the production $Q$ decreases with time.

It can be known from the formula (6-4) that: in the process of water displacing oil, even if the pressure at both ends remains unchanged, the production also changes with time, and the pressure distribution and flow velocity in whole formation also have relationship with time, so different from single-phase seepage, the process of water displacing oil is an unsteady seepage process.

第三区间为$[x, L]$，地层内除了饱和度为$S_{wc}$的束缚水外，其余部分为油所饱和。水为非流动相，油为流动相，在束缚水饱和度下油的相对渗透率为$K_{ro}$，因此油的流量为：

由于地层和液体是不可压缩的，所以必有$Q_{w1} = Q_{w2} = Q_o = Q$，联立式(6-1)、式(6-2)、式(6-3)，可得：

式(6-4)是平面一维活塞式水驱油的产量公式，$K_{rw}/\mu_w$和$K_{ro}/\mu_o$分别称为水和油的流度。

由式(6-4)可见，在地层压力不变的情况下，由于油水界面的位置$x$在水驱油过程中随着时间的变化在不断向前推移，因此，油的产量也是随时间的变化而变化的。当$K_{rw}/\mu_w > K_{ro}/\mu_o$时，产量$Q$将随着时间的延长而增加；反之，当$K_{rw}/\mu_w < K_{ro}/\mu_o$时，产量$Q$将随着时间$t$的延长而减小。

由式(6-4)可以清楚地认识到：在水驱油过程中，即使是两端压力保持不变，产量也随着时间的变化而变化，整个地层的压力分布和渗流速度也与时间有关，所以，与单相渗流不同，水驱油过程是一个不稳定的渗流过程。

The motion law of oil-water interface will be studied next. Assume the porosity of formation is $\phi$, the expression of actual velocity of an arbitrary mass point on the present oil-water interface (the formula of moving velocity of oil-water interface) can be obtained from the formula (6 - 4):

$$v_t = \frac{dx}{dt} = \frac{K(p_0 - p_w)}{\phi \left[ \mu_w x_0 + \frac{\mu_w}{K_{rw}}(x - x_0) + \frac{\mu_o}{K_{ro}}(L - x) \right]} \quad (6-5)$$

After separating the variables in formula (6 - 5) and integrating with respect to $t$ from $0 \to t$ and $x$ from $x_0 \to x$, the following expression can be obtained:

$$t = \frac{\phi}{K(p_e - p_w)} \left\{ \mu_w x_0 (x - x_0) + \frac{\mu_w}{K_{rw}} \frac{1}{2}(x - x_0)^2 + \frac{\mu_o}{2K_{ro}}[(L - x_0)^2 - (L - x)^2] \right\} \quad (6-6)$$

The formula (6 - 6) is the motion law of oil-water interface. The corresponding time $T$ when the oil-water interface reaches bottom (that is $x = L$) is called water breakthrough time, and the following expression can be obtained:

$$T = \frac{\phi}{K(p_e - p_w)} \left[ \mu_w x_0 (L - x_0) + \frac{\mu_o}{2K_{rw}}(L - x_0)^2 + \frac{\mu_o}{2K_{ro}}(L - x_0)^2 \right] \quad (6-7)$$

### 6.1.2 Planar radial piston like displacement of oil by water

If there is a production well in the center of circular formation, the supply radius is $r_e$, supply pressure is $p_e$, well radius is $r_w$. And bottom hole pressure is $p_w$. The distance from the initial oil-water interface to the center of formation is $r_0$ (Figure 6 - 3 Dynamic graph 6 - 1). Assume that after time $t$, the oil-water interface moves to the position where radius is $r$. Because the fluid does radial seepage, the initial oil-water interface and the present oil-water interface are concentric circles, and it's assumed that the pressures on both interfaces are respectively $p_1$ and $p_2$. It's same to the planar one-dimensional piston-like displacement of oil by water, for the water region at $(r_0, r_e)$, the following expression can be obtained:

$$Q_{w1} = \frac{2\pi Kh(p_e - p_1)}{\mu_w \ln \dfrac{r_e}{r_0}}$$

Dynamic graph 6-1 Planar radial piston like displacement of oil by water
动态图 6-1 平面径向活塞式水驱油

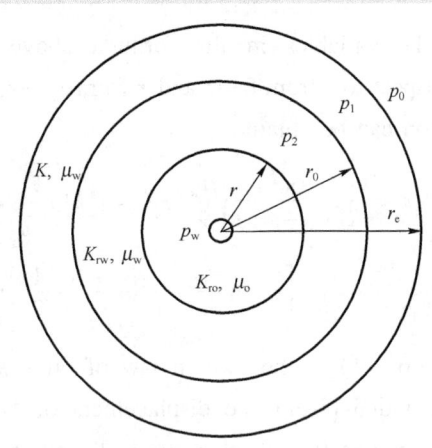

Figure 6 – 3  Planar radial piston like displacement of oil by water
图 6 – 3  平面径向活塞式水驱油

As to the region between ($r$, $r_0$), the following expression can be obtained:

对于($r,r_0$)的区域有:

$$Q_{w2} = \frac{2\pi K K_{rw} h(p_1 - p_2)}{\mu_w \ln \dfrac{r_0}{r}}$$

As to the region between ($r_w$, $r$), the following expression can be obtained:

而对于区域($r_w,r$)有:

$$Q_0 = \frac{2\pi K K_{ro} h(p_2 - p_w)}{\mu_o \ln \dfrac{r}{r_w}}$$

According to the continuity $Q = Q_{w1} = Q_{w2} = Q_0$ can be obtained, after combining the three formulas above, the following expression can be obtained:

根据连续性 $Q = Q_{w1} = Q_{w2} = Q_0$,所以联立以上三式可得:

$$Q = \frac{2\pi Kh(p_e - p_w)}{\mu_w \ln \dfrac{r_e}{r_0} + \dfrac{\mu_w}{K_{rw}} \ln \dfrac{r_0}{r} + \dfrac{\mu_o}{K_{ro}} \ln \dfrac{r}{r_w}} \quad (6-8)$$

The formula (6 – 8) is the production formula of planar radial piston-like displacement of oil by water. The production $Q$ of oil well is related to the position of present oil-water interface $r$. So the planar radial piston-like displacement of oil by water is also unsteady seepage. While as to an arbitrary mass point on the oil-water interface, its true flow velocity is:

式(6 – 8)就是平面径向活塞式水驱油的产量公式。油井产量 $Q$ 与目前油水界面的位置 $r$ 有关。因此,平面径向活塞式水驱油也是不稳定渗流。而对于油水界面上任一液体质点,其真实渗流速度为:

$$v_t = \frac{dx}{dt} = \frac{d(r_e - r)}{dt} = \frac{dr}{dt} = \frac{K(p_e - p_w)}{\phi\left(\mu_w \ln\frac{r_e}{r_0} + \frac{\mu_w}{K_{rw}}\ln\frac{r_0}{r} + \frac{\mu_o}{K_{ro}}\ln\frac{r}{r_w}\right)r}$$

After separating the variables in the formula above and integrating with respect to $t$ from $0 \to t$ and $r$ from $r_0 \to r$, the following expression can be obtained:

将上式分离变量,并将 $t$ 从 $0 \to t$, $r$ 从 $r_0 \to r$ 积分,则:

$$t = \frac{\phi}{K(p_e - p_w)}\left\{\frac{\mu_w}{2}(r_0^2 - r^2)\ln\frac{r_e}{r_0} + \frac{\mu_w}{2K_{rw}}\left[\frac{1}{2}(r_0^2 - r^2) - r^2\ln\frac{r_0}{r}\right]\right.$$
$$\left. + \frac{\mu_o}{2K_{ro}}\left[r_0^2\ln\frac{r_0}{r_w} - r^2\ln\frac{r}{r_w} - \frac{1}{2}(r_0^2 - r^2)\right]\right\} \quad (6-9)$$

The formula (6-9) is the motion law of oil-water interface of planar radial piston-like displacement of oil by water. If the oil-water interface reaches the well bottom that is $r = r_w$, then the corresponding time $t$ is the water breakthrough time marked with $T$, and the following expression also can be obtained:

式(6-9)就是平面径向活塞式水驱油油水界面的运动规律,如果油水界面到达井底,当 $r = r_w$ 时,则相应的 $t$ 就是见水时间,记作 $T$,且:

$$T = \frac{\phi}{K(p_e - p_w)}\left\{\frac{\mu_w}{2}(r_0^2 - r_w^2)\ln\frac{r_e}{r_0} + \frac{\mu_w}{2K_{rw}}\left[\frac{1}{2}(r_0^2 - r_w^2) - r_w^2\ln\frac{r_0}{r_w}\right]\right.$$
$$\left. + \frac{\mu_o}{2K_{ro}}\left[r_0^2\ln\frac{r_0}{r_w} - \frac{1}{2}(r_0^2 - r^2)\right]\right\} \quad (6-10)$$

### 6.1.3　Shrinkage of oil edge

The problem of water displacing oil, even if assumed as piston-like, it is still very difficult to solve in theory. So far, other exact solutions are not yet obtained except that of the planar one-dimensional and planar radial problem. While this problem is of great importance, except numerical calculation, the only way to solve this problem is to assume the mobility ratio of oil and water is equal in order to obtain some approximate ideas.

For example, the oil-water interface in planar infinite formation is a straight line, and there is a production well away from the straight line with distance $a$ as shown in Figure 6-4. And assume $K_{rw}/\mu_w = K_{ro}/\mu_o$ which means that it is homogeneous fluid flow, solve the moving law of an arbitrary liquid mass point on this straight line.

### 6.1.3　含油边缘收缩

水驱油问题,哪怕假定是活塞式的,在理论上也是十分难解决的。到目前为止除了平面一维和平面径向问题有了精确解外,别的精确解一概没有,而这个问题又是非常重要的,除了进行数值计算外,就只有假设油水的流度比相等以获得一些近似概念这条路了。

例如,平面无穷大地层上油水界面为一直线,距该直线 $a$ 处有一口生产井(图 6-4)。设 $K_w/\mu_w = K_o/\mu_o$,实际上这等于假定是均质液体流动,求这条直线上任一液体质点的运动规律。

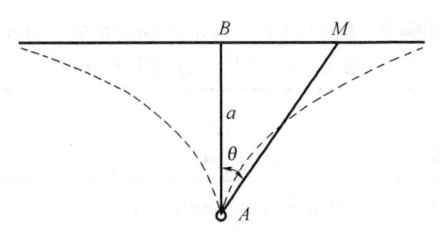

Figure 6 – 4   The contraction graph of linear oil boundary
图 6 – 4   直线含油边缘收缩图

| | |
|---|---|
| Because the mobility is equal, the flow velocity of an arbitrary point $M$ on the straight line is inversely proportional to the distance from the point to the center of well $r$, that is: | 因为流度相等,直线上任一点 $M$ 的渗流速度与该点到井中心的距离 $r$ 成反比,即: |

$$\frac{dx}{dt} = \frac{d(r_e - r)}{dt} = \frac{Q}{2\pi\phi rh} \qquad (6-11)$$

| | |
|---|---|
| Assume the included angle of connection $AM$ from the center of well to point $M$ and the vertical line $AB$ from the center of well to the water-oil interface is $\theta$, after separating the variables in formula (6 – 11) and integrating with respect to $t$ from 0 to $t$ and $r$ from $a/\cos\theta$ to $r$, the following expression can be obtained: | 设井中心到 $M$ 点的连线 $AM$ 与井中心到油水界面的垂线 $AB$ 之间的夹角为 $\theta$,将式(6 – 11)分离变量积分,$t$ 从 $0 \to t$,$r$ 从 $a/\cos\theta \to r$,有: |

$$r^2 = \frac{a^2}{\cos^2\theta} - \frac{Qt}{\pi\phi h} \qquad (6-12)$$

| | |
|---|---|
| What the formula (6 – 12) reflects is just the motion law on the water-oil interface. With same time, the smaller the $\theta$ is, the greater distance it covers. Therefore, the point $B$ reaches the well bottom first. Assume the time that it reaches the well bottom is $T$, because $a \gg r_w$, the following expression can be obtained from the formula (6 – 12): | 式(6 – 12)反映的正是油水界面上的运动规律,经过相同的时间,$\theta$ 越小的点运动的距离越大,因此,$B$ 点率先到达井底,设它到达井底的时间为 $T$,由于 $a \gg r_w$,所以由式(6 – 12)可得: |

$$T \approx \frac{\pi\phi h a^2}{Q} \qquad (6-13)$$

| | |
|---|---|
| $T$ is called the water breakthrough time. When the point $B$ reaches the well bottom, where are the other points on the water-oil interface? After substituting the formula (6 – 13) into the formula (6 – 12), the following expression can be obtained: | $T$ 称为见水时间。当 $B$ 点到达井底时,油水界面上的其余各点的位置在何处呢?将式(6 – 13)代入式(6 – 12)中得: |

$$r^2 = a^2\left(\frac{1}{\cos^2\theta} - 1\right) = a^2\tan^2\theta$$

| | |
|---|---|
| That is: | 即: |

$$r = a|\tan\theta|$$

| | |
|---|---|
| When a different value is chosen for $\theta$, $r$ correspondingly takes different value, the concrete situation can be seen in Table 6 – 1. | 当 $\theta$ 取不同的值时,$r$ 相应地取不同的值,具体情况见表 6 – 1。 |

Table 6-1 The concrete situation of $\theta$ and $r$

表6-1 $\theta$与$r$的对应情况

| $\theta$ | 5° | 30° | 45° | 60° |
|---|---|---|---|---|
| $r$ | $0.087a$ | $\frac{\sqrt{3}}{3}a$ | $a$ | $\sqrt{3}a$ |

It can be seen in Table 6-1, there is only one water-exit point when water breaks through an oil well, while the other points are still on the way to borehole wall. The broken curve in Figure 6-4 is the new oil-water interface. Assume the needed time that the point $\theta = \theta_0$ on the initial oil-water interface reaches bottom is $T_0$, and then the following expression can be obtained from the formula (6-12):

$$T_0 = \frac{\pi \phi h a^2}{Q \cos^2 \theta_0}$$

$\theta_0$ is called flooded angle that water breakthrough on the borehole wall, which will increase with time. Assume that the mobility ratio of oil and water is equal, while the water breakthrough time of different points is not equal and the actual mobility ratio is not equal, then the difference will be bigger.

The movement velocity of liquid mass point moved along the vertical line from well to oil contour is the fastest, which is called the main streamline. While the movement velocity moves along the vertical bisecting line of connection of two wells is the slowest, which is called the middle streamline, as shown in Figure 6-5. Liquid must firstly flow to the bottom along the main streamline, and the closer to the bottom is, the bigger the movement velocity difference is. The situation of line drive flooding and patten flooding is also the same.

由表6-1可看出,油井见水时出水处只有一个点,其余点还未到达井壁,图6-4中的虚线即新的油水界面。设原始油水界面上$\theta = \theta_0$的点到达井底所需时间为$T_0$,则由式(6-12)可得:

$\theta_0$称为水淹角,它是井壁上见水弧的角度,它随时间延长而增大。假定水油流度比相等,不同点见水时间不等,实际流度比不相等,则这一差异将会更大。

液体质点沿井到含油边线的垂直线运动速度最快,称为主流线;而沿两井连线的垂直二等分线运动速度最慢,称为中流线,如图6-5所示。液体必是沿主流线先到井底,且越接近井底,运动速度差异越大。排状注水、面积注水都是如此。

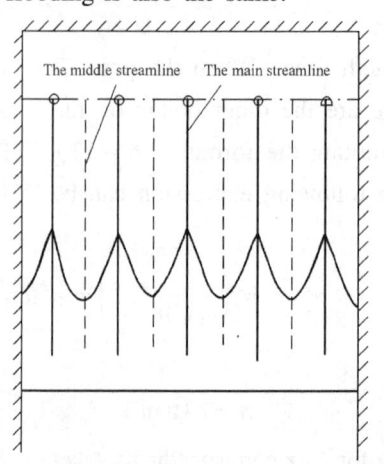

Figure 6-5 The contraction graph of water-oil boundary of linear flooding

图6-5 直线井排含油边缘收缩图

As to oil well, the high permeability zone will appear water breakthrough first, and only the water flooded part of high permeability zone appears breakthrough and the non-flooded part still yields oil. The ratio of flooded area to total area is called sweep efficiency, and the total recovery ratio should be the product of oil displacement efficiency and sweep efficiency.

The sweep efficiency can only be obtained by experimental curve; one part of the oil from oil well is from the water flooded area (displacement of oil by water is non-piston like), and another part is from the non-flooded area, while water is only from water flooded area, so the water content ratio of oil well is:

$$f_w = \frac{Q'_w}{Q'_w + Q'_o + Q''_o} = \frac{f'_w}{1 + \dfrac{Q''_o}{Q'_w + Q'_o}} = \frac{f'_w}{1 + \dfrac{\pi - \theta}{\theta}}$$

Where: $Q'_w$—The production of water in water-flooded area, m³/d;

$Q'_o$—The production of oil in water flooded area, m³/d;

$Q''_o$—The production of oil in non-flooded area, m³/d;

$f'_w$—The water ratio content of water flooded area.

So, changing working system, increasing directions of inflowing water and changing flowing directions can enhance oil recovery to some extent.

## 6.2 Bottom water coning

The structural hydrocarbon reservoir with small inclined angle and big thickness has a very big oil and water contact area. Thus once the oil well is put into production, the oil well will lie in the oil zone with bottom water as shown in Figure 6-6.

Because the pressure near well bottom will decrease when oil is produced, the oil and water contact surface will be deformed and changed to be a mount. The water mount is called the water coning of bottom water. If the water coning breaks through the well, oil and water will be produced simultaneously, and the water coning can even rise to the top of formation, which will make the well produces water only.

Because the problem of bottom water coning is to change contact surface, which is very complicated and until now the exact solution to the problem of vertical well water coning has not been found.

The physical essence of the phenomenon of water coning, the equilibrium condition of water coning and the biggest possible water free production will be briefly introduced next. That is under what condition the water coning will stay immovable and the well will produce pure oil.

The problem of oil flowing to imperfect well when there is bottom water coning remaining immovable will be looked at first.

Assume that the formation is isotropic and the cap formation, bottom formation and the initial oil-water interface are horizontal, that the water coning remains immovable and that it is pure oil that flows into oil well. The coordinate system is set up as shown in Figure 6-6(a), vertical downward is considered as the positive direction of $z$ axis. $h$ is marked as the oil-bearing thickness, $b$ is marked as the drilled thickness and the radius of oil well is $r_w$. The exact solution to the problem of bottom water coning is the solution of Laplace equation meeting the following boundary condition, that is to say:

以下将简要介绍底水锥进现象的物理实质、水锥的平衡条件及油井的最大可能的无水产量,即在何种条件下,水锥稳定不动,井产纯油。

先考察当存在稳定不动的底水水锥时,原油向非完善井流动的问题。

假设地层是各向同性的,并且盖层、底层和初始油水界面是水平的。另外还认为水锥稳定不动,流向油井里的是纯油。按图6-6(a)建立坐标系,$z$轴取垂直向下为正。记含油厚度为$h$,打开厚度为$b$,油井半径为$r_w$。底水锥进问题的精确解即为求解满足下列边界条件的Laplace方程的解,即:

$$\nabla^2 \Phi = 0 \tag{6-14}$$

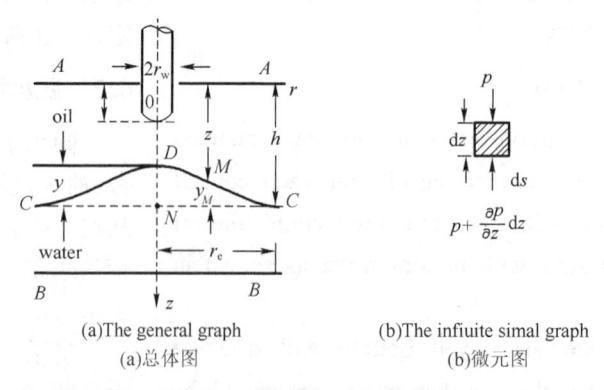

(a)The general graph
(a)总体图

(b)The infiuite simal graph
(b)微元图

Figure 6-6　The schematic of bottom water coning
图6-6　底水锥进示意图

The top surface of formation is impermeable, and the oil and water interface is impermeable to oil. Apparently, its shape and pressure drop have a relationship with the production. It's the quantity we're waiting for.

其中,地层顶面不渗透的;油水界面对于油是不渗透的,显然其形状与压力降和产量大小有关,是待求量。

The difficulty of this problem is that the shape of oil and water interface are unknown, so except the difficulties about solving the Laplace equation, the solving domain is definitely unknown.

The steady state condition of the water mass point on the top of water coning will be clarified first. For this, it is assumed that the pressure distribution function $p$ is known and the bottom pressure is the lowest. On the top of water coning, where $r = 0$, a cylindrical element heigth is $dz$, cross section is $ds$. The pore of element is full of water and this cylinder is considered in the oil region, and acting force on it will be analyzed as shown in Figure 6-6(b).

Assume the pressure on the top of cylinder element is $p(0,z) = p$ and the pressure on the bottom is $p'$, then the following expression can be obtained:

$$p' = p(0, z + dz) = p + \frac{\partial p}{\partial z}dz$$

The upward force pulling the water in the element cylinder is:

$$\phi(p' - p)ds = \phi \frac{\partial p}{\partial z}dzds$$

In the formula above, $\phi$ is porosity. Please pay attention to the fact that the occupied area by liquid is not the total area $ds$, but only a part of it $\phi ds$. The self-weight of water in cylinder element pulls it down again and this gravity force is $\rho_w g \phi ds dz$, in which $\rho_w$ is the density of water. Thus the steady state condition (equilibrium condition) of the water mass point is:

$$\rho_w g \phi ds dz \geqslant \phi \frac{\partial p}{\partial z}dzds \qquad (6-15)$$

Or

$$\rho_w g \geqslant \frac{\partial p}{\partial z} \qquad (6-15')$$

If the formula above is expressed with velocity potential of oil, then the expression will be changed to be:

$$\Phi = \frac{K}{\mu_o}(p - \rho_o gz)$$

$$p = \frac{\mu}{K}\Phi + \rho_o gz$$

In the formulas above, negative sign is taken to the vertically downward direction of $z$ axis. After substituting the two formulas to the equilibrium condition, the following expression can be obtained:

$$\rho_w g \geq \frac{\mu_o}{K}\frac{\partial \Phi}{\partial z} + \rho_o g$$

$$\frac{\partial \Phi}{\partial z} \leq \frac{Kg}{\mu}(\rho_w - \rho_o)$$

Assume that $\rho_w = 1000 \text{kg/m}^3$, then $\frac{\partial p}{\partial z} = 0.00981$ MPa/m. If the pure oil-bearing thickness is assumed to be 10m, the pressure difference of steady state water coning obtained is only 0.0981 MPa, and now the production obtained will be very little. So it is not practical to make the water coning keep balanced.

## 6.3 Theoretical foundation of non-piston like displacement of oil by water

When researching the problem of non-piston like displacement of oil by water, the compressibility of oil and water is not taken into account because the influenced area is not very big and pressure is usually above the saturation pressure. Thus mass conservation and volume conservation are accordant with each other. The net outflow amount of oil and water from the element should be equal to the reductive amount of oil and water in the element in the same time, that is:

$$\text{div } v_w = -\phi \frac{\partial S_w}{\partial t} \quad (6-16)$$

$$\text{div } v_o = -\phi \frac{\partial S_o}{\partial t} \quad (6-17)$$

Assume the seepage of oil and water obeys the Darcy's law, and then the following expression can be obtained:

$$v_w = -\frac{KK_{rw}}{\mu_w}\text{grad } p_w \quad (6-18)$$

$$v_o = -\frac{KK_{ro}}{\mu_o}\text{grad } p_o \quad (6-19)$$

Under the condition that there are only two phases of oil and water, apparently the following equation can be obtained:

$$S_o + S_w = 1 \quad (6-20)$$

Where: $K_{rw}$—The water relative permeability;

$K_{ro}$—The oil relative permeability;

$p_w$—The pressure of water phase;

$p_o$—The pressure of oil phase;

$S_w$—The saturation of water phase;

$S_o$—The saturation of oil phase;

$v_w$—The velocity of water phase;

$v_o$—The velocity of oil phase;

Generally, $K_{rw}$ and $K_{ro}$ are considered as the single valued function of water saturation, which is determined by experiment. The pressure difference of water phase and oil phase at the same point is equal to the capillary pressure of oil and water $p_{cow}$, which is also obtained by experiment and it is the single valued function of water saturation.

$$K_{rw} = K_{rw}(S_w), \quad K_{ro} = K_{ro}(S_o)$$
$$p_{cow} = p_o - p_w \tag{6-21}$$

There are 8 equations above, from which 8 unknown quantities can be solved, and an equation which contains only one unknown function can be obtained with method of elimination. The solution will be derived by one-dimensional oil and water two-phase flow, from which it can be seen under general condition. It is very difficult to solve the differential equation group above, and at present the numerical solution is mainly used, that is, the numerical simulation of reservoir. As to the planar one-dimensional and radial fluid flow, the method of characteristic line can be used to obtain the exact solution.

### 6.3.1 The Buckley-Leverett solution

Starting from the rectangular oil formation as shown in Figure 6-7, research on the theory of non-piston linear water driving oil. The pressure on the supply boundary of the oil formation is $p_e$, the pressure of the drainage channel is $p_w$, the cross section of formation is $A$, the position of initial oil-water interface and the present front respectively are $x_0$ and $x$, and the corresponding pressures are assumed as $p_1$ and $p_2$.

In order to research the motion law of two phases of oil and water efficiently, the main properties must be seized and at the same time, some secondary properties must be neglected, so the following assumptions are made:

(1) The formation is homogeneous, isopachous and isotropic.

(2) The rock and liquid are incompressible.

(3) The relative permeability of oil and water only has relationship with water saturation $S_w$.

(4) The capillary pressure and the action of gravity of oil and water are not taken into account.

Similar to study the piston-like displacement of oil by water, the oil formation shown in Figure 6-7 is divided into three intervals (Dynamic graph 6-2), and in the first and the third interval there is only single phase water and oil flowing, expression as formula (6-1) and formula (6-3).

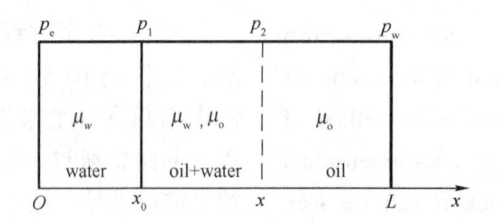

Figure 6-7 Planar one-dimensional non-piston like displacement of oil by water

图 6-7 平面一维非活塞式水驱油

Dynamic graph 6-2 Non-piston like displacement of oil by water

动态图 6-2 非活塞式水驱油

Between the original and present water-oil boundary, water saturation decreases with $x$ increasing. So the flow rate of water is from big to small, while the flow rate of oil is just opposite. Then how do they change with distance and time? Firstly, the equation group of two-phase seepage of oil and water must be simplified by some way to make it one equation containing only one unknown function, and the continuity equations of oil and water respectively are:

$$\frac{\partial v_o}{\partial x} = -\phi \frac{\partial S_o}{\partial t} \quad (6-22)$$

$$\frac{\partial v_w}{\partial x} = -\phi \frac{\partial S_w}{\partial t} \quad (6-23)$$

The saturation of oil and water of an arbitrary point in formation has the following relationship:

$$S_o + S_w = 1 \quad (6-24)$$

Make the formula (6 – 22) and the formula (6 – 23) multiply cross-section area $A$ simultaneously and the following expressions can be obtained:

$$\frac{\partial Q_o}{\partial x} = -\phi A \frac{\partial S_o}{\partial t} \quad (6-25)$$

$$\frac{\partial Q_w}{\partial x} = -\phi A \frac{\partial S_w}{\partial t} \quad (6-26)$$

The following expression can be obtained by adding the formula (6 – 25) to the formula (6 – 26) and utilizing the formula (6 – 24):

$$\frac{\partial(Q_o + Q_w)}{\partial x} = 0 \quad (6-27)$$

The formula (6 – 27) shows that the total flow rate of oil and water ($Q_o + Q_w$) passing through the arbitrary cross section in two-phase area at the same time is a constant marked as $Q$. According to the Darcy's law, at any time $t$ the flow rates of oil and water passing through the arbitrary cross section in two-phase area respectively are:

$$Q_o = -\frac{KK_{ro}}{\mu_o} A \frac{\partial p}{\partial x} \quad (6-28)$$

$$Q_w = -\frac{KK_{rw}}{\mu_w} A \frac{\partial p}{\partial x} \quad (6-29)$$

Therefore, the water content ratio on this cross section (or called fractional flow of water) is:

$$f_w = \frac{Q_w}{Q_w + Q_o} = \frac{K_{rw}/\mu_w}{K_{rw}/\mu_w + K_{ro}/\mu_o} = \frac{1}{1 + \frac{K_{ro}}{K_{rw}} \cdot \frac{\mu_w}{\mu_o}} \quad (6-30)$$

Because the water saturation $S_w$ at different cross section in the two-phase area at time $t$ is different, the effective permeability of oil and water and water content ratio is also different. It can be seen from the formula (6 – 30) that when the viscosity of oil and water keeps unchanged, water ratio $f_w$ is the single valued function of water saturation marked as $f_w(S_w)$.

The following expression can be obtained after adding the formula (6 – 28) to formula (6 – 29):

$$Q = -K\left(\frac{K_{rw}}{\mu_w} + \frac{K_{ro}}{\mu_o}\right)A\frac{\partial p}{\partial x} \qquad (6-31)$$

If the flow resistance in two-phase area at a given time needs to be obtained, at this moment, pressure $p$ is only relevant with $x$. So after integrating the formula (6-31), the following expression can be obtained:

$$\int_{p_1}^{p_2} dp = -\frac{Q}{KA}\int_{x_0}^{x}\frac{dx}{\frac{K_{rw}}{\mu_w}+\frac{K_{ro}}{\mu_o}} \qquad (6-32)$$

It can be seen from the formula (6-32) that the distributing condition of relative permeability of oil and water in two-phase area must be known first to obtain the flow resistance in two-phase area, while the relative permeability of oil and water depends on water saturation only. So the first thing to do is ascertaining the distribution of water saturation $S_w$ in two-phase area at any time $t$, that is, to ascertain the function $S_w(x,t)$.

Because the water flow rate passing through any cross section in the two-phase area at any time $t$ is $Q_w = Qf_w$ and the total liquid $Q$ has no relationship with $x$, the following expression can be obtained by substituting $Q_w = Qf_w$ into the formula (6-26):

$$Q\frac{\partial f_w}{\partial x} = -\phi A\frac{\partial S_w}{\partial t} \qquad (6-33)$$

Because water content ratio $f_w$ is the single valued function of water saturation $S_w$, the expression above can be further written as:

$$Qf'_w(S_w)\frac{\partial S_w}{\partial x} = -\phi A\frac{\partial S_w}{\partial t} \qquad (6-34)$$

The formula (6-34) is the differential equation of saturation distribution in which coefficient $\phi$ and $A$ are both known constants and $Q$ is also a constant at a definite time. It has been talked about before that under the condition that the viscosity ratio is a definite value, if the relative permeability is the known as single valued function of water saturation, then water content ratio is also the single-valued known function of saturation which just appears in the form of experiment curve. The slope $f'_w(S_w)$ of the curve is solved point by point, of course it is the known function of saturation.

The formula (6-34) is the pseudo-linear first order hyperbolic-type partial differential equation which can be solved by the method of characteristics. Because $S_w(x,t)$ is the binary function of coordinate $x$ and time $t$, its total differential is:

$$dS_w = \frac{\partial S_w}{\partial x}dx + \frac{\partial S_w}{\partial t}dt$$

If the motion law is discussed, when $S_w$ is a given value i. e. the motion law of iso-saturation surface $S_w = S_{w1}$. Because the saturation is considered as a constant temporarily, the following expression can be obtained:

$$dS_w = \frac{\partial S_w}{\partial x}dx + \frac{\partial S_w}{\partial t}dt = 0 \qquad (6-35)$$

Please notice that $dx$ in the formula (6-35) shows the moving distance of saturation surface $S_w = S_{w1}$ in time $dt$. The following expression can be obtained from the formula (6-34) and formula (6-35):

$$\frac{dx}{dt} = \frac{Q}{\phi A}f'_w(S_w) \qquad (6-36)$$

Then the formula (6-35) and formula (6-36) are totally equivalent to each other. $dx/dt$ in the left side of formula (6-36) shows that the moving velocity of saturation surface $S_w = S_{w1}$. $S_{w1}$ can be any fixed value, which does not need to be marked specially. Do remember that the formula (6-36) shows the moving velocity when saturation is a definite value. Apparently, the velocity is in direct proportion to the water content ratio rising rate $f'_w$.

Integrate the formula (6-36). $S_w$ should be considered as a constant temporarily, and then only $Q$ may be relevant with time.

$$x - x_0 = \frac{f'_w(S_w)}{\phi A}\int_0^t Qdt \qquad (6-37)$$

In the formula above, $x_0$ is the initial position of a given iso-saturation surface, $x$ is the position of the iso-saturation surface at time $t$, $\int_0^t Qdt$ is the cumulative liquid production or cumulative water injection until time $t$, which is also the total water filtrating into the two-phase area or the total decreased oil in the two-phase area from the time when two-phase area is formed to time $t$.

Although the formula (6-37) is the relational expression of distance and time obtained according to a given saturation that is the motion law of given water saturation, the value of this water saturation $S_w$ is arbitrary. At a given time $t$, the position of different water saturation surface can be obtained by substituting the value of different water saturation into the formula (6-37). Therefore, the formula (6-37) is the distribution formula of saturation.

In the distribution formula of saturation, the water saturation is reflected by the function $f'_w(S_w)$. Because $K_{ro}(S_w)$ or $K_{rw}(S_w)$ is an experimental curve, which is different for different formation, $f_w(S_w)$ and $f'_w(S_w)$ can be only expressed with the form of curve and $S_w(x,t)$ cannot be directly written in the analytical form. Then how to ascertain the distribution of saturation in two-phase area?

Generally, both the viscosity of oil and water and the relative permeability curve can be measured in laboratory. What the Figure 6-8 shows is the relative permeability curves measured by experiment. By substituting the measured viscosity of oil and water $\mu_o$ and $\mu_w$, and the relative permeability data of oil and water under different water saturation into the formula (6-30), then the corresponding water content ratio $f_w(S_w)$ can be obtained. Consider $f_w(S_w)$ as the $y$-coordinate, $S_w$ as the $x$-coordinate, and the relation curve of water content ratio and water saturation as shown in Figure 6-9 can be obtained with the point-describing method, which is a monotone increasing curve. Then on this curve, make the tangent line of the corresponding point at different water saturation, and the slope of the tangent line is the corresponding $f'_w(S_w)$ at the water saturation. So the relation curve of $f'_w(S_w)$ and $S_w$ can be made as shown in Figure 6-10, which is a curve with peak value.

渗入两相区的总水量或两相区内减少的总油量。

式(6-37)虽然是针对某一饱和度而得到的距离和时间的关系式,即某含水饱和度的运动规律,但这个含水饱和度$S_w$的值是任意的。对于给定的时刻$t$,可以将不同的含水饱和度值代入式(6-37)求出不同的含水饱和度面所在的位置。因此,式(6-37)就是饱和度分布公式。

在饱和度分布公式中,含水饱和度是通过函数$f'_w(S_w)$反映出来的。由于$K_{ro}(S_w)$、$K_{rw}(S_w)$是实验曲线,因油层而异,所以$f_w(S_w)$和$f'_w(S_w)$也只好用曲线形式表达,$S_w(x,t)$不可能直接写成解析形式。那么,怎样才能确定两相区内饱和度的分布呢?

通常,油水的黏度以及相对渗透率曲线都可以在实验室进行测定。图6-8就是实验测定的相对渗透率曲线。将实验测定的油水黏度$\mu_o$、$\mu_w$及不同含水饱和度下的油水相对渗透率数据代入式(6-30)就可以得到相应的含水率$f_w(S_w)$。以$f_w(S_w)$为纵坐标、$S_w$为横坐标,用描点法就可以作出图6-9所示的含水率与含水饱和度的关系曲线,它是一条单调上升的曲线。然后在这条关系曲线上,对不同含水饱和值所对应的点作切线,这切线的斜率就是与该含水饱和度相对应的$f'_w(S_w)$,于是,可以作出图6-10所示的$f'_w(S_w)$与$S_w$的关系曲线,它是一条有峰值的曲线。

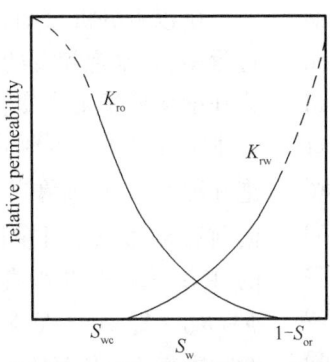

Figure 6-8  The relative permeability of oil and wate
图 6-8  油水相对渗透率曲线

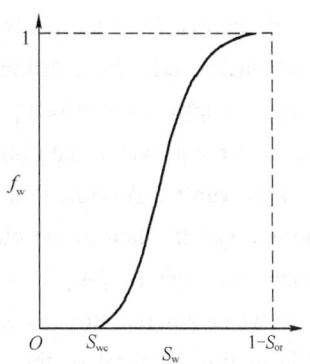

Figure 6-9  The water content ratio curve
图 6-9  含水率曲线

From the Figure 6-10, the corresponding $f'_w(S_w)$ at different water saturation $S_w$ can be obtained, and the position of the saturation surface at any time can also be obtained by substituting the value of $f'_w(S_w)$ into the distribution formula of saturation (6-37). What the Figure 6-11 shows is the distribution curve of water saturation in two-phase area at a certain time. It can be seen from Figure 6-11 that water saturation $S_w$ is a double-valued function of position $x$ apparently. Because it does not distinguish that water displace oil or oil displace water under studying oil-water two phase fluid flow. If it is water displacing oil, the water saturation $S_w$ decreases with $x$ increasing, while if it is oil displacing water, it is just the opposite. Therefore, the upper line of Figure 6-11 represents water displacing oil, and the lower curve is oil displacing water. Because the researching object is water displacing oil, the useful one is the upper branch, and the terminal point of the upper branch is the water-oil front. The upstream of front (the side with high pressure) is the oil-water flowing simultaneously and the downstream of front (the side with lower pressure) is the single-phase oil flowing. However, the location of the terminal point of the upper branch needs to be further discussed.

由图 6-10 可以求得与不同含水饱和度 $S_w$ 相对应的$f'_w(S_w)$的值,并将它们代入饱和度分布式(6-37)即可求得任一时刻这些饱和度面所在的位置。图 6-11 就是某时刻两相区内含水饱和度的分布曲线。由图 6-11 可看出,含水饱和度 $S_w$ 是位置 $x$ 的双值函数,显然是因为在研究油水两相同时渗流时没有区分是水驱油还是油驱水。如果是水驱油,则含水饱和度 $S_w$ 随着 $x$ 增大而下降,若是油驱水结果则相反。所以,图 6-11 曲线的上支相当于水驱油,而下支则是油驱水。由于研究的是水驱油,因此,有用的是上支,而上支的终点就是油水前缘。前缘的上游(压力高的一侧)是油水两相同时流动,前缘的下游(压力低的一侧)是单相油的流动。但是,上支终点的位置还有待于进一步讨论。

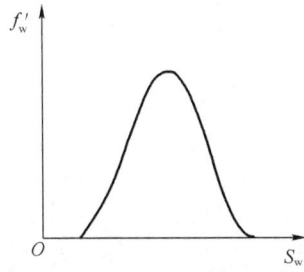

Figure 6-10  The derivative curve of water content ratio
图 6-10  含水率导数的曲线

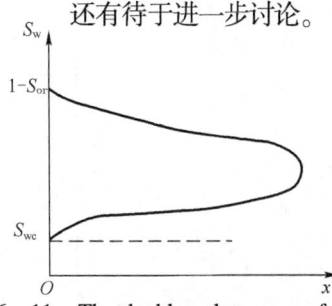

Figure 6-11  The double value curve of saturation
图 6-11  饱和度与位置的双值曲线

It can be known from the analysis above, in the course of water displacing oil, the distribution of water saturation must change abruptly in the position of front, which will change from frontal water saturation $S_w$ to initial water saturation $S_{wc}$ abruptly. Assume that the front moves forward $dx$ in time interval $dt$, then in the element it's length $dx$ the water amount increases to $\phi A(S_{wf} - S_{wc})dx$ which must be equal to the water amount flowing into the unit in the same time, while in time interval dt the extra water flowing into the element is $Qf_{wf}dt$. Make the two equal and the following expression can be obtained:

$$\left.\frac{dx}{dt}\right|_{x=x_f} = \frac{Qf_{wf}}{\phi A(S_{wf} - S_{wc})} \qquad (6-38)$$

The left side of the formula (6 - 38) shows the velocity of frontal water saturation $S_{wf}$. Compared with the formula (6 - 36), the following expression can be obtained:

$$f'_{wf} = \frac{f_{wf}}{S_{wf} - S_{wc}} \qquad (6-39)$$

This is the formula of frontal water saturation. Because $f_w(S_{wc}) = 0$, the right side of the formula (6 - 39) shows the slope of the connection of two points $(S_{wc}, 0)$ and $(S_{wf}, f_{wf})$ on the plot of water content ratio, and the left side is the slope of tangent line at point $(S_{wf}, f_{wf})$. in order to make the left is equal to the right, the only thing to do is to make tangent line to water content ratio curve from the point $(S_{wc}, 0)$ and the corresponding water saturation of the tangential point is the water saturation of water-oil displacement front, and the corresponding water content ratio is the frontal water content ratio $f_{wf}$ and then $f'_{wf}$ can be obtained by substituting it into the formula (6 - 39) as shown in Figure 6 - 12.

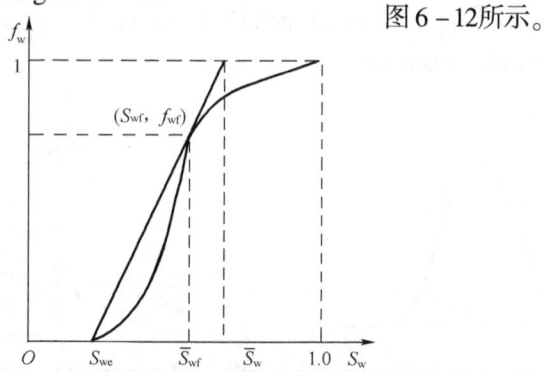

Figure 6 - 12  The tangent method of ascertaining frontal saturation
图 6 - 12  确定前缘饱和度的切线法

Another way can be also adopted to obtain frontal water saturation $S_{wf}$. According to the law of conservation of mass, the cumulative water injection in time $t$ should be equal to the additional water amount in formation at the same time.

还可以通过另一途径求前缘含水饱和度 $S_{wf}$,根据质量守恒原理,$t$ 时间内的累积注水量应等于地层内在同一时间中水量的增加。

$$\int_0^t Q \mathrm{d}t = A\phi \int_{x_0}^{x_f} (S_w - S_{wc}) \mathrm{d}x \qquad (6-40)$$

At this moment, because $t$ is a constant, the relationship of integrand and coordinate can be obtained from the formula (6-37):

此时,由于 $t$ 是固定值,由式(6-37)可得被积函数与坐标的关系为:

$$\mathrm{d}x = \frac{\mathrm{d}f'_w}{\phi A} \int_0^t Q \mathrm{d}t \qquad (6-41)$$

By substituting the formula (6-41) into the formula (6-40) and pay attention to the interval of $x$ is $x_0 \to x_f$, $S_w$ is $(1-S_{or}) \to S_{wf}$, then the following expression can be obtained:

将式(6-41)代入式(6-40)中,并注意到 $x$ 由 $x_0 \to x_f$,$S_w$ 则由 $(1-S_{or}) \to S_{wf}$,则有:

$$\int_{1-S_{or}}^{S_{wf}} (S_w - S_{wc}) \mathrm{d}f'_w = 1 \qquad (6-42)$$

What the formula (6-42) shows is that just the area of curvilinear trapezoid $ABEF$ equals 1 shown in Figure 6-13. It is not difficult to verify that:

式(6-42)表示的正是图6-13中的曲边梯形面积 $ABEF$,且值为1。不难验证:

$$\int_{S_{wc}}^{1-S_{or}} f'_w \mathrm{d}S_w = 1 \qquad (6-43)$$

What the formula above shows is that the area of curvilinear trapezoid $ABCDF$ shown in Figure 6-13 also equals 1. So it can be seen that if only moving the straight line parallel with the $S_w$ axis on the relational figure of $S_w$ with $f'_w$, and just making the area of $BCDB$ equal to the area of $FDEF$, then the intersection point of this straight line with the upper branch is the frontal water saturation $S_{wf}$.

而这一公式表示的是图6-13中的曲线 $ABCDF$ 所包围的面积也为1。于是可以看出,只要在 $S_w$ 与 $f'_w$ 的关系图上移动与 $S_w$ 轴平行的直线,如果它正好使面积 $BCDB$ 等于面积 $FDEF$,则此直线与上支曲线的交点就是前缘饱和度 $S_{wf}$。

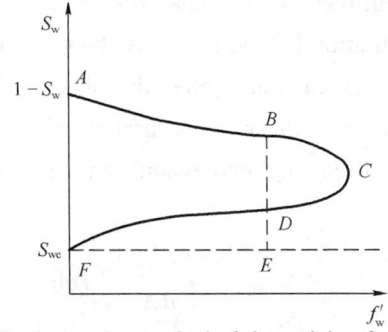

Figure 6-13  The integration method of determining frontal saturation
图6-13  确定前缘饱和度的积分法

Assume the average water saturation of the two-phase area is $\bar{S}_w$, according to the law of conservation of mass, the amount of injected water is equal to the additional water in time $t$ that is:

$$\int_0^t Q \mathrm{d}t = (\bar{S}_w - S_{wc}) \phi A (x_f - x_0) \qquad (6-44)$$

It can be known from the formula (6-37) that:

$$x_f - x_0 = \frac{f'_{wf}}{\phi A} \int_0^t Q \mathrm{d}t$$

Then the following expression can be obtained by substituting the expression above into the formula (6-44):

$$f'_{wf} = \frac{1}{\bar{S}_w - S_{wc}} \qquad (6-45)$$

The formula (6-45) shows: making the tangent line to water content ratio curve on the point $(S_{wc}, 0)$ and extending it to intersect with the horizontal line at water content ratio 100%, the abscissa of intersection point is the average water saturation $\bar{S}_w$ of the two-phase area as shown in Figure 6-12.

When the water-oil front reaches the delivery end, and if the volume factor is not taken into account, the non-water displacement efficiency is:

$$E_b = \frac{\bar{S}_w - S_{wc}}{1 - S_{wc}} \qquad (6-46)$$

After the water-oil front reaches the delivery end, the oil well starts water breakthrough and enters the long-period water content ratio production. Then, how will the water content ratio at the delivery end? Apparently, whether it is before or after water breakthrough, the distribution of water saturation in two-phase area both meet the formula (6-37), so the distribution curve of saturation can be made, as shown in Figure 6-14. Assume at a given time $t$ after the oil-water front reaches the delivery end, the water saturation of delivery end is $S_{wL}$, then the following expression can be obtained:

$$L - x_0 = \frac{f'_w(S_{wL})}{\phi A} \int_0^t Q \mathrm{d}t \qquad (6-47)$$

Make

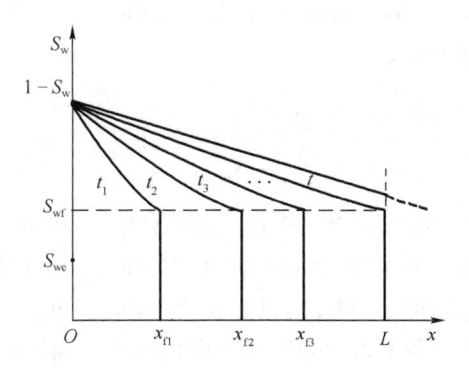

Figure 6 – 14 The distribution of water saturation

图 6 – 14 含水饱和度分布

$$V_D = \frac{\int_0^t Q dt}{\phi A (L - x_0)}$$

In the formula above, $V_D$ is the ratio of cumulative water injection to the pore volume of oil-bearing area, which is called injection multiple. Because the cumulative water injection can be obtained by measuring, the value of $V_D$ at different time also can be obtained, and then $f'_w(S_{wL})$ can be obtained by substituting $V_D$ into the formula (6 – 47), then the water saturation of delivery end can be obtained from the relation curve of $f'_w$ and $S_w$, and the relation curve of $S_{wL}$ with $t$ can also be further obtained.

After the delivery end starts to break through water, the average water saturation in the oil-water two-phase area will also change constantly. Because this water saturation is very useful in knowing the water-flood efficiency, it is of great importance to study it. The average water saturation in the oil-bearing area after water breakthrough is:

式中, $V_D$ 是累积注水量与含油区的孔隙体积之比, 称为注水倍数。由于累积注水量是可以计量得到的, 所以, 不同时刻 $V_D$ 的值也容易算出, 代入式(6 – 47) 就可以得到 $f'_w(S_{wL})$, 然后从 $f'_w$ 与 $S_w$ 的关系曲线上求出出口处的含水饱和度, 还可以进一步获得 $S_{wL}$ 与 $t$ 的关系曲线。

当出口端见水后, 油水两相区中的平均含水饱和度也将不断变化。由于这一饱和度对于认识水驱效率十分有用, 因此, 研究它十分必要, 而见水后含油区中的平均含水饱和度为:

$$\overline{S}_w = \frac{\int_{x_0}^L S_w dx}{L - x_0}$$

By substituting the formula (6 – 41) and the formula (6 – 47) into the expression above, and changing the integration limits $x\ (x_0 \to L)$ to $S_w[(1 - S_{or}) \to S_{wL}]$, then the following expression can be obtained:

将式(6 – 41)和式(6 – 47)代入上式, 并将积分变量 $x(x_0 \to L)$ 换成 $S_w[(1 - S_{or}) \to S_{wL}]$, 则:

$$\overline{S}_w = \frac{1}{f_{wL}} \int_{1 - S_{or}}^{S_{wL}} S_w df'_w$$

After integrating by parts to the expression above, the following expression can be obtained:

$$f'_{wL} = \frac{1 - f_{wL}}{\overline{S}_w - S_{wL}} \quad (6-48)$$

The geometry meaning of the formula (6-48) is also very clear, and the right side of the formula is the slope of the connection of point $(S_{wL}, f_{wL})$ and point $(\overline{S}_w, 1)$. If the aim is to make it equal to $f'_{wL}$, the only thing to do is making the tangent line at the point $(S_{wL}, f_{wL})$ (Figure 6-15), and prolong it to intersect with the horizontal line at water ratio 100%, the abscissa of intersection point is $\overline{S}_w$. Thus, the corresponding water displacement efficiency can be obtained according to the average water saturation $\overline{S}_w$ at different time:

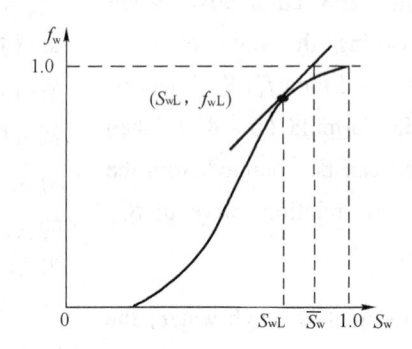

$$E = \frac{\overline{S}_w - S_{wc}}{1 - S_{wc}} \quad (6-49)$$

Figure 6-15  Calculate the average water saturation in the oil-bearing area after water breakthrough

图 6-15  求见水后的平均含水饱和度

### 6.3.2 The pressure and production on oil-water two-phase seepage

It has been discussed previously that the flow resistance in two-phase area can be expressed by the formula (6-32); according to the definition of water content, the following expression can be obtained:

$$\frac{1}{\dfrac{K_{rw}}{\mu_w} + \dfrac{K_{ro}}{\mu_0}} = \frac{\mu_w f_w}{K_{rw}}$$

By substituting the formula above and the formula (6-41) into the formula (6-33) and changing the integration variable $x$ into $f'_w$ and integration limits $x_0 \to x_f$ to $0 \to f'_{wL}$, and the following expression can be obtained:

$$p_1 - p_2 = \frac{\mu_w Q}{KA\phi} \int_0^t Q dt \int_0^{f'_{wf}} \frac{f_L}{K_{rw}} df'_w$$

Because:

$$\frac{\int_0^t Q dt}{\phi A} = \frac{x_f - x_0}{f'_{wf}}$$

So:

$$p_1 - p_2 = \frac{\mu_w Q(x_f - x_0)}{KA} \frac{1}{f'_{wf}} \int_0^{f'_w} \frac{f_w}{K_{rw}} df'_w$$

Make $a = \frac{1}{f'_{wf}} \int_0^{f'_w} \frac{f_w}{K_{rw}} df'_w$, in which $a$ is called the resistance factor of two-phase area, and the resistance formula of two-phase area is:

$$p_1 - p_2 = \frac{\mu_w Q(x_f - x_0) a}{KA} \quad (6-50a)$$

After the water-oil front reaches the delivery end, the resistance factor of two-phase area $a^*$ can be obtained with the same method:

$$a^* = \frac{1}{f'_{wL}} \int_0^{f'_{wL}} \frac{f_w}{K_{rw}} df'_w$$

While the resistance formula of two-phase area is:

$$p_1 - p_2 = \frac{\mu_w Q(L - x_0) a^*}{KA} \quad (6-50b)$$

When the viscosity ratio of oil and water is a definite value, the resistance factor $a$ is relevant with the relative permeability curve. Before the water-oil front reaches the delivery end, $a$ is a constant, while after the water-oil front reaches the delivery end, $a$ will changed constantly. Whether $a$ changes or not, it can be calculated by numerical method, that is making a curve considered the derivative of water content ratio $f_w$ to water saturation $S_w$ as the X-axis and $\frac{f_w}{K_{rw}}$ as the Y-axis (Figure 6-16), the shadow area under the curve is $\int_0^{f'_{wf}} \frac{f_w}{K_{rw}} df'_w$, and the resistance factor $a$ can be finally

— 249 —

obtained that $\int_0^{f'_{wf}} \frac{f_w}{K_{rw}} df'_w$ divided by derivative $f'_{wf}$ of water content ratio $f_w$ at the frontal water saturation $S_{wf}$.

In the same way, the shadow area under the curve as shown in Figure 6-17 is just the value of $\int_0^{f'_{wL}} \frac{f_w}{K_{rw}} df'_w$. After dividing it by $f'_{wL}$, the resistance factor $a^*$ of two-phase area after water breakthrough can be obtained.

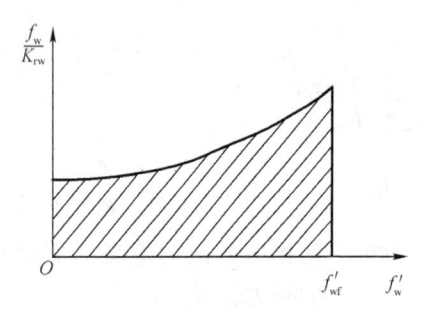

Figure 6-16　The relation curve of $f_w / K_{rw}$ with $f'_w$ (I)

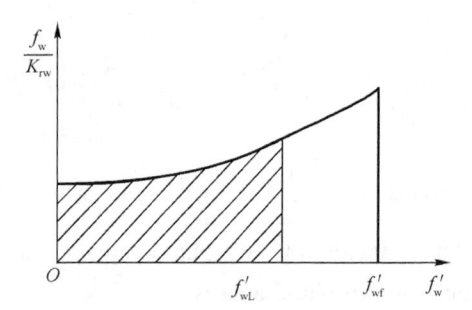

Figure 6-17　The relation curve of $f_w / K_{rw}$ with $f'_w$ (II)

Because rock and fluid are incompressible, as to the rectangular reservoir shown in Figure 6-17, before the water-oil saturation reaches the well bottom:

$$Q = \frac{KA(p_e - p_1)}{\mu_w x_0} = \frac{KA(p_1 - p_2)}{a\mu_w(x_f - x_0)} = \frac{KK_{ro}A(p_2 - p_w)}{\mu_o(L - x_f)}$$

After arranging the expression above:

$$Q = \frac{KA(p_e - p_w)}{\mu_w x_0 + a\mu_w(x_f - x_0) + \frac{\mu_o(L - x_f)}{K_{ro}}} \quad (6-51)$$

After the oil-water front reaches the delivery end, the following expression can be obtained:

$$Q = \frac{KA(p_e - p_w)}{\mu_w x_0 + a^* \mu_w(L - x_0)} \quad (6-52)$$

Where: $L$—The length of rectangular reservoir, m;

　　　$A$—The cross section area of rectangular reservoir, m²;

$x_0$—The initial oil – water front, m;
$x_f$—The present oil – water front, m;

The formula (6 – 51) and the formula (6 – 52) respectively are the formula productions before and after water breakthrough of planar one-dimensional non-piston displacement of oil by water.

When producing pressure drop maintains unchanged, the liquid producing capacity is not a constant but a variable. This is because before the oil-water front reaches the delivery end, $x_f$ is a variable. After the oil-water front reaches the delivery end, $a^*$ is a variable, and oil production $Q_o$ has the following relationship with the liquid production:

$$Q_o = (1 - f_{wL})Q \qquad (6-53)$$

The water content ratio of delivery end $f_{wL}$ will increase constantly with production, and the oil production will decrease under the condition that the producing capacity is invariable. So if aiming at making the oil production unchanged, the only way is to increase liquid producing capacity.

### 6.3.3 Planar radial non-piston displacement of oil by water

Assume there is a production well with radius $r_w$ in the circular formation, the supply radius is $r_e$, the pressure on the supply boundary line is $p_e$, and the radius of initial oil-water interface is $r_0$, the position of oil-water front is $r_f$ at time $t$ and it is non-piston like displacement of oil by water (Figure 6 – 18). The assuming conditions of formation and fluid are the same as that of planar one-dimensional non-piston like displacement of oil by water.

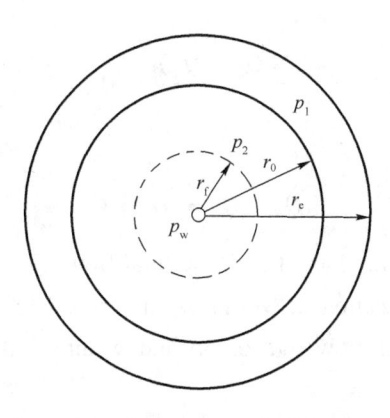

Figure 6 – 18  Non-piston like displacement of oil by water in circular reservoir

In the same way, when studying planar one-dimensional non-piston like displacement of oil by water, the key point of studying planar radial non-piston like displacement of oil by water is to know the distribution of saturation in the two-phase area and the changing law of production and resistance. And the starting point of resolving these problems is still the continuity equation of oil and water:

$$\frac{1}{r}\frac{\partial}{\partial r}(rv_o) = -\phi\frac{\partial S_o}{\partial t} \qquad (6-54)$$

$$\frac{1}{r}\frac{\partial}{\partial r}(rv_w) = -\phi\frac{\partial S_w}{\partial t} \qquad (6-55)$$

Because formation is homogeneous and isopachous, if the two right sides of the formula (6-54) and formula (6-55) are multiplied by $2\pi h$ simultaneously and use the relationship between flow rate and flow velocity, the formula (6-54) and formula (6-55) are changed to be:

$$\frac{1}{r}\frac{\partial Q_o}{\partial r} = -2\pi\phi h\frac{\partial S_o}{\partial t} \qquad (6-56)$$

$$\frac{1}{r}\frac{\partial Q_w}{\partial r} = -2\pi\phi h\frac{\partial S_w}{\partial t} \qquad (6-57)$$

In the formulas above, $Q_o$ and $Q_w$ respectively represent the flow rate of oil and water on the section with radius $r$. While saturations of oil and water have the following relationship on any cross section or in any element:

$$S_o + S_w = 1$$

Therefore, add the formula (6-56) to the formula (6-57), and use the expression above, and the following expression can be obtained:

$$\frac{\partial(Q_o + Q_w)}{\partial r} = 0$$

That is:

$$Q_o + Q_w = Q = C \qquad (6-58)$$

The formula (6-58) shows at any time $t$ the total flow rate passing through any cross section in two-phase area is a constant marked by $Q$. And the flow rate of oil and water respectively is:

$$Q_o = 2\pi rh \frac{KK_{ro}}{\mu_o}\frac{\partial p}{\partial r}$$

$$Q_w = 2\pi rh \frac{KK_{rw}}{\mu_w}\frac{\partial p}{\partial r}$$

So, at this moment the water content ratio on the cross section is:

$$f_w = \frac{Q_w}{Q_w + Q_o} = \frac{K_{rw}/\mu_w}{K_{rw}/\mu_w + K_{ro}/\mu_o}$$

By substituting $Q_w = f_w Q$ into the formula (6-57), the following expression can be obtained:

$$\frac{1}{r}Qf'_w\frac{\partial S_w}{\partial r} = -2\pi\phi h\frac{\partial S_w}{\partial t} \qquad (6-59)$$

This is the differential equation of distribution of water saturation in the two-phase area of planar radial non-piston like displacement of oil by water, and the method of characteristic line is also used to get the solution. Then as to any iso-saturation surface, the following expression can be obtained:

$$\frac{\partial S_w}{\partial r}dr + \frac{\partial S_w}{\partial t}dt = 0 \qquad (6-60)$$

After combining the formula (6-59) and the formula (6-60), the following expression can be obtained:

$$\frac{dr}{dt} = \frac{Qf'_w(S_w)}{2\pi r\phi h} \qquad (6-61)$$

Due to the opposite direction between the radial radius $r$ and the oil-water front $r_f$, if $r$ is still used to represent the position of the oil-water front, then:

$$\frac{dr_f}{dt} = -\frac{Qf'_w(S_{wf})}{2\pi rh\varphi} \qquad (6-61')$$

In the formula above, $\frac{dr}{dt}$ is the velocity of any saturation surface, which is in direct proportion to $f'_w(S_w)$. After separating the variables of the formula (6-61) and integrating, the following expression can be obtained:

$$r_o^2 - r_f^2 = \frac{f'_w(S_w)}{\pi\phi h}\int_0^t Q dt \qquad (6-62)$$

The formula (6-62) shows the distribution of water saturation in the two-phase area of planar radial non-piston like displacement of oil by water.

From the formula (6-62), the distribution curve of water saturation $S_w$ in the two-phase area at any time is a double-valued function of $r$, which apparently is not fit for the actual situation. It can be known from the analysis that water saturation changes abruptly at the location of the oil-water front $r_f$, that is, decreasing abruptly from the frontal water saturation $S_{wf}$ to the bound water saturation $S_{wc}$. Assume the oil-water front moves forward $-dr$ in time $dt$, and in the element with thickness $dr$ the incremental amount of water is $2\pi r_f \phi h (S_{wf} - S_{wc}) dr$, which is definitely equal to the additional amount of water flowing into the element in the same time, which is $Qf_{wf}dt$.

Make the two equal and the following expression can be obtained:

$$\left. \frac{dr}{dt} \right|_{r=r_f} = -\frac{Qf_{wf}}{2\pi r_f \phi h (S_{wf} - S_{wc})} \qquad (6-63)$$

The left side of the formula (6-63) shows the velocity of frontal saturation, which coincides the formula (6-61). Make the two equal, and the following expression can be obtained:

$$f'_{wf} = \frac{f_{wf}}{S_{wf} - S_{wc}} \qquad (6-64)$$

The formula (6-64) is the totally same as the formula (6-39), so it can be used as the way to obtain the frontal saturation as shown in Figure 6-12.

Assume the average water saturation is $\overline{S}_w$ in the two-phase area at time $t$, according to the law of conservation of mass, the cumulative water injection in time $t$ should be equal to the incremental water in the two-phase area that is:

$$(\overline{S}_w - S_{wc})\pi(r_0^2 - r_f^2)\phi h = \int_0^t Q dt$$

By substituting the formula (6-62) into the expression above, the following expression can be obtained:

$$f'_{wf} = \frac{1}{\overline{S}_w - S_{wc}} \qquad (6-65)$$

The formula (6-65) is also entirely the same as the formula (6-45). Assume after water breakthrough of oil well, the water saturation of well bottom is $S_{wL}$ and water content ratio is $f_{wL}$, let:

$$V_D = \frac{\int_0^t Q dt}{\pi \phi h (r_o^2 - r_w^2)}$$

By substituting the expression into the formula (6-62), the following expression can be obtained:

$$V_D = \frac{1}{f'_{wL}}$$

In the formula above, $V_D$ is called the injection multiple, $f'_{wL}$ can be obtained, and then $S_{wL}$ can be obtained from the relation curve of $f'_w$ and $S_w$ according to it. Like the planar one-dimensional non-piston like displacement of oil by water, after water breakthrough in oil well, the average water saturation in the two-phase area meets the formula (6-48), that is:

$$f'_{wL} = \frac{1 - f_{wL}}{\overline{S}_w - S_{wL}} \tag{6-66}$$

According to the average water saturation $\overline{S}_w$ in the two-phase area after water breakthrough, the water displacement efficiency of different time can also be calculated.

As to any time $t$, because at this moment the pressure $p$ in the two-phase area is only relevant to position $r$, the total flow rate passing through the any cross section in the two-phase area is:

$$Q = 2\pi K h r \left( \frac{K_{ro}}{\mu_o} + \frac{K_{rw}}{\mu_w} \right) \frac{dp}{dr} \tag{6-67}$$

Separate the variables of formula (6-67) and integrate from the initial oil-bearing boundary $r_0$ to oil-water front $r_f$, then the following expression can be obtained:

$$p_0 - p_f = \frac{Q}{2\pi K h} \int_{r_0}^{r_f} \frac{1}{\frac{K_{rw}}{\mu_w} + \frac{K_{ro}}{\mu_o}} \frac{dp}{r} = \frac{Q\mu_w}{2\pi K h} \int_{r_0}^{r_f} \frac{f_w}{K_{rw}} \frac{dp}{r}$$

Compared with the one-dimensional linear problem, the resistance factor $a^*$ of two-phase area is no longer a constant.

### Exercises

6-1  Please try to derive the continuity equation of oil-water two-phase seepage.

6-2  Assume water made planar one-dimensional non-piston like displacement of oil by water, please try to derive the differential equation of saturation distribution of non-piston like displacement of oil by water starting from the element, and further derive the distribution formula of saturation.

6-3 Please try to find the basic data needed when solving the seepage law of non-piston like displacement of oil by water.

6-4 Assume there is non-piston like displacement of oil by water of one-dimensional horizontal formation, and that the data of phase permeability of oil-water combined flowing is shown in Table 6-2.

**Table 6-2 The data of phase permeability curve**

| $S_w$ | $K_{rw}$ | $K_{ro}$ | $S_w$ | $K_{rw}$ | $K_{ro}$ |
|---|---|---|---|---|---|
| 0.200 | 0.000 | 0.800 | 0.550 | 0.100 | 0.120 |
| 0.250 | 0.002 | 0.610 | 0.600 | 0.132 | 0.081 |
| 0.300 | 0.009 | 0.470 | 0.650 | 0.170 | 0.050 |
| 0.350 | 0.020 | 0.370 | 0.700 | 0.208 | 0.027 |
| 0.400 | 0.038 | 0.285 | 0.750 | 0.251 | 0.010 |
| 0.045 | 0.051 | 0.220 | 0.800 | 0.300 | 0.000 |
| 0.500 | 0.075 | 0.163 | | | |

(1) If the volume factor of oil is $B_o = 1.30$ and that of water is $B_w = 1.0$, the irreducible water saturation is $S_{wc} = 0.2$, please respectively calculate the water content ratio and cumulative recovery under the three situations shown in Table 6-3 when water breaks through.

**Table 6-3 The viscosity of oil and water under three situations**

| situation | 1 | 2 | 3 |
|---|---|---|---|
| $\mu_o$, mPa·s | 50.0 | 5.0 | 0.4 |
| $\mu_w$, mPa·s | 0.5 | 0.5 | 1.0 |

(2) If the water injection per day is $Q_j = 160 \text{m}^3/\text{d}$, the irreducible water saturation is $S_{wc} = 0.2$, the residual oil saturation is $S_{or} = 0.2$, the width is $B = 100\text{m}$, $h = 12\text{m}$, $\mu_o = 5.0\text{mPa·s}$, $\mu_w = 0.5\text{mPa·s}$, $\phi = 0.18$ and the distance of injection well row and the production well row is 600m, please find the time of water breakthrough.

(3) Please try to obtain the saturation distribution when time $t = 100\text{d}$, 200d and 400d and make the figure (on computer).

6-5　Assume there is a production well in the center of circular reservoir with sufficient edge water energy; please try to prove that the resistance factor at after water breakthrough in oil well can be expressed by the following expression:

6-5　设边水能量充足的圆形油藏中心有一口生产井,试证油井见水后的阻力系数为:

$$a^* = \frac{1}{f'_{wL}} \int_0^{f'_{wL}} \frac{f_w}{K_{rw}} df'_w$$

Chapter 6 Study Guide
第六章学习指南

# 7 Seepage theory of oil-gas two phases (dissolved gas drive)

油气两相渗流理论（溶解气驱）

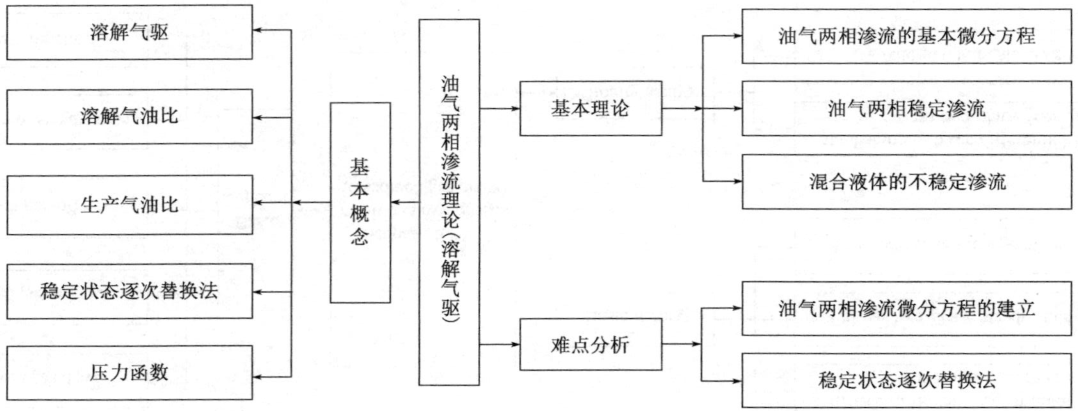

# 第七章知识图谱

```
                    油气两相渗流理论(溶解气驱)
                              │
              ┌───────────────┼───────────────┐
              ↓               ↓               ↓
           基本理论                          难点分析
    ┌────────┼────────┐              ┌──────┴──────┐
    ↓        ↓        ↓              ↓             ↓
油气两相   油气两相   混合液体的    油气两相渗流   稳定状态
渗流的基   稳定渗流   不稳定渗流    微分方程的     逐次替换法
本微分方程                          建立

                        基本概念
        ┌──────┬──────┼──────┬──────┐
        ↓      ↓      ↓      ↓      ↓
     溶解气  溶解气  生产气  稳定状态  压力函数
      驱    油比    油比   逐次替换法
```

```
                          ┌─────────────────────────────────────┐
                          │ Basic differential equation of      │
                          │ oil-gas two-phase seepage           │
                          ├─────────────────────────────────────┤
                          │ Oil-gas two-phase steady state seepage │
                          ├─────────────────────────────────────┤
                          │ Unsteady state seepage of grassy fluid │
                          └─────────────────────────────────────┘
                                          ▲
                                   ┌─────────────┐
                                   │ Basic theory│
                                   └─────────────┘
                                          ▲
┌──────────────────────────────┐
│ Seepage theory               │
│ of oil-gas two phases        │──── Difficulty analysis ──▶ ┌──────────────────────────────────────┐
│ (dissolved gas drive)        │                              │ Establishing of basic differential   │
└──────────────────────────────┘                              │ equation of oil-gas two-phase seepage│
                  │                                           ├──────────────────────────────────────┤
                  ▼                                           │ Alternation of steady state          │
           ┌─────────────┐                                    └──────────────────────────────────────┘
           │ Basic concept│
           └─────────────┘
                  │
   ┌────────┬─────┼─────┬──────────┐
   ▼        ▼     ▼     ▼          ▼
Dissolved Dissolved Production Alternation Pressure
gas drive gas-oil   gas-oil    of steady   function
          ratio     ratio      state
```

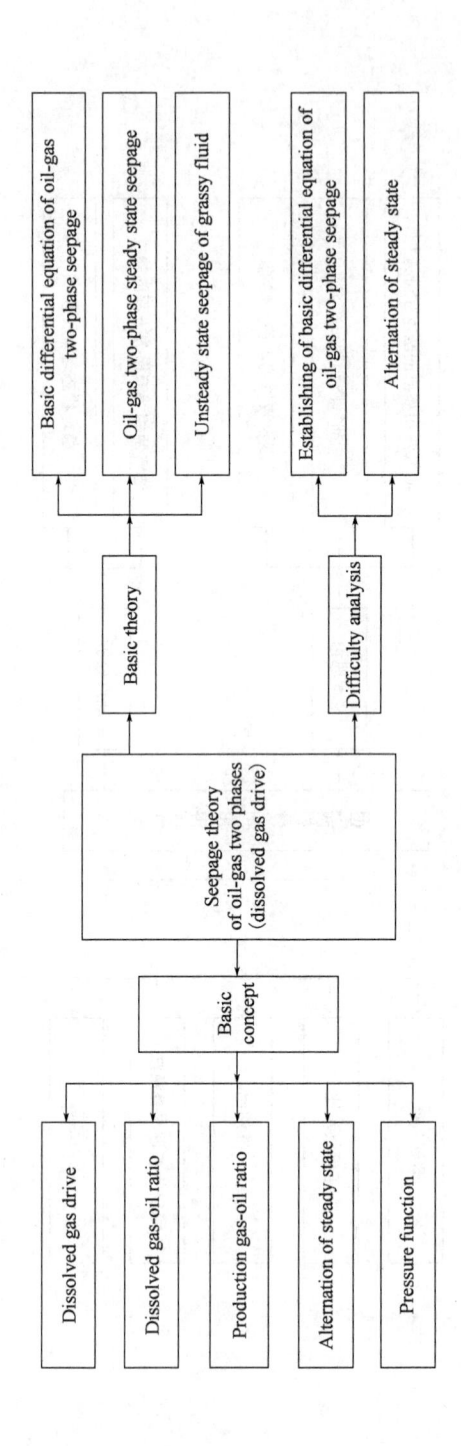

Chapter 7 Knowledge Graph

There is also lots of natural gas dissolved in oil in any reservoir. No matter if reservoir is opened or closed, when the bottom hole pressure of oil well is lower than the saturation pressure, in definite area near bottom hole the initial natural gas dissolved in oil before will separate from oil, which will cause oil-gas two-phase seepage in this area. With the decrease of formation pressure, the two-phase area will expand to the whole formation. And when the formation pressure is lower than the saturation pressure, the whole reservoir will form the oil-gas two-phase seepage.

The dissolved gas drive is a kind of drive mode of the least recovery ratio which is lower than 10%. However, it is the least cost because it totally depends on the natural energy to develop. The theory calculation proves that the recovery ratio will not decrease when formation pressure is 20% lower than saturation pressure, while that is not approved in reality.

## 7.1 Basic differential equation of oil-gas two-phases seepage

In order to solve any seepage problem, a corresponding basic differential equation of the problem, the mathematical model must be set up first. And for setting up the basic differential equation of oil-gas two-phase seepage, take the element (which is the parallelepiped under the cartesian coordinate), and then set up the continuity equation according to the conservation law of mass. The difference is when studying the oil-gas two-phases seepage, the volume of oil and gas in the flowing process is conversed to be the volume under the standard conditions. Thus the mass conservation can be replaced with the volume conservation.

Assume taking any element in the two-phase seepages flow area (Figure 7-1), the flow velocity of oil is $v_o$ when passing through the center of the element point $M$, the oil volume factor is $B_o$, and the flow velocity of free gas is $v_g$, the gas volume factor is $B_g$. Let's assume that the oil and gas flow into and out of the hexahedron along the direction of $x$, $y$ and $z$, then the component of flow velocity along the direction $x$ is $\dfrac{v_{ox}}{B_o}$ when the oil phase is conversed to be the standard conditions on point $M$.

任何一个油藏,原油中总是溶有相当多的天然气。无论油藏是开启还是封闭的,当油井的井底压力低于饱和压力时,在井底附近的一定范围内原来溶解在油中的天然气将分离出来,在该区域内形成油气两相渗流,随着地层压力的进一步降低,两相区有可能扩大到整个地层。当地层压力低于饱和压力时,全油藏将形成油气两相渗流。

溶解气驱是采收率最低的一种驱动方式,采收率一般低于10%。然而,由于它完全依靠天然能量进行开采,因此,成本较低。理论计算证明,地层压力比饱和压力低20%左右不会降低采收率,实际往往并非如此。

## 7.1 油气两相渗流的基本微分方程

要解决任何一种渗流问题,首先必须建立与该问题相对应的基本微分方程式,即数学模型,建立油气两相渗流的基本微分方程,也要从取单元体(在直角坐标系下就是平行六面体)开始,然后根据质量守恒定律建立连续性方程。所不同的是在研究油气两相渗流时,将渗流过程中的油气体积都换算到地面标准条件下的体积,这样,质量守恒就可以用体积守恒来代替。

在油气两相渗流区内任取一微小的六面体(图7-1),其中心$M$点处原油的流速为$v_0$,体积系数为$B_o$,自由气的流速为$v_g$,体积系数为$B_g$。不妨假定油气沿$x$、$y$、$z$方向流入和流出六面体,则$M$点油相换算到标准条件下的渗流速度在$x$方向上的分量为$\dfrac{v_{ox}}{B_o}$。

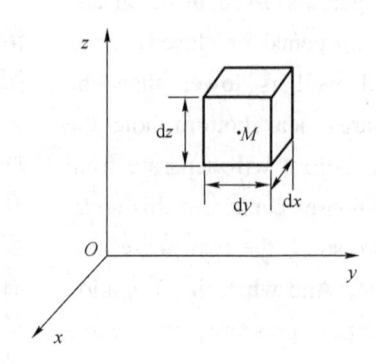

Figure 7 – 1  The element in two-phase area
图 7 – 1  两相区微元体示意图

Because the length of side $\mathrm{d}x$ of the element taken is very small, so the change of velocity within this area can be considered linear. The standard volume of oil flowing into the hexahedron from back in time $\mathrm{d}t$ along the direction $x$ is:

因为所取的六面体的边长 $\mathrm{d}x$ 很小，可以认为速度在这个范围内按线性变化，$\mathrm{d}t$ 时间内沿 $x$ 方向从后面流入六面体内的原油标准体积为：

$$\left[\frac{v_{ox}}{B_o} - \frac{1}{2}\frac{\partial}{\partial x}\left(\frac{v_{ox}}{B_o}\right)\mathrm{d}x\right]\mathrm{d}y\mathrm{d}z\mathrm{d}t$$

With the same reason, the standard volume of oil flowing out of the hexahedron from front in time $\mathrm{d}t$ along the direction $x$ is:

同理，$\mathrm{d}t$ 时间内沿 $x$ 方向从前面流出六面体的原油标准体积为：

$$\left[\frac{v_{ox}}{B_o} + \frac{1}{2}\frac{\partial}{\partial x}\left(\frac{v_{ox}}{B_o}\right)\mathrm{d}x\right]\mathrm{d}y\mathrm{d}z\mathrm{d}t$$

Therefore, the difference between standard volume of oil flowing into and out of the hexahedron in time $\mathrm{d}t$ along the direction $x$ is:

从而求得 $\mathrm{d}t$ 时间内沿 $x$ 方向流入和流出六面体的原油标准体积之差为：

$$-\frac{\partial}{\partial x}\left(\frac{v_{ox}}{B_o}\right)\mathrm{d}x\mathrm{d}y\mathrm{d}z\mathrm{d}t$$

With the same reason, the differences between standard volume of oil flowing into and out of the hexahedron in time $\mathrm{d}t$ along the direction $y$ and $z$ respectively are:

同理，可求得 $\mathrm{d}t$ 时间内沿 $y$ 和 $z$ 方向流入及流出六面体的原油标准体积之差为：

$$-\frac{\partial}{\partial y}\left(\frac{v_{oy}}{B_o}\right)\mathrm{d}x\mathrm{d}y\mathrm{d}z\mathrm{d}t$$

$$-\frac{\partial}{\partial z}\left(\frac{v_{oz}}{B_o}\right)\mathrm{d}x\mathrm{d}y\mathrm{d}z\mathrm{d}t$$

So the total difference between standard volume of oil flowing into and out of the hexahedron in time $\mathrm{d}t$ along the direction $x$, $y$ and $z$ is:

所以，$\mathrm{d}t$ 时间内沿 $x$、$y$、$z$ 三个方向流入和流出六面体的总的原油标准体积之差为：

$$-\left[\frac{\partial}{\partial x}\left(\frac{v_{ox}}{B_o}\right)+\frac{\partial}{\partial y}\left(\frac{v_{oy}}{B_o}\right)+\frac{\partial}{\partial z}\left(\frac{v_{oz}}{B_o}\right)\right]\mathrm{d}x\mathrm{d}y\mathrm{d}z\mathrm{d}t \quad (7-1)$$

While the increment of standard volume of oil phase in hexahedron in time $\mathrm{d}t$ is:

$$\frac{\partial}{\partial t}\left(\frac{S_o\phi}{B_o}\right)\mathrm{d}x\mathrm{d}y\mathrm{d}z\mathrm{d}t \quad (7-2)$$

Where: $S_o$—The saturation of oil phase;
$\phi$—The porosity, if the elasticity of rock is not taken into account, $\phi$ is a constant.

Apparently, the formula (7-1) and the formula (7-2) should be equal to each other, so the following expression can be obtained:

$$\frac{\partial}{\partial x}\left(\frac{v_{ox}}{B_o}\right)+\frac{\partial}{\partial y}\left(\frac{v_{oy}}{B_o}\right)+\frac{\partial}{\partial z}\left(\frac{v_{oz}}{B_o}\right)=-\frac{\partial}{\partial t}\left(\frac{S_o\phi}{B_o}\right) \quad (7-3)$$

The formula (7-3) just is the continuity equation of oil phase.

As to the gas phase, not only the free gas needs to be considered, but also the gas dissolved in oil. Along the direction $x$ at the position of point $M$, the flow velocity of free gas under the standard conditions is $\frac{v_g}{B_g}$, while along the direction $x$ the flow velocity of dissolved gas under the standard conditions is $R_s\frac{v_{ox}}{B_o}$ in which $R_s$ is the dissolved gas-oil ratio. Same with the way of studying the seepage law of oil phase, under the standard conditions, the volume difference of gas (free gas and dissolved gas) flowing into and out of the hexahedron in time $\mathrm{d}t$ along the direction $x$ is:

$$-\frac{\partial}{\partial x}\left(\frac{v_{gx}}{B_g}+R_s\frac{v_{ox}}{B_o}\right)\mathrm{d}x\mathrm{d}y\mathrm{d}z\mathrm{d}t$$

Along the direction $y$ and $z$ respectively are:

$$-\frac{\partial}{\partial y}\left(\frac{v_{gy}}{B_g}+R_s\frac{v_{oy}}{B_o}\right)\mathrm{d}x\mathrm{d}y\mathrm{d}z\mathrm{d}t$$

$$-\frac{\partial}{\partial z}\left(\frac{v_{gz}}{B_g}+R_s\frac{v_{oz}}{B_o}\right)\mathrm{d}x\mathrm{d}y\mathrm{d}z\mathrm{d}t$$

So the total difference of standard volume of gas flowing into and out of the hexahedron in time $\mathrm{d}t$ is:

$$-\left[\nabla\left(\frac{R_s v_o}{B_o} + \frac{v_g}{B_g}\right)\right]dxdydzdt \tag{7-4}$$

While the increment of standard volume of gas in hexahedron in time d$t$ is:

$$\frac{\partial}{\partial t}\left(\frac{\phi S_o R_s}{B_o} + \frac{\phi S_g}{B_g}\right)dxdydzdt \tag{7-5}$$

Apparently, the formula (7-4) and the formula (7-5) should be equal to each other, so the following expression can be obtained:

$$\nabla\left(\frac{R_s v_o}{B_o} + \frac{v_g}{B_g}\right) = -\frac{\partial}{\partial t}\left(\frac{R_s \phi S_o}{B_o} + \frac{\phi S_g}{B_g}\right) \tag{7-6}$$

The formula (7-6) is the continuity equation of gas phase when oil-gas two-phase seepage.

If the oil-gas seepage obeys the Darcy's law, then the motion equation of oil phase is:

$$v_o = -\frac{K_o}{\mu_o}\nabla p_o \tag{7-7}$$

And the motion equation of gas phase is:

$$v_g = -\frac{K_g}{\mu_g}\nabla p_g \tag{7-8}$$

Where: $K_o$—The effective permeability of oil, $10^{-3}\mu m^2$;
$K_g$—The effective permeability of gas, $10^{-3}\mu m^2$;
$\mu_o$—The viscosity of oil, mPa·s;
$\mu_g$—The viscosity of gas, mPa·s;
$p_o$—The pressure of oil phase, MPa;
$p_g$—The pressure of gas phase, MPa.

The pressure of oil phase and the pressure of gas phase have the following relationship:

$$p_{cgo} = p_g - p_o \tag{7-9}$$

In the formula above, $p_{cgo}$ is called the capillary pressure of oil and gas, if it is neglected, the formula (7-7) and the formula (7-8) can be written as the following forms:

$$v_o = -\frac{K_o}{\mu_o}\nabla p \tag{7-10}$$

$$v_g = -\frac{K_g}{\mu_g}\nabla p \tag{7-11}$$

The basic differential equation of oil-gas seepage can be finally obtained by respectively substituting the motion equation of oil phase (7-10) and the equation of gas phase (7-11) into the formula (7-3) and the formula (7-6):

$$\nabla\left(\frac{K_o}{\mu_o B_o}\nabla p\right) = \frac{\partial}{\partial t}\left(\frac{\phi S_o}{B_o}\right) \tag{7-12}$$

$$\nabla\left[\left(\frac{R_s K_o}{\mu_o B_o}+\frac{K_g}{\mu_g B_g}\right)\nabla p\right] = \frac{\partial}{\partial t}\left[\phi\left(\frac{R_s S_o}{B_o}+\frac{S_g}{B_g}\right)\right] \tag{7-13}$$

Although it is of great difficulty to obtain the analytic solution of the problem of oil-gas two-phase seepage starting from the partial differential formula (7-12) and the formula (7-13), this problem can definitely be solved through numerical calculation, the key point of doing this is obtaining precise and reliable initial data.

## 7.2 Oil-gas two-phase steady seepage

The steady state condition of oil-gas two-phase seepage is same with that of single-phase seepage, that is how much oil and gas produced from well. Besides, the same amount of oil and gas must be supplemented into formation. While as to actual reservoir, it is definitely hard to realize. Under the mode of dissolved gas drive, the oil-gas two-phase seepage process is unsteady state, the production and pressure of oil well change along with time. However, in every short time of the whole course, seepage can be approximately considered as steady state, and then the whole unsteady state course can be considered as the superposition of numerous steady state courses. When a small time interval is taken, the changing law of production (bottom hole pressure) successively obtained according to the steady state basically fits for the actual situation, and this method is called method of alternation of steady state. Therefore, the oil-gas two-phase unsteady state seepage law should be studied first in order to research oil-gas two-phase unsteady state seepage. In this section, only the planar radial and planar one-dimensional seepage of them mixture of gas and liquid will be discussed. For the convenience, the following assumptions will be made:

(1) The formation is horizontal, homogeneous, isopachous and isotropic, the oil-gas differentiation caused by gravity is not taken into account.

(2) The natural gas distributes uniformly in oil, there is no condition of forming secondary gas cap.

(3) The interfacial tension of oil and gas is not taken into account.

(4) The oil-gas system always remains the thermodynamic equilibrium state. It is that the natural gas escapes from oil instantaneously, and there is no existence of supersaturated state.

(5) The relative permeability of oil and gas is the single valued function of saturation. In fact, the viscosity ratio and interfacial tension of oil and gas have influence on the relative permeability.

Assume there is a production well with radius $r_w$ in the center of circular reservoir with supply radius $r_e$ and supply pressure $p_e \leqslant p_b$, the bottom pressure is $p_w$ ($p_w < p_b$), the formation is homogeneous, isopachous, isotropic and incompressible. The dissolved gas - oil ratio, viscosity of oil and gas, the changing data of volume factor with pressure (PVT) are given on (Figure 7-2) and the relative permeability curves (Figure 7-3) are given. So the flow rate of oil and gas under standard conditions passing through the cross section from the center of well with distance $r$ is:

$$Q_o = \frac{2\pi rhKK_{ro}}{\mu_o B_o}\frac{dp}{dr} \qquad (7-14)$$

$$Q_g = \frac{2\pi rhKK_{rg}}{\mu_g B_g}\frac{dp}{dr} + Q_o R_s \qquad (7-15)$$

Where: $K$—The absolute permeability, $10^{-3}\mu m^2$;

$K_{ro}$—The relative permeability of oil;

$K_{rg}$—The relative permeability of gas;

$B_o$—The volume factor of oil;

$B_g$—The volume factor of gas;

$R_s$—The dissolved gas - oil ratio.

（1）地层是水平、均质、等厚、各向同性的,不考虑重力引起的油气分异。

（2）天然气在油中是均匀分布的,不存在形成次生气顶的条件。

（3）不考虑油气间的界面张力。

（4）油气系统一直处于热力学平衡状态,即天然气从油中瞬时逸出,不存在脱气滞后所造成的过饱和状态。

（5）油气的相对渗透率是饱和度的单值函数。事实上,油气黏度比、界面张力等因素对相对渗透率是有影响的。

设供给半径为 $r_e$、供给压力 $p_e \leqslant p_b$ 的圆形油气藏,其中心有一口半径为 $r_w$ 的生产井,井底压力为 $p_w$ ($p_w < p_b$),地层是均质、等厚、各向同性且不可压缩的。溶解气油比、油气黏度,体积系数随压力的变化数据(高压物性资料)图 7-2 及相对渗透率曲线图 7-3 都是已经给定的。则通过距离井中心 $r$ 远处的断面的油气标准状况下的流量分别为:

式中: $K$ = 绝对渗透率, $10^{-3}\mu m^2$;

$K_{ro}$ = 油的相对渗透率;

$K_{rg}$ = 气的相对渗透率;

$B_o$ = 油的体积系数;

$B_g$ = 气的体积系数;

$R_s$ = 溶解气油比。

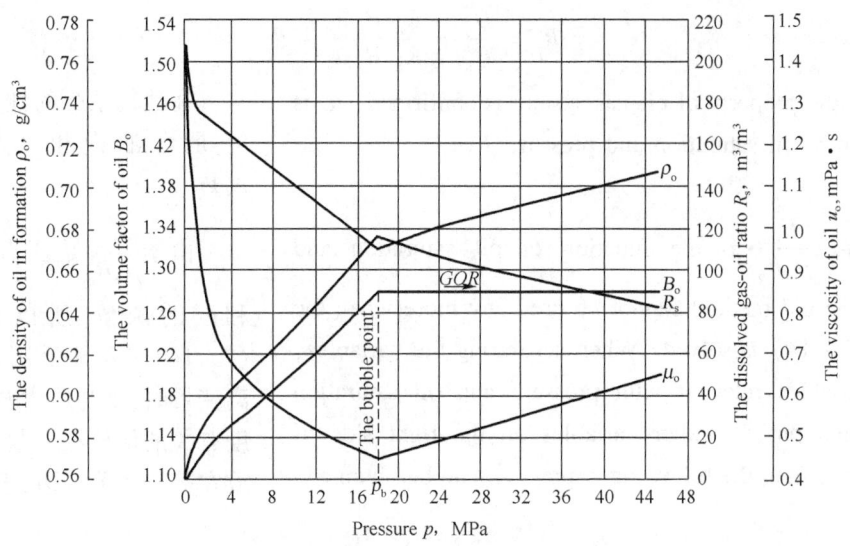

Figure 7-2  The PVT curve of oil
图 7-2  原油的高压物性曲线

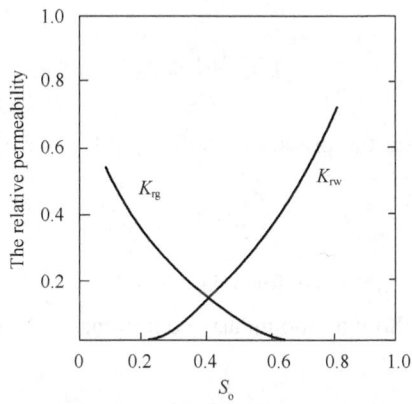

Figure 7-3  The relative permeability curve of oil and gas
图 7-3  油气相对渗透率曲线

Because the relative permeability of oil and gas in the formula (7-14) and the formula (7-15) is the single valued function of saturation, and the oil saturation is the function of pressure, they all are functions of pressure, and the formula (7-14) and the formula (7-15) all are non-linear equations.

Because seepage is steady state, so the pressure of every point in formation, the flow rate of oil and gas passing through every cross section and the gas-oil ratio are all irrelevant with time.

In actual production, the oil production and the ratio of gas production to oil production usually need to be calculated, which is called production gas-oil ratio marked with $R_p$:

式(7-14)和式(7-15)中的油气相对渗透率是含油饱和度的单值函数,而含油饱和度又是压力的函数,所以它们都是压力的函数,且式(7-14)和式(7-15)都是非线性方程。

由于渗流是稳定的,所以地层各点的压力、通过各截面的油气流量以及气油比都与时间无关。

实际工作中常要计算油的产量及天然气与原油的产量之比,即称为生产气油比,用 $R_p$ 表示:

$$R_p = \frac{Q_g}{Q_o} = \frac{K_{rg}}{K_{ro}} \frac{B_o \mu_o}{B_g \mu_g} + R_s \qquad (7-16)$$

When the property of oil-gas system is confirmed, $R_p$ is the function of oil saturation and pressure.

Because $\frac{K_{ro}}{\mu_o B_o}$ is the function of oil saturation and pressure, the old method cannot be used any more when the equation (7-14) is solved. When separating the variables, put the variables relevant with pressure and oil saturation together, and put the other variables on the right side of equality sign, then the following expression can be obtained:

$$\frac{K_{ro}}{\mu_o B_o} dp = \frac{Q_o}{2\pi Kh} \frac{dr}{r} \qquad (7-17)$$

Let:

$$H = \int \frac{K_{ro}}{\mu_o B_o} dp \qquad (7-18)$$

In the formula above, $H$ is called the pressure function, and:

$$dH = \frac{K_{ro}}{\mu_o B_o} dp \qquad (7-19)$$

Substitute the formula (7-19) into the formula (7-17) and integrate, and take the following boundary conditions into account:

$$H\big|_{r=r_e} = H_e; \quad H\big|_{r=r_w} = H_w$$

Then the production of oil of planar radial steady seepage of gassy oil can be obtained:

$$Q_o = \frac{2\pi Kh(H_e - H_w)}{\ln \frac{r_e}{r_w}} \qquad (7-20)$$

The formula (7-20) is very similar to the Dupuit formula, and the main difference $(p_e - p_w)/\mu_o$ is replaced with $H_e - H_w$, so the key point of this problem is how to obtain the pressure function $H$. Because the volume factor and viscosity of oil and gas are all function of pressure, while the relative permeability of oil and gas is the function of oil saturation, so the relationship of oil saturation and pressure must be found out. According to the formula (7-16) the following expression can be obtained:

$$\frac{K_{rg}}{K_{ro}} = \frac{R_p - R_s}{\frac{\mu_o B_o}{\mu_g B_g}} \tag{7-21}$$

In the formula above, $R_p$ is a constant and the variables left all are relevant with pressure. According to the $p, V, T$ data and the production gas-oil ratio $R_p$, the relationship of $\frac{K_{rg}}{K_{ro}}$ with pressure $p$ can be ascertained (Figure 7-4).

式(7-21)中的 $R_p$ 是常数，其余各变量都与压力有关。根据高压物性资料和生产气油比 $R_p$，就可以确定 $\frac{K_{rg}}{K_{ro}}$ 与压力 $p$ 之间的关系(图7-4)。

Otherwise, according to the relative permeability curve (Figure 7-3), the relationship of $\frac{K_{rg}}{K_{ro}}$ with oil saturation $S_o$ can be also obtained (Figure 7-5):

另外，根据相对渗透率曲线(图7-3)可以求出 $\frac{K_{rg}}{K_{ro}}$ 与含油饱和度 $S_o$ 的关系(图7-5)。

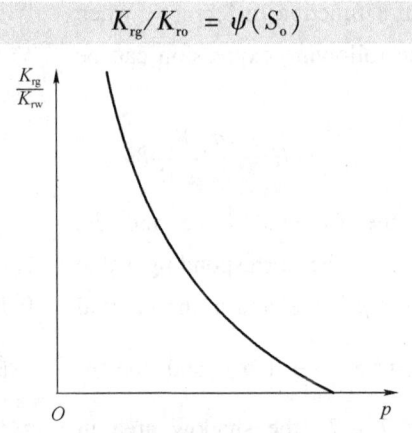

$$K_{rg}/K_{ro} = \psi(S_o)$$

Figure 7-4  The relationship between gas-oil relative permeability ratio and pressure
图7-4  气油相渗透率比值与压力关系

According to the Figure 7-4 and the Figure 7-5, the relationship between pressure and oil saturation can be easily obtained (Figure 7-6).

根据图7-4和图7-5，易求得压力和含油饱和度的关系(图7-6)。

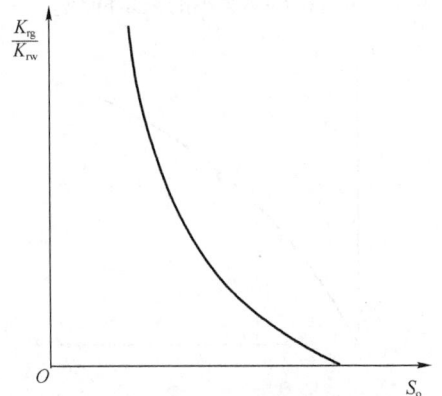

Figure 7-5  The relationship between gas-oil relative permeability ratio and oil saturation
图7-5  气油相渗透率比与含油饱和度关系

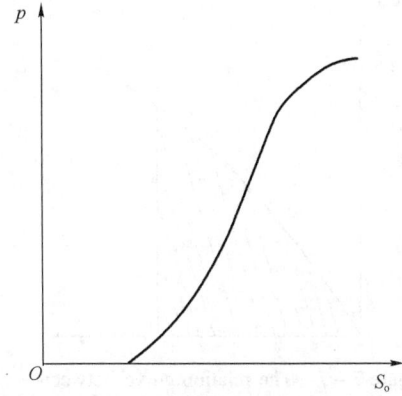

Figure 7-6  The relationship of pressure and oil saturation
图7-6  压力与含油饱和度关系

After the preliminary work above, the relationship between pressure function and pressure p can be further ascertained. Give an arbitrarily pressure p, according to the p, V, T data and the Figure 7 - 4 ~ Figure 7 - 6, the corresponding value of $\frac{K_{ro}}{\mu_o B_o}$ can be obtained, so a point can be further obtained on the figure considering pressure p as the X-axis and $\frac{K_{ro}}{\mu_o B_o}$ as the Y - axis. Arbitrarily give another pressure, another point can be also obtained on the figure, and finally a relation curve of $\frac{K_{ro}}{\mu_o B_o} - p$ can be obtained by repeating it like this in Figure 7 – 7. When used it usually appears with the form of pressure function difference, when calculating pressure function the following expression can be considered correct:

$$H = \int_0^p \frac{K_{ro}}{\mu_o B_o} dp$$

The difference between the formula above and the formula (7 - 18) is a constant, so the corresponding value of pressure function H on pressure p is the area in the interval [0, p] bellow the relation curve $\frac{K_{ro}}{\mu_o B_o}(p)$ and the p-coordinate as shown in Figure 7 – 7, the shadow area in Figure 7 – 7. Consider pressure p as the x-axis and pressure function as the y-axis, and draw the value of pressure function under different pressure obtained on the figure, then the relation curve as shown in Figure 7 – 8 can be finally obtained.

经过以上准备工作,可以进一步确定压力函数与压力 p 之间的关系。任给一个压力 p,根据高压物性资料和图 7 - 4 至图 7 - 6,可以求出相应的 $\frac{K_{ro}}{\mu_o B_o}$ 的值,从而在以压力 p 为横坐标、以 $\frac{K_{ro}}{\mu_o B_o}$ 为纵坐标的图上得到一个点。再任给一个压力,又可在图上得到另一个点。这样重复进行下去就可以得到一条 $\frac{K_{ro}}{\mu_o B_o}$ —p 的关系曲线,如图 7 - 7 所示。由于实用时通常以压力函数的差值出现,所以在计算压力函数时不妨认为:

上式与式(7 - 18)之间只相差一个常数,所以与压力 p 相对应的压力函数值 H 就是图 7 - 7 所示的 $\frac{K_{ro}}{\mu_o B_o}(p)$ 的关系曲线与 p 坐标轴之间的在区间[0,p]内的面积,即图 7 - 7 所示的阴影部分。以压力 p 为横坐标、压力函数为纵坐标,将计算得到的不同压力下的压力函数值点在图上,可以得到图 7 - 8 所示的关系曲线。

Figure 7 - 7  The relation curve between $\frac{K_{rw}}{\mu_o B_o}$ and pressure p

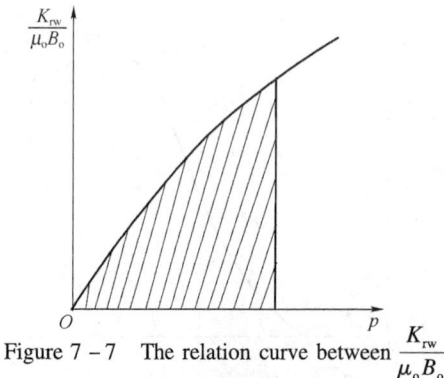

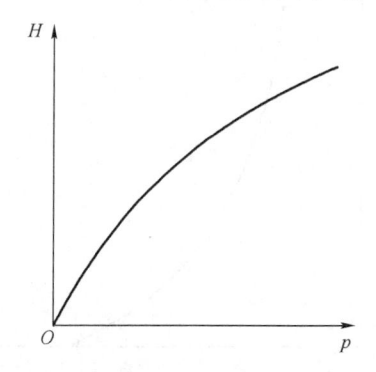

图 7 - 7  $\frac{K_{rw}}{\mu_o B_o}$ 与 p 的关系曲线

Figure 7 - 8  The relation curve between value of pressure function H with pressure p

图 7 - 8  压力函数值 H 与压力 p 的关系曲线

If we start from the basic differential equation of oil-gas seepage, as to the steady seepage, the formula (7 – 12) and the formula (7 – 13) can be simplified as the following forms:

如果从油气渗流的基本微分方程出发，对于稳定渗流，式(7 – 12)和式(7 – 13)可以简化为：

$$\nabla\left(\frac{K_{ro}}{\mu_o B_o}\nabla p\right) = 0 \qquad (7-22)$$

$$\nabla\left[\left(\frac{R_s K_{ro}}{\mu_o B_o} + \frac{K_{rg}}{\mu_g B_g}\right)\nabla p\right] = 0 \qquad (7-23)$$

If pressure function is introduced, the formula (7 – 22) can be written as:

引入压力函数，式(7 – 22)可写成：

$$\nabla^2 H = 0 \qquad (7-24)$$

Obviously, after the pressure function is introduced, the basic differential equation of gassy oil seepage meets the Laplace equation. The relation between production and pressure can be also obtained by solving the formula (7 – 24) under definite boundary condition. If the formula (7 – 24) is changed to be the form of polar coordinates, then the following expression can be obtained:

显然引入压力函数后，混气原油渗流时的基本微分方程满足拉普拉斯方程。在一定的边界条件下求解方程(7 – 24)也可得产量与压力的关系式。若将式(7 – 24)改成极坐标形式，则：

$$\frac{1}{r}\frac{d}{dr}\left(r\frac{dH}{dr}\right) = 0$$

It has been known that the boundary conditions are:

已知边界条件：

$$H\big|_{r=r_e} = H_e; \quad H\big|_{r=r_w} = H_w$$

The solution of this problem is:

则该问题的解为：

$$H = H_e - \frac{H_e - H_w}{\ln\frac{r_e}{r_w}}\ln\frac{r_e}{r} \qquad (7-25)$$

$$Q_o = 2\pi rh\frac{KK_{ro}}{\mu_o B_o}\frac{dp}{dr} = 2\pi Krh\frac{dH}{dr}$$

The formula (7 – 20) can be obtained by derivating $r$ in formula (7 – 25) and then substituting the expression obtained into the formula above.

将式(7 – 25)对 $r$ 求导数并代入上式，可得式(7 – 20)。

As to gas, the following expression can be obtained by the equation (7 – 23):

对于气体，由方程(7 – 23)可得：

$$\nabla\left[\frac{K_{ro}}{\mu_o B_o}\left(R_s + \frac{K_{rg}}{K_{ro}}\frac{\mu_o B_o}{\mu_g B_g}\right)\nabla p\right] = 0$$

That is:

即：

$$\nabla\left(\frac{K_{ro}}{\mu_o B_o}R_p \nabla p\right) = 0$$

After expanding the expression above, the following expression can be obtained:

将上式展开得：

$$R_p \nabla \left( \frac{K_{ro}}{\mu_o B_o} \nabla p \right) + \frac{K_{ro}}{\mu_o B_o} \nabla p \cdot \nabla R_p = 0$$

| After considering the formula (7-22), the following expression can be obtained: | 考虑式(7-22),于是有: |
|---|---|

$$\frac{K_{ro}}{\mu_o B_o} \nabla p \cdot \nabla R_p = 0 \tag{7-26}$$

| Apparently: | 显然: |
|---|---|

$$\frac{K_{ro}}{\mu_o B_o} \nabla p = v_o$$

| So: | 于是只有: |
|---|---|

$$\nabla R_p = 0$$

The expressions above show that along the streamlines, the gas-oil ratio remains unchanged, that is to say, the gas-oil ratio passing through every cross section is equal when gassy oil makes planar radial steady seepage. In fact long term steady state production is impossible, therefore, the law of unsteady seepage of gassy oil needs to be further discussed.

When conducting a steady well testing under the condition of dissolved gas drive, the relationship between oil production $Q_o$ and producing pressure drop $\Delta p_w$ is a curve (Figure 7-9), so $\Delta p_w$ needs to be conversed to be $\Delta H_w = H_e - H_w$. When production is low, $Q_o$ and $\Delta H_w$ show a linear relationship (Figure 7-10), and according to the formula (7-20) the productivity index $J_H$ can be obtained which is:

这说明沿着流线,气油比保持不变,也就是说混气原油作平面径向稳定渗流时通过各断面的气油比相等。实际上长期的稳定生产是不可能的。因此,有必要进一步讨论混气石油不稳定渗流规律。

在溶解气驱条件下进行稳定试井时,获得的产油量 $Q_o$ 与生产压差 $\Delta p_w$ 的关系曲线是一条曲线(图7-9),因而需将 $\Delta p_w$ 换算成 $\Delta H_w = H_e - H_w$。当产量较低时,$Q_o$ 与 $\Delta H_w$ 呈线性关系(图7-10),且根据式(7-20)可得产油指数 $J_H$ 为:

$$J_H = \frac{2\pi Kh}{\ln \frac{r_e}{r_w}}$$

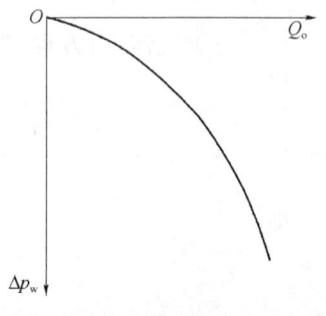

Figure 7-9 The relation curve of oil between production $Q_o$ and producing pressure drop $\Delta p_w$
图7-9 产油量 $Q_o$ 与生产压差 $\Delta p_w$ 的关系曲线

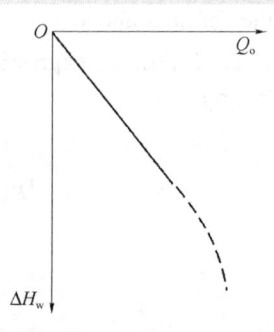

Figure 7-10 The relation curve of oil production $Q_o$ between pressure function drop $\Delta H_w$
图7-10 产油量 $Q_o$ 与拟生产压差 $\Delta H_w$ 的关系曲线

Thus, the formation permeability $K$ near well bottom can be obtained, while the production under unit producing pressure drop is the productivity index, not a constant.

## 7.3　Unsteady seepage of gassy fluid

What happens in reservoir during actual production is unsteady seepage, the pressure and oil saturation of every point in formation will definitely decrease with the increasing of cumulative production, and also the production and gas-oil ratio will definitely change constantly. Same as the elastic unsteady seepage of homogeneous fluid, the unsteady seepage of gassy fluid can be divided into transient period (the first period) and the pseudo-steady state period (the second period), and the trasient period between the two periods is usually not taken into account. The time of transient period is very short which is also not taken into account, so once an oil well is put into production, it is generally considered to be on the pseudo-steady state period.

One of the methods that solve the unsteady seepage of gassy fluid is successive iteration at steady state. This kind of method assumes that the unsteady seepage is consisted of a series of pseudo-steady state seepage, so it is therefore an approximation method.

Assume the pore volume of reservoir is $V_p$, when reservoir pressure drops to $p$, the standard volume of oil left in formation $N_o$ is:

$$N_o = \frac{S_o V_p}{B_o} \qquad (7-27)$$

While the standard volume of residual gas left (free gas and dissolved gas) $G$ is:

$$G = \frac{R_s V_p S_o}{B_o} + \frac{(1 - S_o - S_{wc}) V_p}{B_g} \qquad (7-28)$$

Where: $S_o$—The average oil saturation in formation under any formation pressure;

　　　$S_{wc}$—The irreducible water saturation.

On the right side of the formula (7-28), the first item is the volume of residual dissolved gas, and the second item is the volume of free gas. Taking the derivative of the formula (7-27) with respect to time $t$, and the formation is

7.3　混气液体的不稳定渗流

在实际生产过程中油藏中发生的都是不稳定流,地层内各点的压力以及含油饱和度必然随着累积产量的增加而下降,产量和气油比必然是不断变化的。像均质液体弹性不稳定渗流一样,混气液的不稳定流也可以分为传播期(第一时期)和拟稳定期(第二时期),一般不考虑二者之间的过渡期。传播期的时间很短,通常也不予考虑,即认为油井一投产就进入拟稳定期。

解决混气液体不稳定渗流的方法之一是稳定状态逐次替换法,这种方法假设不稳定渗流是由一系列拟稳定流所组成的,因而是一种近似方法。

设油藏的孔隙体积为 $V_p$,当油藏压力降到 $p$ 时,剩余在地层中的原油的标准体积 $N_o$ 为:

而剩余的气体(自由气体和溶解气)的标准体积 $G$ 为:

式中: $S_o$ = 任一地层压力下地层中的平均含油饱和度;

　　　$S_{wc}$ = 束缚水饱和度。

式(7-28)等号右边的第一项为剩余溶解气的体积,第二项则是自由气的体积。将式(7-27)对时间 $t$ 求导,并认为地

considered as incompressible (compared with that of oil and gas, the compressibility of formation can be neglected), then the following expression can be obtained:

$$Q_o = -\frac{dN_o}{dt} = -V_p \frac{d}{dt}\left(\frac{S_o}{B_o}\right)$$

In the formula above, $Q$ is the production of oil.

After separating the variables in the formula above and integrating, the following expression can be obtained:

$$V_p\left(\frac{S_{oi}}{B_{oi}} - \frac{S_o}{B_o}\right) = \int_0^t Q dt \qquad (7-29)$$

In the formula above, $B_{oi}$ and $S_{oi}$ respectively are the volume factor and average initial oil saturation under initial pressure (generally is the saturation pressure). In order to obtain the integration on the right of the formula (7-29), the relationship of $Q$ and $t$ must be known first, and also the relationship of pressure function $H$ and $t$, and relationship of pressure $p$ and $t$. And at last the relationship between pressure and saturation must be known. Therefore, after respectively differential the formula (7-27) and the formula (7-28) with respect to pressure $p$ the following expression can be obtained:

$$\frac{dN_o}{dp} = V_p\left(\frac{1}{B_o}\frac{dS_o}{dp} - \frac{S_o}{B_o^2}\frac{dB_o}{dp}\right) \qquad (7-30)$$

$$\frac{dG}{dp} = V_p\left[\frac{R_s}{B_o}\frac{dS_o}{dp} + \frac{S_o}{B_o}\frac{dR_s}{dp} - \frac{R_s S_o}{B_o^2}\frac{dB_o}{dp} - \frac{(1-S_o-S_{wc})}{B_g^2}\frac{dB_g}{dp} - \frac{1}{B_g}\frac{dS_o}{dp}\right] \qquad (7-31)$$

The production gas-oil ratio $R_p$ can be obtained with the formula (7-31) divided by the formula (7-30):

$$R_p = \frac{dG}{dN_o} = \frac{\frac{R_s}{B_o}\frac{dS_o}{dp} + \frac{S_o}{B_o}\frac{dR_s}{dp} - \frac{R_s S_o}{B_o^2}\frac{dB_o}{dp} - \frac{1}{B_g}\frac{dS_o}{dp} - \frac{1}{B_g^2}(1-S_o-S_{wc})\frac{dB_g}{dp}}{\frac{1}{B_o}\frac{dS_o}{dp} - \frac{S_o}{B_o^2}\frac{dB_o}{dp}} \qquad (7-32)$$

The formula (7-32) and the formula (7-16) should be equal, and then the relationship between formation pressure $p$ and oil saturation $S_o$ is:

$$\frac{dS_o}{dp} = \frac{\frac{B_g S_o}{B_o}\frac{dR_s}{dp} - \frac{1}{B_g}(1-S_{wc}-S_o)\frac{dB_g}{dp} + \frac{K_{rg}}{K_{ro}}\frac{\mu_o S_o}{\mu_g B_o}\frac{dB_o}{dp}}{1 + \frac{K_{rg}}{K_{ro}}\frac{\mu_o}{\mu_g}} \qquad (7-33)$$

The formula (7-33) is a nonlinear ordinary differential

equation which can only be solved by approximation method. For the convenience of calculation, the pressure functions in the formula (7-33) will be expressed with the following formula:

$$X(p) = \frac{B_g}{B_o}\frac{dR_s}{dp}$$

$$Y(p) = \frac{\mu_o}{\mu_o B_o}\frac{dB_o}{dp}$$

$$Z(p) = -\frac{1}{B_g}\frac{dB_g}{dp}$$

Because $X(p)$, $Y(p)$ and $Z(p)$ are all functions of pressure $p$, and the formula (7-33) is nonlinear, the method of iteration can be used to solve the equation. When explaining it, replace the differentiation with difference, so the formula (7-33) can be written as the following formula:

$$\frac{\Delta S_o}{\Delta p} = \frac{S_o X(p) + \frac{K_{rg}}{K_{ro}} S_o Y(p) + (1 - S_o - S_{wc}) Z(p)}{1 + \frac{K_{rg}}{K_{ro}}\frac{\mu_o}{\mu_g}} \quad (7-34)$$

Where: $\Delta p$—The changing amount of average formation pressure, MPa;

$\Delta S$—The changing amount of average formation saturation.

The relationship between average formation pressure and average oil saturation can be calculated with the formula (7-34), and the concrete steps are shown as follows:

(1) Give the accuracy $\varepsilon$ of calculation.

(2) Assume a pressure $p_1$ which is lower $\Delta p_1$ than the initial formation pressure $p_{oi}$ (generally is the saturation pressure), and assume the formation average oil saturation is $S_1$ when the average formation pressure drops to $p_1$ and give the initial conjecture $S_1 = S_{oi}$.

(3) Calculate $\bar{p}_1 = \frac{p_{oi} + p_1}{2}$, and further calculate the pressure functions $X(\bar{p}_1)$、$Y(\bar{p}_1)$ and $Z(\bar{p}_1)$.

(4) Make $k = k + 1$, calculate $\bar{S}_1 = \frac{S_{oi} + S_1}{2}$, $K_{ro}(\bar{S}_1)$ and $K_{rg}(\bar{S}_1)$, and then substitute them into the formula

(7-34) to obtain $\Delta S^{(k)}$.

(5) If $k < 2$, turn to the step (7); otherwise turn to the step (6).

(6) If $|\Delta S_o^{(k)} - \Delta S_o^{(k-1)}| < \varepsilon$, turn to the step (8), otherwise turn to the step (7);

(7) Calculate the average oil saturation $S_1 = S_{oi} - \Delta S_o^{(k)}$ and then turn to the step (4).

(8) Calculate the formation average oil saturation $S_1 = S_{oi} - \Delta S_o^{(k)}$ when formation average pressure drops to $p_1$.

(9) Assume an average formation pressure $p_2$ which is lower than $p_1$, at this moment, $S_{oi} = S_1$ and $p_{oi} = p_1$ can be considered, while $p_2$ is amount to the $p_1$ in the last step, repeat the step (2) ~ step (8), the average oil saturation $S_2$ of oil layer when average formation pressure is $p_2$ can be ascertained.

The relation curve between average formation pressure and average oil saturation can be calculated by repeating the steps above (Figure 7-11).

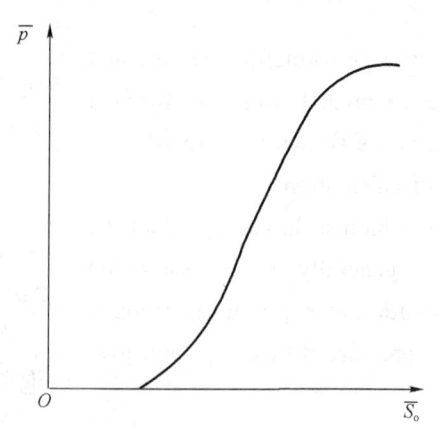

Figure 7-11  The curve of average formation pressure with average oil saturation
图 7-11  平均地层压力与平均含油饱和度的关系曲线

With the Figure 7-11, the $p, V, T$ data and the relative permeability curve, the relation curve between pressure function $H$ and the formation pressure can be obtained.

In the flow of gassy oil, in every pressure interval the seepage can be considered as steady state, the average pressure and the average oil saturation in interval $[p_j, p_{j+1}]$ respectively are: $\bar{p} = \dfrac{p_j + p_{j+1}}{2}$, $\bar{S}_o = \dfrac{S_j + S_{j+1}}{2}$, and at this moment, the relationship between oil production $Q_o$ and pressure difference is:

$$Q_o = \dfrac{2\pi Kh(\bar{H} - H_w)}{\ln \dfrac{r_e}{r_w} - \dfrac{1}{2}}$$

The oil production $Q_o$ can be further obtained because both $\bar{p}$ and $p_w$ are known and $\bar{H} - H_w$ can be solved from Figure 7-8. The average production gas-oil ratio in this interval is:

$$R_p = \dfrac{K_{rg}(\bar{S}_o)\mu_o(\bar{p})\bar{B}_o(\bar{p})}{K_{ro}(\bar{S}_o)\mu_g(\bar{p})B_g(\bar{p})} + R_s(\bar{p})$$

Likewise, the relationship among $Q_o$, $R_p$ and the average oil saturation in every interval can be calculated as shown in Figure 7-12.

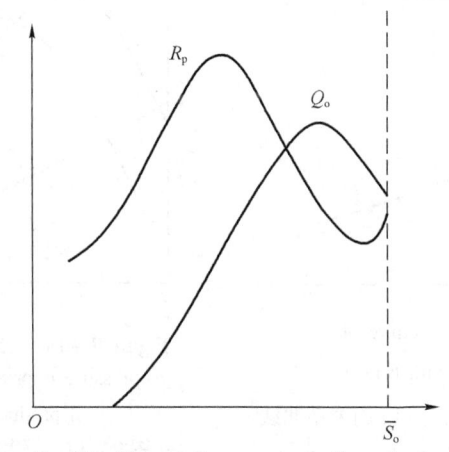

Figure 7-12  The relation curve of oil production $Q_o$, production gas-oil ratio $R_p$ with the oil saturation $\bar{S}_o$

图 7-12  产油量 $Q_o$ 和生产气油比 $R_p$ 与平均含油饱和度 $\bar{S}_o$ 的关系曲线

The changing law of average formation pressure, oil saturation and production gas-oil ratio with time can be further obtained according to the formula (7-29). Because in every interval the seepage of gassy oil can be considered as steady state, and the oil production $Q_o$ is a constant, the required time of the average formation pressure drops from $p_{j-1}$ to $p_j$ can be obtained from the formula (7 – 29) which is:

$$\Delta t_j = \frac{V_p}{Q_j}\left(\frac{S_{j-1}}{B_{j-1}} - \frac{S_j}{B_j}\right)$$

When $j = 1$, $S_{j-1}$ is just the initial oil saturation $S_{oi}$, and $B_{j-1}$ is the volume factor of oil $B_{oi}$ under initial formation pressure, while the required time of average formation pressure drops from $p_{oi}$ to $p_j$ is:

$$t = \sum_{k=1}^{j} \Delta t_k \tag{7 – 36}$$

当 $j = 1$ 时，$S_{j-1}$ 就是原始含油饱和度 $S_{oi}$，$B_{j-1}$ 为原始地层压力下的原油体积系数 $B_{oi}$，而平均地层压力从 $p_{oi}$ 下降到 $p_j$ 所需的时间为：

(7 – 35)

$p_j$ 所需的时间为：

The relation curve of average oil saturation with time $t$ can be obtained by formula (7 – 35) and the formula (7 – 36) as shown in Figure 7 – 13, and the relation curve of $\bar{p}$, $R_p$, $Q_o$ with time $t$ can be also obtained with the known relationship of $\bar{p}$, $R_p$, $Q_o$ and $\bar{S}_o$ as shown in Figure 7 – 14. From the former analysis, the economical ultimate production time, the possible production downtime and the final recovery efficiency etc. can be inferred.

由式(7 – 35)和式(7 – 36)可求出平均含油饱和度与时间 $t$ 的关系曲线(图 7 – 13)，再利用已知的 $\bar{p}$、$R_p$、$Q_o$ 与 $\bar{S}_o$ 的关系，可求出 $\bar{p}$、$R_p$、$Q_o$ 与 $t$ 的关系曲线(图 7 – 14)，从而可以推测经济极限开采时间、油井可能停产的时间以及最终采收率等。

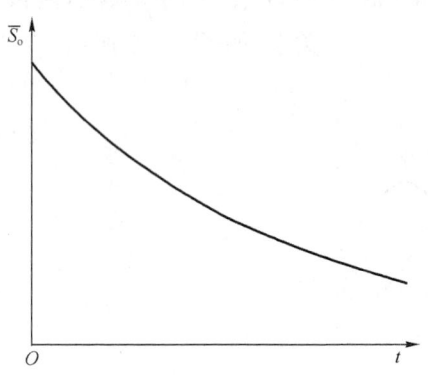

Figure 7 – 13　The relation curve of average oil saturation $\bar{S}_o$ with time $t$

图 7 – 13　平均含油饱和度 $\bar{S}_o$ 与时间 $t$ 的关系曲线

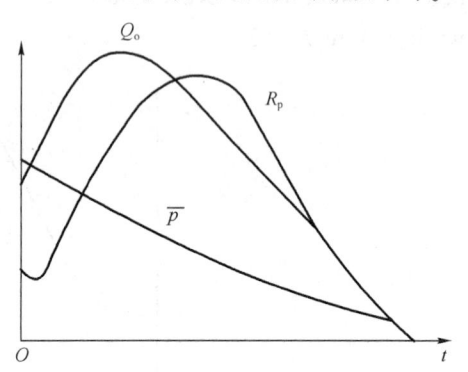

Figure 7 – 14　The relation curve of average pressure $\bar{p}$, production gas-oil ratio $R_p$, oil production $Q_o$ with time $t$

图 7 – 14　平均压力 $\bar{p}$、生产气油比 $R_p$、产油量 $Q_o$ 与时间 $t$ 的关系曲线

In the dissolved gas drive, the gas-oil ratio will rise rapidly after dropping at first, and later will drop rapidly after reaching the extreme value. The corresponding production of oil and gas will also increase rapidly, and the time that oil production reaches the extreme is earlier than that of the gas-oil ratio reaches the extreme, and then the oil production decreases rapidly. As to the average formation pressure it always drops with time. The theory calculation

溶解气驱开发方式下，气油比开始下降以后迅速上升，达到极值后又迅速下降。相应的油产量也迅速上升，油产量达到极值的时间早于气油比达到极值的时间，然后迅速下降。至于平均地层压力则一直随时间降低。理论计算表明开始生产时，气油

shows when the production starts, the gas-oil ratio can drop because slight gas just escapes from the oil in disperse state and cannot flow continuously, and has displacing function, so when the formation pressure does not drop too much, the oil recovery can be enhanced. However, in actual reservoir development, the declining stage of production gas-oil ratio cannot be seen. Otherwise, it can be seen in the Figure 7-13 that the oil recovery of dissolved gas drive is very low.

## Exercises

7-1 Please try to derive the production formula of one-dimensional isothermal steady state seepage of dissolved gas drive, and write out the solving steps of pressure function $H$.

7-2 Assume the saturation pressure of a given reservoir is $p_b = 17.0$ MPa, and now assume the reservoir produces under the saturation pressure; it has the known data in Table 7-1 and Table 7-2. Please respectively make the relation curve of pressure function $H(p)$ with pressure $p$ when $R_p = 96.270$ m$^3$/m$^3$ and 960 m$^3$/m$^3$.

**Table 7-1  The production data**

| $p$, MPa | $\mu_o$, mPa·s | $\mu_w$, mPa·s | $B_o$ | $B_g$ | $R_s$, m$^3$/m$^3$ |
|---|---|---|---|---|---|
| 1.7 | 2.372 | 0.01328 | 1.080 | 0.049 | 6.858 |
| 3.4 | 2.116 | 0.01424 | 1.096 | 0.025 | 40.567 |
| 5.1 | 1.901 | 0.01516 | 1.116 | 0.0164 | 22.857 |
| 6.8 | 1.736 | 0.01576 | 1.139 | 0.0113 | 31.619 |
| 8.5 | 1.584 | 0.91652 | 1.161 | 0.00913 | 40.567 |
| 10.2 | 1.464 | 0.01712 | 1.185 | 0.00742 | 49.905 |
| 11.9 | 1.352 | 0.01780 | 1.207 | 0.00625 | 59.048 |
| 13.6 | 1.280 | 0.01848 | 1.235 | 0.00546 | 69.714 |
| 15.3 | 1.220 | 0.01892 | 1.268 | 0.00479 | 81.904 |
| 17.0 | 1.192 | 0.01936 | 1.305 | 0.00440 | 96.0 |

**Table 7-2  The data of relative permeability**

| $S_o$ | 0.10 | 0.20 | 0.25 | 0.30 | 0.35 | 0.40 | 0.45 | 0.50 | 0.55 | 0.50 | 0.65 | 0.70 | 0.75 | 0.80 |
|---|---|---|---|---|---|---|---|---|---|---|---|---|---|---|
| $K_{ro}$ | 0.00 | 0.00 | 0.0175 | 0.0456 | 0.0842 | 0.116 | 0.175 | 0.218 | 0.281 | 0.340 | 0.421 | 0.496 | 0.596 | 0.705 |
| $K_{rg}$ | 0.50 | 0.351 | 0.267 | 0.193 | 0.140 | 0.0982 | 0.0632 | 0.0438 | 0.0263 | 0.0088 | 0.0042 | 0.00 | 0.00 | 0.00 |

# 8 Seepage in dual-medium

双重介质中的渗流

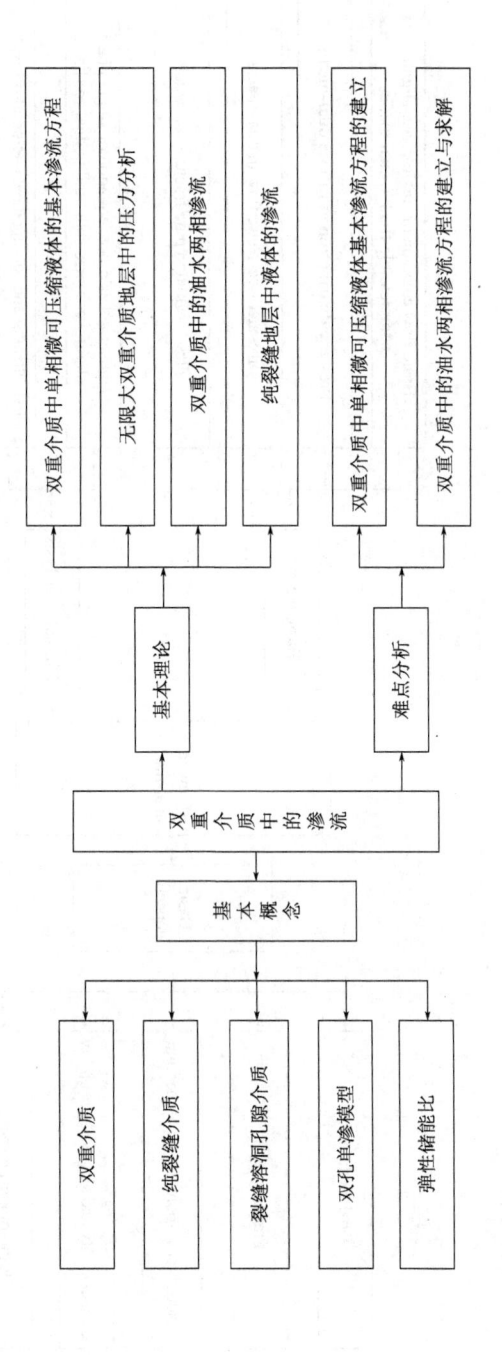

## 第八章知识图谱

```
                    ┌─ 双重介质中单相微可压缩液体的基本渗流方程
                    ├─ 无限大双重介质地层中的压力分析
          ┌─ 基本理论 ─┼─ 双重介质中的油水两相渗流
          │         └─ 纯裂缝地层中液体的渗流
双重介质中的渗流 ─┤
          │         ┌─ 双重介质中单相微可压缩液体基本渗流方程的建立
          └─ 难点分析 ─┴─ 双重介质中的油水两相渗流方程的建立与求解
          │
          └─ 基本概念 ─┬─ 双重介质
                    ├─ 纯裂缝介质
                    ├─ 裂缝溶洞孔隙介质
                    ├─ 双孔单渗模型
                    └─ 弹性储能比
```

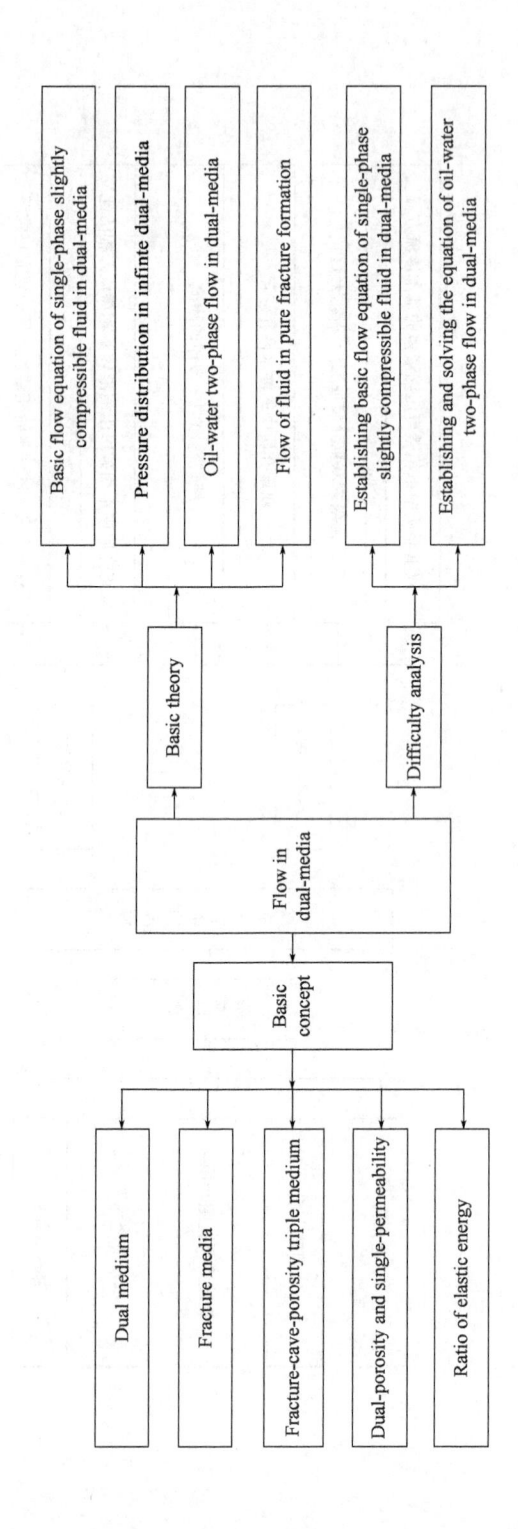

## Chapter 8 Knowledge Graph

```
Flow in dual-media
├── Basic concept
│   ├── Dual medium
│   ├── Fracture media
│   ├── Fracture-cave-porosity triple medium
│   ├── Dual-porosity and single-permeability
│   └── Ratio of elastic energy
├── Basic theory
│   ├── Basic flow equation of single-phase slightly compressible fluid in dual-media
│   ├── Pressure distribution in infinite dual-media
│   ├── Oil-water two-phase flow in dual-media
│   └── Flow of fluid in pure fracture formation
└── Difficulty analysis
    ├── Establishing basic flow equation of single-phase slightly compressible fluid in dual-media
    └── Establishing and solving the equation of oil-water two-phase flow in dual-media
```

With the progress of exploration and development of reservoirs in our country, the fractured carbonate rock gradually becomes the main reservoir of oil and gas. According to the statistics, about 35% ~ 50% of the oil reserves in the world is in the fractured carbonate rock reservoir and the tight sandstone reservoir also has many fractures. The fluid flow in the fracture-pore oil formation is not same with that in the pores, so as to research the flow of fluid in the fracture – pore oil formation is of great actual value and theory significance.

## 8.1 Seepage characteristics in dual-medium

The fractured reservoir (fractured media) mainly includes: the pure fractured media, the fracture – pore media and the fracture-cave-pore media.

The fracture – pore media is also called the dual – medium which is composed of rock with pores and the fractures cutting up the rocks, the oil, gas and water mainly are stored in the pores and the fractures are the main flow paths of oil, gas and water (Figure 8 – 1). Because the pores of rocks of the dual-medium are controlled by the action of sedimentation and diagenesis, so the porosity is big and the permeability is very small. However, as to the secondary fracture joint, because the fracture volume occupies very small part of the formation volume, the porosity of it is very small, while the permeability is very big. The distinctiveness of the structure of dual – medium decides the flow characteristics of fluid in dual – medium. The fracture permeability in dual-medium is dozens of or hundreds of times bigger than that of matrix. So after the oil well is put into production, oil in fractures will start to flow to the well first, which will cause pressure disequilibrium between matrix and fractures, under the action of the pressure difference, the fluid stored in the pores of matrix will flow into the fractures first and then flow to the oil well through the factures. The flow from the matrix to the fractures is called "cross flow". Between the fractures and the pores of matrix exists the exchange of fluid which will form two interlaced hydrodynamic fields, and each of the hydrodynamic field is with different pressure and flow velocity, so when researching the flow problem of fluid in the dual-medium, the changing characteristics and the interaction of the two hydrodynamic potential of fractures and matrix should be analyzed mainly.

压力和流速,因而,研究流体在双重介质中的渗流问题时,应着重分析裂缝和岩块两个水动力场的变化特征与相互影响。

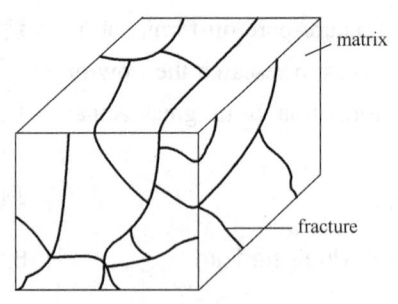

Figure 8 – 1　The rock-fracture system
图 8 – 1　岩块与裂缝系统

Because the compressibility of fractures is much bigger than that of pores in matrix, so when pressure in fractures drops, the porosity and permeability will also change with pressure, and the change of permeability is bigger than porosity. Many researchers show the relationship between porosity and permeability of fractures and pressure is nonlinear. When the pressure does not change too much, the change of permeability of matrix with pressure can be neglected. Because fractures have the feature of directional distribution, so the dual-medium has high anisotropy and the flow resistance in every direction is different. Therefore, the permeability in dual-medium doesn't only change with direction but also with the direction of measured plane, it is a second-rank tensor.

由于裂缝的压缩性比岩块孔隙的压缩性大很多倍,裂缝中压力下降时,孔隙度与渗透率都将随压力变化,渗透率的变化又比孔隙度大。很多研究表明,裂缝孔隙度与渗透率和压力的关系是非线性的。在压力变化不大时,可以忽略岩块渗透率随压力的变化。由于裂缝具有定向分布的特点,因而双重介质又具有高度的各向异性,各方向的渗流阻力不一样。因此,双重介质中渗透率不仅随方向变化,而且还和测量面的方向有关,是一个二阶张量。

The pure fractured media is the oil reservoir in which oil only stores in the fractures and the matrix has no permeable ability. In this kind of oil reservoir, fractures usually develop very well, so the additional inertial resistance is easy to happen when the flow velocity is big, thereby the Darcy's law is destroyed. In the pure fractured media, the permeability of fractures also has the features of anisotropy and changes with pressure.

纯裂缝介质是指那些只有裂缝储油,岩块无渗透能力的油层,这种油层中一般裂缝比较发育,在渗流速度较大时易产生附加的惯性阻力,从而破坏达西定律。纯裂缝介质中裂缝渗透率也具有各向异性,并随压力而变化。

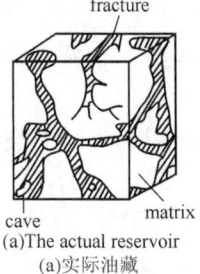

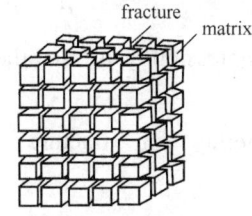

Recent years, the flow problem of fracture-cave-pore multiple media is proposed which needs to be further studied. It has been known to all that in the carbonate formation, the big and small caves always coexist with fracture system, and the matrix flow maybe happen when fluid in the fracture system flowing through the caves. Therefore, the research on the flow of multiple media that is taking the influence of caves into account is of great actual and theoretical significance.

## 8.2 Basic seepage equation of single-phase slightly compressible fluid in dual-medium

The flow of single-phase slightly compressible fluid in dual-medium is a more complex problem. Баренблат т of the Soviet Union and Warren of America and others proposed the flow model of slightly compressible fluid in dual-medium, and simplified the rock-fracture system as shown in Figure 8-1, the simplified model shown in Figure 8-2, and made the following assumptions in one infinitesimal element:

(1) The primary pore media (the matrix) is homogeneous and isotropic, and is consisted of equal parallel arrayed rectangular matrix.

(2) The secondary fractures are consisted of set of continuous, evenly and orthogonal systems, the space and width in each fracture system simulates the degree of anisotropy of fracture.

(3) Between the primary pore and the secondary fractures cross flow happens, but the liquid flow to the well through fractures only, and inflow through the pore system is neglected. This model is called dual-porosity and single-permeability model, and consider the flow of fluid in the fractures obeys the Darcy's law:

(a) The actual reservoir
(b) The model reservoir

Figure 8-2　The simplified model

$$v_f = -\frac{K_f}{\mu} \nabla p_f \qquad (8-1)$$

The permeability of matrix system segregated and surrounded by fractures is usually several quantity degrees smaller than that of fracture system, therefore, flow in the matrix system can be neglected and matrix can be considered as a evenly distributed source that releases liquid into the fractures—fluid channeling. Assume the liquid exchanges between the media systems is linear quasi-steady state, and then the fluid channeling rate $q$ is in direct proportion to the pressure difference between the media systems that are:

$$q = \frac{\alpha \rho K_m (p_m - p_f)}{\mu} \qquad (8-2)$$

Where: $\alpha$—The exchange factor of liquid between systems is also called matrix shape factor and is relevant with the geometric shape and characters of matrix and the concentration of fractures, its dimension is the reciprocal of area, $cm^{-2}$;

$p_f, p_m$—The pressure in the flow field of fracture system and matrix system, MPa;

$K_f, K_m$—The permeability of the same point in the fracture system and matrix, $\mu m^2$;

$v_f$—The average flow velocity in the fracture system, cm/s;

$\rho$—The density of liquid, g/cm³.

The equation of state of liquid that is the relationship between density and pressure is:

$$\rho = \rho_o e^{C_o(p-p_o)} \approx \rho_o [1 + C_o(p_f - p_o)] \qquad (8-3)$$

Where: $C_o$—The compressibility of oil (the liquid phase).

The compressibility of fracture system can be expressed as:

$$C_f = \frac{1}{\phi_f} \frac{\partial \phi_f}{\partial p_f} \qquad (8-4)$$

The compressibility of pore system can be expressed as:

$$C_m = \frac{1}{\phi_m}\frac{\partial \phi_m}{\partial p_m} \qquad (8-5)$$

Where: $C_f, C_m$—The compressibility of fracture system and pore system, $\text{MPa}^{-1}$.

$\phi_f, \phi_m$—The porosity of fracture system and pore system.

If the gravity, the anisotropy of dual-medium are not taken into account and the flow is isothermal, take an element from the dual-medium (Figure 8-1) and turn it to be the simplified model as shown in Figure 8-2. According to the law of conservation of mass, the continuity equation in every media system can be written out as:

$$\frac{\partial (\rho \phi_f)}{\partial t} = -\nabla(\rho \cdot v_f) + q \qquad (8-6)$$

$$\frac{\partial (\rho \phi_m)}{\partial t} + q = 0 \qquad (8-7)$$

The formula (8-6) is the continuity equation of liquid flow in the fracture system, the formula (8-7) shows the continuity of fluid channeling from the pore system to the fracture system, and there is no flow in pore system.

Expand the left side of the formula (8-6), and then the following expression can be obtained:

$$\frac{\partial (\phi_f \rho)}{\partial t} = \phi_f \frac{\partial \rho}{\partial t} + \rho \frac{\partial \phi_f}{\partial t} = \phi_f C_o \rho_o \frac{\partial p_f}{\partial t} + \phi_f C_f \rho_o \frac{\partial p_f}{\partial t} + \phi_f C_o C_f \rho_o (p_f - p_o) \frac{\partial p_f}{\partial t}$$

On the right side of the formula above, the third item is a second-order mini value which can be omitted, and then the following expression can be obtained:

$$\frac{\partial (\phi_f \rho)}{\partial t} = \phi_f C_o \rho_o \frac{\partial p_f}{\partial t} + \phi_f C_f \rho_o \frac{\partial p_f}{\partial t} = \phi_f (C_o + C_f) \rho_o \frac{\partial p_f}{\partial t}$$

Assume $C_{tf} = C_o + C_f$, then the following expression can be obtained:

$$\frac{\partial (\phi_f \rho)}{\partial t} = \phi_f C_{tf} \rho_o \frac{\partial p_f}{\partial t} \qquad (8-8)$$

Likewise, the following expression can be obtained:

$$\frac{\partial(\phi_m \rho)}{\partial t} = \phi_m C_{tm} \rho_o \frac{\partial p_m}{\partial t} \qquad (8-9)$$

Where: $C_{tf}$, $C_{tm}$—The total compressibility coefficient of fracture system and pore system, $\text{MPa}^{-1}$.

After substituting the formula (8-1), the formula (8-8) and the formula (8-9) into the formula (8-6) and the formula (8-7), the following expression can be obtained:

$$\phi_f \rho_o C_{tf} \frac{\partial p_f}{\partial t} = \frac{K_f}{\mu} \rho \nabla^2 p_f + \frac{K_f}{\mu} \rho_o C_o (\nabla p_f)^2 + q$$

Assume the pressure gradient is not big, then $(\nabla p_1)^2$ can be neglected. In fact, except the area near the well, this assumption can be accepted. Simplify the expression above to be the following form:

$$\phi_f \rho_o C_{tf} \frac{\partial p_f}{\partial t} = \frac{K_f}{\mu} \rho \nabla^2 p_f + q \qquad (8-10)$$

If it is the problem of axisymmetric flow, in the polar coordinates the following expression can be obtained:

$$\phi_f \rho_o C_{tf} \frac{\partial p_f}{\partial t} = \frac{K_f}{\mu} \rho \left( \frac{\partial^2 p_f}{\partial r^2} + \frac{1}{r} \frac{\partial p_f}{\partial r} \right) + q \qquad (8-11)$$

With the same method to process the formula (8-7), the following expression can be obtained:

$$\phi_m \rho_o C_{tm} \frac{\partial p_m}{\partial t} + q = 0$$

Approximately take $\rho \approx \rho_o$, and substitute the formula (8-2) into the expression above and the formula (8-11), then the following expression can be obtained:

$$\begin{cases} \phi_f C_{tf} \dfrac{\partial p_f}{\partial t} = \dfrac{K_f}{\mu} \left( \dfrac{\partial^2 p_f}{\partial r^2} + \dfrac{1}{r} \dfrac{\partial p_f}{\partial r} \right) + \dfrac{\alpha K_m}{\mu} (p_m - p_f) \\ \phi_m C_{tm} \dfrac{\partial p_m}{\partial t} + \dfrac{\alpha K_m}{\mu} (p_m - p_f) = 0 \end{cases} \qquad (8-12)$$

In the early 1960s, Баренблат т of the Soviet Union and others proposed that when the single-phase slightly compressible fluid flows in the dual-medium, there is also the coexistence of flow along the fractures and in the matrix, and the liquid exchange between the matrix and the fractures. On the other hand, flow in the fracture system and the pore system both obey the Darcy's law, then the following expression can be obtained:

$$\begin{cases} v_{\mathrm{f}} = \dfrac{-K_{\mathrm{f}}}{\mu} \nabla p_{\mathrm{f}} \\ v_{\mathrm{m}} = \dfrac{-K_{\mathrm{m}}}{\mu} \nabla p_{\mathrm{m}} \end{cases} \quad (8-13)$$

Where: $v_{\mathrm{m}}$—The average flow velocity in the pore system, m/s.

After that the continuity of each system can be obtained as:

式中：$v_{\mathrm{m}}$ = 孔隙系统中的平均渗流速度，m/s。

进而可分别写出各系统的连续性关系为：

$$\begin{cases} \dfrac{\partial(\rho\phi_{\mathrm{f}})}{\partial t} = -\nabla(\rho v_{\mathrm{f}}) + q \\ \dfrac{\partial(\rho\phi_{\mathrm{m}})}{\partial t} = -\nabla(\rho v_{\mathrm{m}}) - q \end{cases} \quad (8-14)$$

Because the fluid flows from the pore media to the fracture media, it is inflow to the fracture media, while to the pore media it is outflow, so the sign of $q$ is different. Likewise, assume the liquid exchange between the systems is linear quasi-steady state as the formula (8-2) shown. With the same method above, the flow equations of the fracture system and the pore system can be set up:

由于液体从孔隙介质流向裂缝介质，对裂缝介质来说是流入，而对孔隙介质来说是流出，所以 $q$ 符号不同。同样设系统间液量交换是线性拟稳定状态，如式(8-2)所示。用与上面相同的方法，可建立裂缝系统与孔隙系统的渗流方程为：

$$\begin{cases} C_{\mathrm{tf}}\phi_{\mathrm{f}}\mu \dfrac{\partial p_{\mathrm{f}}}{\partial t} = K_{\mathrm{f}}\nabla^{2}p_{\mathrm{f}} + \alpha K_{\mathrm{m}}(p_{\mathrm{m}} - p_{\mathrm{f}}) \\ C_{\mathrm{tm}}\phi_{\mathrm{m}}\mu \dfrac{\partial p_{\mathrm{m}}}{\partial t} = K_{\mathrm{m}}\nabla^{2}p_{\mathrm{m}} - \alpha K_{\mathrm{m}}(p_{\mathrm{m}} - p_{\mathrm{f}}) \end{cases} \quad (8-15)$$

In 1970s, when Boulton and others researched the unsteady flow underground water in the fractured water-bearing formation, considering the fluid flow from the pore system to the fracture system is elastic unsteady state, so the improved flow equation of dual-medium and the corresponding solutions are proposed. Recent years, big caves are taken into account besides the matrix and the fracture system, so the flow problem in the triple medium is further proposed and the flow model of triple medium is set up.

20世纪70年代Boulton等在研究地下水在裂缝性含水层中不稳定渗流时，进一步考虑液体从孔隙系统向裂缝渗流是弹性不稳定的，从而提出改进的双重介质渗流方程及相应的解。近年来考虑到除孔隙岩块、裂缝系统外，还存在大的溶洞，因此，进一步提出三重介质中的渗流问题，并建立了三重介质的渗流模型。

## 8.3 Pressure distribution in infinite dual-medium

In infinite dual-medium, the whole thickness of formation is vertically drilled and put into production with constant production aiming to obtain the propagation law of pressure in the elastic unsteady state at first stage.

The definite conditions are:

$$p_f(r,0) = p_i \qquad (8-16)$$

$$r \frac{\partial p_f}{\partial r}\bigg|_{r=r_w} = \frac{Q_o \mu B_o}{2\pi K_f h} \qquad (8-17)$$

$$p_f(\infty,t) = p_i$$

Now the differential formula (8 – 11) will be solved later.

Under the definite conditions of formula (8 – 16) and formula (8 – 17), the approximate formula of well bottom hole pressure when time is not very short given by J. E. Warren by the method of Laplace transformation which is:

$$p_{wf} = p_i - \frac{Q_o \mu B_o}{4\pi K_f h}\left\{\ln t_D + \text{Ei}\left[-\frac{\lambda t_D}{\omega(1-\omega)}\right] - \text{Ei}\left[-\frac{\lambda t_D}{(1-\omega)}\right] + 0.809 + 2S\right\} \qquad (8-18)$$

$$t_D = \frac{K_f t}{(\phi_f C_{tf} + \phi_m C_{tm})\mu r_w^2} \qquad \omega = \frac{C_{tf}\phi_f}{C_{tf}\phi_f + C_{tm}\phi_m} \qquad \lambda = \frac{\alpha K_m}{K_f} r_w^2$$

In the formula above: $\alpha$ is the shape factor of matrix, $\lambda$ is called the cross flow factor which is a dimensionless parameter. The magnitude of $\lambda$ decides the ability of cross flow from the matrix to the fractures. The value of $\lambda$ will increase when the permeability of matrix or the density of fractures is big, which will make the flow velocity from the matrix to the fractures increase. $\omega$ is called elastic storativity ratio which is the ratio of elastic energy of fracture to the total (fractures and matrix) elastic energy, the bigger the ratio of the fracture porosity to the total porosity is, the bigger the elastic energy of fracture and the value of $\omega$ are.

According to the properties of the function $\text{Ei}(x)$, when time is short, the function $\text{Ei}(x)$ can be approximately expressed with logarithmic function. As to the condition of general reservoirs, when $\frac{\lambda t_D}{\omega(1-\omega)} \leq \frac{1}{400}$, the formula (8 – 18) can be simplified as:

$$p_{wf} = p_i - \frac{Q_o \mu B_o}{4\pi K_f h}(\ln t_D + 0.809 + 2S - \ln\omega) \qquad (8-19)$$

When time is long and $\frac{\lambda t_D}{\omega} > 3$, the function Ei is very close to zero, then the following formula can be obtained:

当时间较长,且 $\frac{\lambda t_D}{\omega} > 3$ 时,Ei 函数非常接近零,则:

$$p_{wf} = p_i - \frac{Q_o \mu B_o}{4\pi K_f h}(\ln t_D + 0.809 + 2S) \qquad (8-20)$$

After drawing the relation curve of pressure drop with the logarithm of time $\ln t$ (Figure 8-3), two straight lines with same slope will be obtained, and $K_f h$ can be further obtained by the slope of straight line. The ratio of elastic energy $\omega$ can be calculated with the latitude $\Delta p$ of the two straight lines, that is:

绘制压力降与时间对数值 $\ln t$ 的关系曲线(图 8-3),得到斜率相等的两条直线,用直线斜率求出 $K_f h$。由两条直线的纵距 $\Delta p$ 可以计算弹性储能比 $\omega$,即:

$$\ln\omega = -\frac{\Delta p}{i} \qquad (8-21)$$

In the formula above, $i$ is the slope of straight line. If the compressibility and the porosity of matrix are known and take $C_f \approx C_o$, then the porosity of fractures $\phi_f$ can be calculated.

式中,$i$ 为直线斜率。如果已知岩块压缩系数和孔隙度,并取 $C_f \approx C_o$,可以算出裂缝孔隙度 $\phi_f$。

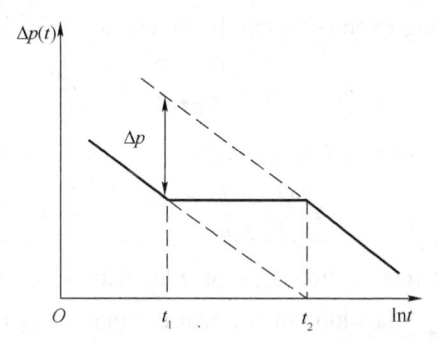

Figure 8-3  The relation curve of pressure drop with logarithm of time
图 8-3  压降与时间对数关系曲线

The bottom pressure buildup formula can be obtained by the formula (8-18) and the Duhamel principle. Assume the well is shut down after producing with constant production $q_o$ for time $t$, and the shut-down time is $\Delta t$. Then the buildup pressure $p_{ws}$ is:

利用式(8-18)和 Duhamel 原理,可以求得关井后井底压力恢复公式,设该井以常产量 $q_o$ 生产 $t$ 时后关井,关井时间为 $\Delta t$,则恢复压力 $p_{ws}$ 为:

$$p_{ws} = p_i + \frac{Q_{sco}\mu B_o}{4\pi K_f h}\left\{\ln\frac{\Delta t}{t + \Delta t} + \text{Ei}\left[-\frac{\lambda(t_D + \Delta t_D)}{\omega(1-\omega)}\right] - \text{Ei}\left[-\frac{\lambda \Delta t_D}{\omega(1-\omega)}\right] \right.$$
$$\left. - \text{Ei}\left[-\frac{\lambda(t_D + \Delta t_D)}{1-\omega}\right] + \text{Ei}\left(-\frac{\lambda \Delta t_D}{1-\omega}\right)\right\} \qquad (8-22)$$

Generally, $t \gg \Delta t$, that is the producing time is very long and the well-testing time is short, it is very possible in actual production

一般 $t \gg \Delta t$,即投产时间很长,试井时间很短,这在生产实践中

practice, because the production of dual–medium reservoir is always very high, so the pressure builds up fast.

In the early stage of shut down when $\Delta t$ is very small, the third and the fifth item in the vinculum of the formula (8–22) both can be replaced by approximate formulas they are:

$$-\text{Ei}\left[-\frac{\lambda \Delta t_D}{\omega(1-\omega)}\right] = \ln\frac{\omega(1-\omega)}{\lambda \Delta t_D} - 0.5772 \quad (8-23)$$

$$\text{Ei}\left[-\frac{\lambda \Delta t_D}{\omega(1-\omega)}\right] = \ln\frac{\lambda \Delta t_D}{1-\omega} + 0.5772 \quad (8-24)$$

$t$ is very long, the relevant function Ei is close to zero, so the early law of pressure buildup is:

$$p_{ws} = p_i + \frac{Q_{sco}\mu B_o}{4\pi K_f h}\left(\ln\frac{\Delta t}{t+\Delta t} + \ln\omega\right) = p_i - \frac{Q_{sco}\mu B_o}{4\pi K_f h}\left(\ln\frac{t+\Delta t}{\Delta t} + \ln\frac{1}{\omega}\right) \quad (8-25)$$

To the later stage of pressure buildup, the value of $\Delta t$ is very big, then all the items of Ei in the formula (8–22) are close to zero, then the following expression can be obtained:

$$p_{ws} = p_i - \frac{Q_{sco}\mu B_o}{4\pi K_f h}\ln\frac{t+\Delta t}{t} \quad (8-26)$$

The slope of later straight line section $l_2$ is obtained:

$$i_2 = \frac{Q_{sco}\mu B_o}{4\pi K_f h} \quad (8-27)$$

Apparently, the early straight line section $l_1$ is parallel with the later straight line section $l_2$, the slope of the two is equal that is $i_1 = i_2$. The latitude $D_p$ between $l_1$ and $l_1$ is:

$$D_p = i\ln\frac{1}{\omega} \quad (8-28)$$

$$i = i_1 \text{ or } i = i_2$$

From the preparation above, the ratio of elastic energy of dual–medium $\omega$ can be obtained which is:

$$\omega = e^{-\frac{D_p}{i}} \quad (8-29)$$

Lots of data show that the two parallel semi-log straight lines may not appear which depends on the conditions of wells and the properties of medium. When the two parallel semi-log straight lines appear, the first semi-log straight line can last for several hours.

What the Figure 8–4 shows is the typical oil well pressure buildup curve in dual–medium. The end time of the early straight line is $\Delta t_1$, the start time of the later straight line is $\Delta t_2$, $(\Delta t_2 - \Delta t_1)$ is the transition time. It can be seen from the figure, the flow can be divided into four phases, in which the phase of wellbore storage is not easy to see in actual production practice.

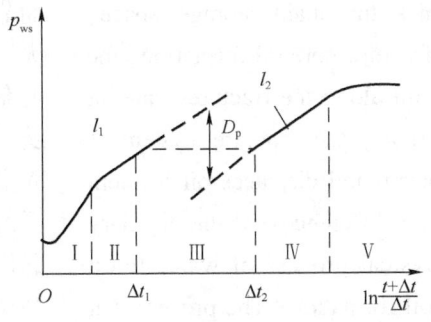

Figure 8-4 The typical oil well pressure buildup curve in dual-medium

图 8-4 双重介质油井压力恢复典型曲线

Ⅰ—The phase of wellbore storage; Ⅱ—The phase of early straight line; Ⅲ—The transitional phase;
Ⅳ—The phase of later straight line; Ⅴ—The phase of boundary reflection

The analysis above is solved under the assumptions that the ratio of elastic energy $\omega$ and the cross flow factor $\lambda$ are constant, while the actual condition is not like this, especially when the formation pressure drops under the saturation pressure. $\omega$ and $\lambda$ not only depend on the properties of matrix, but also are relevant with the properties of fluid, therefore, $\lambda$ measured in the same well changes with time.

What needs to be figured out is that the pressure behavior of multiple-medium reservoir in which the permeability of every layer is apparently different with others is similar to the dual-medium reservoir.

If the pressure drop stops in the transitional phase that is the production time is short, the pressure buildup dynamic in the dual-medium reservoir is always similar to the dynamic state in the homogeneous pore reservoir with constant pressure boundary.

## 8.4 Oil-water two-phase seepage in dual-medium

The mechanism of water displacing oil in multiple-medium has been introduced in the chapter 6. In 1964, Bare proposed the mathematical model of oil-water two-phase flow in fracture-pore media, its main characteristic is that it still includes two continuous flow fields of matrix and fractures. The permeability of fracture is much bigger than that of matrix which plays the main role on channel. However, the porosity of matrix is much bigger than that of fracture system which becomes the main storage space. When injecting water into the fracture-pore oil formation, the injected water firstly displaces oil along the fractures, and at the same time the water entering the fracture system is imbied into the surrounding matrix and displaces oil in them because of the action of pressure difference and the capillary pressure. This is the basic physical process of water flood recovery in the fracture-pore oil formation. The process that the injected water is imbied into the matrix and displaces oil in them because of the action of capillary pressure is called the action of imbition, it is an important problem to express the imbition correctly.

If the flow in matrix system is neglected and the matrix system is considered as the source supplying oil to the fracture system, the oil and water are incompressible, and the action of capillary pressure in fractures is neglected, then the flow equation group describing the fracture system is:

Bare 提出裂缝—孔隙介质油水两相渗流的数学模型,其主要特点仍然是存在岩块系统和裂缝系统两个连续的渗流场,裂缝系统的渗透率远大于岩块系统的渗透率,起着主要通道的作用,而岩块系统的孔隙度又显著地大于裂缝系统的孔隙度,成为主要储集空间。当向亲水的裂缝—孔隙油层注水时,注入水首先沿着裂缝驱油,同时进入裂缝系统的水因压力差和毛管压力的作用被吸入周围的岩块系统,并从其中置换出油。这就是亲水的裂缝—孔隙油层中注水采油的基本物理过程,注入水因毛管压力被吸入岩块孔隙置换出油称为渗吸作用,正确表达渗吸油量是一个重要的问题。

如果忽略岩块系统中的流动,把岩块系统看成向裂缝系统补给油量的源,油和水是不混溶和不可压缩的,忽略裂缝中的毛管压力作用,描述裂缝系统中的渗流方程组为:

$$\begin{cases} \boldsymbol{v}_{wf} = -\dfrac{K_f K_{wf}}{\mu_w} \nabla(p_f + \rho_w gz) \\[4pt] \boldsymbol{v}_{of} = -\dfrac{K_f K_{of}}{\mu_o} \nabla(p_f + \rho_o gz) \\[4pt] \phi_f \dfrac{\partial S_{wf}}{\partial t} + \nabla \boldsymbol{v}_{wf} = q_w \\[4pt] \phi_f \dfrac{\partial S_{of}}{\partial t} + \nabla \boldsymbol{v}_{of} = q_o \\[4pt] S_{wf} + S_{of} = 1 \\[4pt] q_w + q_o = 0 \end{cases} \quad (8-30)$$

As to the matrix system, the following expression can be obtained:

对岩块系统则有:

$$\phi_m \frac{\partial S_{wm}}{\partial t} + q_w = 0 \qquad (8-31)$$

Where: $v_{wf}$, $v_{of}$—The flow velocity of water and oil in fractures, m/s;

$K_{wf}$, $K_{of}$—The relative permeability of water and oil in fractures;

$K_f$—The absolute permeability of fractures, $10^{-3}\mu m^2$;

$S_{wf}$, $S_{of}$—The saturation of water and oil in fractures;

$S_{wm}$, $S_{om}$—The saturation of water and oil in matrix;

$\rho_w$, $\rho_o$—The density of water and oil, kg/m³;

$q_w$—The water absorbed into the matrix system, m³;

$q_o$—The displaced oil from the matrix system, m³;

$z$—The upward vertical coordinate, m.

To ascertain the imbibition equation $Q$ between the two systems is the key point to solve the problem of oil displacing oil in dual-medium. According to the result of Aronofsky imbibition experiment, that is putting the hydrophilic oil-bearing matrix into the pure water from time $t = 0$, because of the imbibition, to time $t$ the cumulative oil displaced from the matrix is:

$$Q_o(t) = R(1 - e^{-\lambda t}) \qquad (8-32)$$

Where: $R$—The final cumulative oil from the unit matrix under the action of imbibition;

$\lambda$—A constant expressing the intensity of imbibition, and its dimension is the reciprocal of time.

$R$ and $\lambda$ are the constant that is relevant to the properties of oil, water and matrix and can be measured by experiment.

The intensity of imbibition solved is:

$$q_o(t) = \frac{dQ_o(t)}{dt} = R\lambda e^{-\lambda t} \qquad (8-33)$$

When water is injected, the arbitrary point in formation $q_o$ is the function of saturation $S_{wf}(x,y,z,t)$ in the fracture system, water will flow into the smallest fractures and then gradually expand to the bigger fractures.

What needs to be especially noticed is: the automatic water imbition and oil displaced from the matrix to the fractures are not only relevant with time, but also with the water saturation $S_{wf}$ in fractures. The experiment by Aronofsky was ran by putting the matrix directly into the pure water, while in the process of water displacing oil in dual-medium, the water saturation in the fractures contacting with matrix is changing with time, there is not pure water around the matrix but the oil-water mixture. Then the intensity of imbibition should be expressed as:

$$q_o(t) = R\lambda S_{wf}(t_0) e^{-\lambda(t-t_0)}$$

In the formula above, $(t-t_0)$ is the contact time on matrix with fluid in the fractures with water saturation $S_{wf}(t_0)$.

In the actual process of oil displacing oil, the saturation of arbitrary point in fractures changes with time, the imbibition consists of two parts: one part is $S_{wf}(t_0)$, the action time is $(t-t_0)$; another part is $S_{wf}(t_1) - S_{wf}(t_0)$, the action time is $(t-t_1)$, so when $t > t_1$ the intensity of imbibition is changed to be:

$$\begin{aligned} q_o(t) &= R\lambda\{S_{wf}(t_0) e^{-\lambda(t-t_0)} + [S_{wf}(t_1) - S_{wf}(t_0)] e^{-\lambda(t-t_1)} + \cdots \\ &\quad + [S_{wf}(t_n) - S_{wf}(t_{n-1})] e^{-\lambda(t-t_{n-1})}\} \\ &= R\lambda\{S_{wf}(t_0)[e^{-\lambda(t-t_0)} - e^{-\lambda(t-t_1)}] + S_{wf}(t_1)[e^{-\lambda(t-t_1)} - e^{-\lambda(t-t_2)}] + \cdots \\ &\quad + S_{wf}(t_{n-2})[e^{-\lambda(t-t_{n-2})} - e^{-\lambda(t-t_{n-1})}] + S_{wf}(t_{n-1}) e^{-\lambda(t-t_{n-1})}\} \end{aligned}$$

According to the differential mean value theorem, the following expression can be obtained:

$$e^{-\lambda(t-t_{i-1})} - e^{-\lambda(t-t_i)} = -\frac{d}{d\tau} e^{-\lambda(t-\tau)}\bigg|_{\tau=\tau_{i-1}+\theta\Delta\tau_i} \Delta\tau_i \quad (0 \leq \theta \leq 1; i=1,2,\cdots,n)$$

And so the following expression can be obtained:

$$q_o(t) = R\lambda\left\{\sum_{i=0}^{n-1} -S_{wf}(\tau_i)\left[\frac{d}{d\tau}e^{-\lambda(t-\tau)}\bigg|_{\tau=\tau_i+\theta\Delta\tau_i}\right]\Delta\tau_i + S_{wf}(t_{n-1}) e^{-\lambda(t-t_{n-1})}\right\}$$

Make $n\to\infty$ and $\|\Delta\tau_i\|\to 0$, then the following expression can be obtained:

$$q_o = R\lambda\left[S_{wf} - \lambda\int_0^t S_{wf}(x,\tau)e^{-\lambda(t-\tau)}d\tau\right] \qquad (8-34)$$

The formula (8-30) can be written as the following form under the horizontal one-dimensional condition:

$$v_{wf} = -\frac{K_f K_{wf}}{\mu_w}\frac{\partial p_f}{\partial x} \qquad (8-35)$$

$$v_{of} = -\frac{K_f K_{of}}{\mu_o}\frac{\partial p_f}{\partial x} \qquad (8-36)$$

$$\phi_f\frac{\partial S_{wf}}{\partial t} + \frac{\partial}{\partial x}v_{wf} = q_w \qquad (8-37)$$

$$\phi_f\frac{\partial S_{of}}{\partial t} + \frac{\partial}{\partial x}v_{of} = q_o \qquad (8-38)$$

$$S_{wf} + S_{of} = 1 \qquad (8-39)$$

$$q_w + q_o = 0 \qquad (8-40)$$

$$\phi_m\frac{\partial S_{wm}}{\partial t} + q_w = 0 \qquad (8-41)$$

After adding the formula (8-37) to the formula (8-38) and considering the formula (8-39) and the formula (8-40), then the following expression can be obtained:

$$\frac{\partial v_{wf}}{\partial x} + \frac{\partial v_{of}}{\partial x} = 0$$

And so the following expression can be obtained:

$$v_{wf} + v_{of} = v(t) \qquad (8-42)$$

In the formula above, $v(t)$ is the flow velocity of injected water.

After you add the formula (8-35) to the formula (8-36) and consider the formula (8-42), then the following expression can be obtained:

$$\frac{\partial p_f}{\partial x} = -\frac{v(t)}{K_f\left(\dfrac{K_{wf}}{\mu_w} + \dfrac{K_{of}}{\mu_o}\right)} \qquad (8-43)$$

After you substitute the expression above into the formula (8-35), the following expression can be obtained:

$$v_{wf} = v(t)f_{wf}(t) \qquad (8-44)$$

$$f_{wf}(S_{wt}) = \frac{1}{1 + \dfrac{\mu_w}{\mu_o}\dfrac{K_{of}}{K_{wf}}} \qquad (8-45)$$

After substituting the formula (8-34) and the formula (8-44) into the formula (8-37), the basic equation of two-phase displacement in double-porous media can be obtained which is:

$$v(t)f'_{wf}S_{wf}\frac{\partial S_{wf}}{\partial x} + \phi_f \frac{\partial S_{wf}}{\partial t} + R\lambda\left[S_{wf} - \lambda\int_0^t S_{wf}(x,\tau)e^{-\lambda(t-\tau)}d\tau\right] = 0 \qquad (8-46)$$

When $R=0$ or $\lambda=0$, there is no imbibition, the equation (8-46) is changed to be the Buckley-Leverett equation in the general porous media. The equation (8-46) is a pseudo-linear hyperbolic equation which is difficult to obtain its precise analytic solution. In some relevant document, only the numerical solution and the approximate analytic solution are given.

Now the approximate solution of formula (8-46) is introduced. Firstly, make the $S_{wf}(x,\tau)$ under the integral sign in the formula (8-34) to be equal to $S_{wf}(x,\tau)$ and 0 respectively, then the following expressions can be obtained:

$$q_o^{(1)} = R\lambda S_{wf}e^{-\lambda t} \qquad (8-47)$$
$$q_o^{(2)} = R\lambda S_{wf} \qquad (8-48)$$

For the convenience of analysis, transform the formula (8-46) to be an equation with dimensionless parameters that is:

$$v_D(t_D)f'_{wf}(S_{wf})\frac{\partial S_{wf}}{\partial x_D} + \phi_f \frac{\partial S_{wf}}{\partial t_D} + R\left[S_{wf} - \int_0^{t_D} S_{wf}(x_D,\tau_D)e^{-(t_D-\tau_D)}d\tau_D\right] = 0 \qquad (8-49)$$

In the formula above, $v_D(t_D) = \dfrac{v(t_D)}{L\lambda}, x_D = \dfrac{x}{L}$, and $t_D = \lambda t$, $L$ is the length of linear double-porosity media.

Substitute the formula (8-47) and the formula (8-48) respectively into the formula (8-49), the following expressions can be obtained:

$$v_D(t_D)f'_{wf}(S_{wf})\frac{\partial S_{wf}}{\partial x_D} + \phi_f \frac{\partial S_{wf}}{\partial t_D} + RS_{wf}e^{-t_D} = 0 \qquad (8-50)$$

$$v_D(t_D)f'_{wf}(S_{wf})\frac{\partial S_{wf}}{\partial x_D} + \phi_f \frac{\partial S_{wf}}{\partial t_D} + RS_{wf} = 0 \qquad (8-51)$$

Apparently, the solution obtained from the equation (8-50) reduces the action of imbibition which is called the approximate solution of loss-imbibition, while the solution obtained from the equation (8-51) exaggerates the action of imbibition which is called the approximate solution of exceed-imbibition, and the real solution is between the two approximate solutions.

Because $x_D(t_D)$ is the function of $t$ in $S_{wf}(x_D, t_D)$:

$$\frac{dS_{wf}}{dt_D} = \frac{\partial S_{wf}}{\partial x_D}\frac{dx_D}{dt_D} + \frac{\partial S_{wf}}{\partial t_D} \qquad (8-52)$$

The equation of characteristic line is:

$$\frac{dx_D}{dt_D} = \frac{v_D(t_D)f'_{wf}(S_{wf})}{\phi_f} \qquad (8-53)$$

According to the formula (8 – 46), along the characteristic line is the moving velocity of an arbitrary saturation:

$$\frac{dS_{wf}}{dt_D} = \frac{R}{\phi_f}\left[e^{-t_D}\int_0^{t_D} S_{wf}(x_D, \tau_D)e^{-\tau_D}d\tau_D - S_{wf}\right] \qquad (8-54)$$

According to the formula (8 – 50), the saturation equation on the characteristic line for loss-imbibition is:

$$\frac{dS_{wf}}{dt_D} = \frac{R}{\phi_f}S_{wf}e^{-t_D} \qquad (8-55)$$

Its solution is:

$$S_{wf} = S_{wf0}\exp\left[\frac{R}{\phi_f}(e^{-t_D} - e^{-t_{D0}})\right] \qquad (8-56)$$

In the formula above, $S_{wf0}$ is the corresponding water saturation in fractures under the initial condition $t_{D0}$.

Because on the characteristic line $S_{wf}$ is only the function of $t_D$, the equation (8 – 53) of characteristic line obtained is:

$$x_D - x_{D0} = \frac{1}{\phi_f}\int_{t_{D0}}^{t_D} v_D(t_D)f'_{wf}(S_{wf})dt_D \qquad (8-57)$$

Because in the fractures the capillary effect is negligible, the relative permeability curve of fracture system can be taken as the diagonal line, that is:

$$K_{wf}(S_{wf}) = S_{wf};\ K_{of}(S_{wf}) = 1 - S_{wf}$$

Because:

$$f_{wf}(S_{wf}) = \frac{1}{1 - m\left(1 - \dfrac{1}{S_{wf}}\right)};\ m = \frac{\mu_w}{\mu_o}$$

$$f'_{wf}(S_{wf}) = \frac{\mu}{\left[(1-m)S_{wf} + \mu\right]^2} \qquad (8-58)$$

The following expression can be obtained:

$$x_D - x_{D0} = \frac{m}{\phi_f}\int_{t_{D0}}^{t_D}\frac{v_D(t_D)}{\left[(1-\mu)S_{wf0}\exp\dfrac{R}{\phi_f}(e^{-t_D} - e^{-t_{D0}}) + m\right]^2}dt_D \qquad (8-59)$$

Likewise, the solution of exceed-imbibition can be obtained as:

$$x_D - x_{D0} = \frac{\mu}{\phi_f} \int_{t_{D0}}^{t_D} \frac{v_D(t_D)}{\left[(1-m)S_{wf0}\exp\frac{R}{\phi_f}(t_D - t_{D0}) + m\right]^2} dt_D \quad (8-60)$$

When the injecting velocity keeps unchanged and $v_D(t_D) = v_{D0}$, the formula (8-60) can be changed to be:

$$x_D - x_{D0} = \frac{v_{D0}}{Rm}\left\{\ln\frac{(1-m)S_{wf} + m}{S_{wf}} + (1-m)\left[\frac{S_{wf}}{(1-m)S_{wf} + m} - 1\right]\right\} \quad (8-61)$$

$$S_{wf} \geq \exp\left(-\frac{R}{\phi_f}t_D\right)$$

It can be known from the analysis above, when injecting water into the hydrophilic fracture-pore formation to displace oil, the action of imbibition to water entering the fractures from the matrix system is the basic character of water displacing oil. The process of injecting water to displace oil in the hydrophilic fracture-pore formation can be divided into three stages:

(1) The initial strong stage, at this moment, the displacement between water and oil still has no time to proceed, the behavior in the dual-medium is like the pure fracture media, the distribution of saturation in the fracture system and other development indexes are all close to the solution of equation Buckley-Leverett. Because the porosity of fracture system is very small, and the frontal water saturation is zero and the front velocity is very fast, therefore the delivery end of formation will water breakthrough quickly.

(2) The stage of vigorous imbibition, at this moment imbibition starts to work sufficiently, the oil supply from matrix increases, and the saturation in fracture system will gradually attain the pseudo-steady state along the direction of displacement. The increase of water-oil ratio gradually becomes flattened, and the state is totally different from the process of Buckley-Leveret, and the stage is destroyed after lasting for a long time. Most recoverable crude oil is produced in this stage.

(3) The stage of subsidence. The action of imbibition gradually subsides and water displacing oil starts to enter the later phase in which the process which is similar to the Buckley-Leveret process.

As for the dual-medium, the water free oil production period is very short, but because of the action of imbibition, the increase of water-oil ratio is relatively slow. Different from the pore media, in dual-medium recovery degree of the formation and the water-oil ratio and so on are all relevant with the water injection rate, the smaller the rate is, the better the indexes is. It can be known from the case study, the approximate analytic solution of loss-imbibition is a fairly good approximation of the exact solution which can be absolutely used in the reservoir engineering calculation.

What needs to be pointed out is: the model of intensity of imbibition obtained by Aronofsky only suits for matrix with small volume that is the imbibition is mainly because of the result of capillary pressure. If the matrix is of big volume, besides the influence of capillary pressure there is also the affect of gravity. As to the problem that comprehensively considers the influence of capillary pressure and gravity, please refer to the relevant papers.

## 8.5 Seepage of fluid in pure fracture formation

The movement of fluid in the pure fractures can be described with the binomial expression because of its high velocity which is:

$$\frac{\partial p}{\partial r} = \frac{\mu}{K_f} v_f + \beta \frac{\mu}{K_f} v_f^2 \qquad (8-62)$$

Where: $p$— The pressure of fluid, Pa;
$r$— The radial radius, m;
$\beta$— A constant;
$K_f$— The permeability of fractures, m$^2$;
$\mu$— The viscosity of fluid, Pa·s.

The first item on the right side of the formula (8-62) shows the loss pressure overcoming the frictional resistance between the fracture media and fluid, the second item shows the lost pressure overcoming the inertial resistance of fluid. If the flow velocity is very small, then the second item can be neglected. At this moment, the flow obeys the Darcy's law. If the flow velocity of fluid is very big in fractures, compared with the inertia force, the viscosity force can be omitted, and then the relationship between pressure drop and the velocity is square relationship.

The formula (8-62) can be also changed to be the form which is similar to the Darcy's law, which is:

$$v_f = B_1 \frac{dp}{dr} \quad (8-63)$$

$$B_1 = \frac{\sqrt{1 + 4\beta \frac{K_f}{\mu} \frac{dp}{dr}} - 1}{2\beta \frac{dp}{dr}} \quad (8-64)$$

Or change it to be the parabolic relation, which is:

$$(2\beta v_f + 1)^2 = 4K_f \frac{dp}{dr} \frac{\beta \rho}{\mu} + 1 \quad (8-65)$$

Compare the formula (8-63) with the Darcy's law, and it can be seen that when fluid flows in the fractures, the coefficient $\frac{K_f}{\mu}$ in the Darcy's law is replaced with $B_1$. The Figure 8-5 shows the relationship of both sides of the formula (8-65).

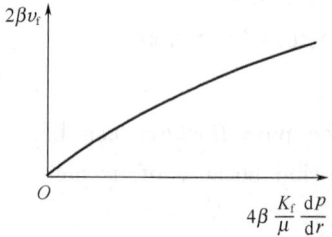

Figure 8-5 The relation curve of $4\beta \frac{K_f}{\mu} \frac{dp}{dr}$ with $2\beta v_f$

### 8.5.1 The steady state flow towards the well of fluid

The continuity relation of steady state flow of incompressible fluid should meet the following expression:

$$\nabla \cdot v = 0, \operatorname{div} v = 0 \quad (8-66)$$

The flow equation can be obtained after substituting the formula (8-63) into the formula (8-66) and the obtained expression is expressed with polar coordinates, which are:

$$\frac{d}{dr}\left[r\left(\sqrt{1 + 4\beta \frac{K_f}{\mu} \frac{dp}{dr}} - 1\right)\right] = 0 \quad (8-67)$$

The boundary condition is:

$$p|_{r=r_w} = p_w, p|_{r=r_e} = p_e$$

According to the formula (8-67), when the production well is working, the pressure of any point in formation is:

$$p(r) = C_2 + C_1 \frac{\mu}{2\beta K_f}\ln r - C_1^2 \frac{\mu}{4\beta K_f}\frac{1}{r} \qquad (8-68)$$

When the injection well is injecting, the pressure of any point in formation is: 注水井注入时，地层中任一点压力为：

$$p(r) = C_2 - C_1 \frac{\mu}{2\beta K_f}\ln r + C_1^2 \frac{\mu}{4\beta K_f}\frac{1}{r} \qquad (8-69)$$

In the formulas above, $C_1$ and $C_2$ are both the integral constants. $C_2$ can be removed by the boundary condition, then the following expression can be obtained: 式中，$C_1$、$C_2$ 为积分常数。由边界条件可消去 $C_2$，求得：

$$\Delta p = p_e - p_w = C_1 \frac{\mu}{2\beta K_f}\ln \frac{r_e}{r_w} + C_1 \frac{\mu}{4\beta K_f}\left(\frac{1}{r_w} - \frac{1}{r_e}\right) \qquad (8-70)$$

Therefore, the production of well $Q_o$ can be obtained: 从而求得井产量 $Q_o$ 为：

$$Q_o = 2\pi r_w h v_f \bigg|_{r=r_w} = \frac{\pi h}{\beta}\left[r\left(\sqrt{1+4\beta\frac{K_f}{\mu}\frac{dp}{dr}}-1\right)\right]_{r=r_w} \qquad (8-71)$$

$$\Delta p = p_e - p_w = Q_o \frac{\mu}{2\pi h K_f}\ln \frac{r_e}{r_w} + Q_o^2 \frac{\mu\beta}{(2\pi h)^2 K_f}\left(\frac{1}{r_w} - \frac{1}{r_e}\right) \qquad (8-72)$$

The formula (8-72) can be simplified as the form below: 将式(8-72)简化成：

$$\Delta p = AQ_o + BQ_o^2 \qquad (8-73)$$

The relation curve of $\Delta p$ with $Q_o$ in the formula (8-73) is shown in the Figure 8-6. For the convenience of analysis, change the formula (8-73) to the form below: 式(8-73)中 $\Delta p$ 与 $Q_o$ 的关系曲线如图 8-6 所示，为了分析方便，将式(8-73)改写为：

$$\frac{\Delta p}{Q_o} = A + BQ_o \qquad (8-74)$$

As shown in the Figure 8-7, $\frac{\Delta p}{Q_o}$ with $Q_o$ is a linear relationship, so the corresponding $A$ and $B$ can be obtained. Compared with the formula (8-72), $K_f h_1$ can be calculated from the value of $A$ and $\beta$ can be calculated from the value of $B$. 如图 8-7 所示，$\frac{\Delta p}{Q_o}$ 与 $Q_o$ 为直线关系，从而可求出相应的 $A$ 与 $B$。对照式(8-72)，由 $A$ 值可计算出 $K_f h_1$，由 $B$ 值可计算出 $\beta$。

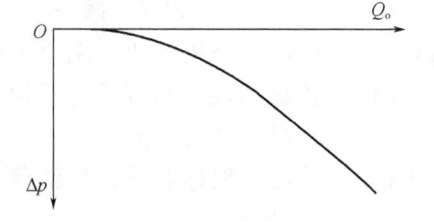

Figure 8-6　The relation curve of $\Delta p$ with $Q_o$
图 8-6　$\Delta p$ 与 $Q_o$ 的关系曲线

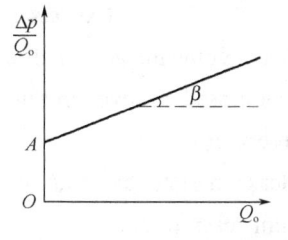

Figure 8-7　The relation curve of $\Delta p/Q_o$ with $Q_o$
图 8-7　$\Delta p/Q_o$ 与 $Q_o$ 的关系曲线

## 8.5.2　The unsteady state flow towards the well of fluid　8.5.2　液体向井的不稳定渗流

The continuity relation of fluid under the axisymmetric condition should meet the following expression: 轴对称条件下液体的连续性关系应满足：

$$\frac{\partial(\phi_f\rho)}{\partial t} + \frac{1}{r}\frac{\partial}{\partial r}(r\rho v_f) = 0 \tag{8-75}$$

Therefore, the flow equation of fluid obtained is: 因此,求得液流的渗流方程为:

$$\frac{1}{r}\frac{\partial}{\partial r}\left[\frac{r}{2\beta}\left(\sqrt{1+4\beta\frac{K_f}{\mu}\frac{\partial p}{\partial r}}-1\right)\right] = C_t\frac{\partial p}{\partial t} \tag{8-76}$$

$$C_t = C_f + C_{Lf}$$

Where: $C_t$—The total compressibility, Pa$^{-1}$;
$\phi_f$—The porosity of fractures;
$C_{Lf}$—The compressibility of liquid in fracture, Pa$^{-1}$;
$C_f$—The compressibility of fracture, Pa$^{-1}$.

式中: $C_t$ = 综合压缩系数, Pa$^{-1}$;
$\phi_f$ = 裂缝孔隙度;
$C_{Lf}$ = 裂缝系统中液体压缩系数, Pa$^{-1}$;
$C_f$ = 裂缝系统压缩系数, Pa$^{-1}$;

This writing style is same with the definition of compressibility of the Soviet Union $C_\phi = \frac{1}{V}\frac{dV_p}{dp}$ ($V_p$ is the volume of pores), while the definition of compressibility of Europe and America is $C_f = \frac{1}{V_p}\frac{dV_p}{dp}$, the difference of the two definitions is a multiple $\phi$.

这个写法同苏联压缩系数的定义 $C_\phi = \frac{1}{V}\frac{dV_p}{dp}$ ($V_p$ 为孔隙体积), 而欧美压缩系数的定义 $C_f = \frac{1}{V_p}\frac{dV_p}{dp}$, 二者相差 $\phi$ 倍。

If the oil production is constant that is $Q_o$ = constant, then the pressure of any point in formation obtained is:

若保持原油产量恒定,即 $Q_o$ 为常数,求得地层中任一点压力为:

$$p(r,t) = p_e + \frac{Q_o\mu}{4\pi K_f}\exp\left(\frac{r_w^2}{L^2}\right)\text{Ei}\left(-\frac{r^2}{L^2}\right) + \frac{\beta\mu Q_o^2}{4\pi^2 K_f r}\exp\left(\frac{r_w^2 - r^2}{L^2}\right)$$
$$+ \frac{\beta\mu Q_o^2}{K_f L(2\pi)^{\frac{3}{2}}}\exp\frac{2r_w^2}{L^2}\left[1 - \Phi\left(\frac{r\sqrt{2}}{L}\right)\right] \tag{8-77}$$

$$L = \frac{\sqrt{\beta^2 Q_o^2 + 128\pi K t} - \beta Q_o}{4\sqrt{2\pi}} \tag{8-78}$$

In the formula above, $\Phi(x)$ is the probability integral. 式中, $\Phi(x)$ 为概率积分。

**Exercises** 习 题

8-1 Please compare and analyze the features of flow of reservoir composed of two production layers and the dual-medium reservoir.

8-1 分析对比由两个生产层组成的油藏和双重介质油藏生产时的渗流特点。

8-2 Please narrate the features of pressure drop of the dual-medium reservoir.

8-2 叙述双重介质油藏的压力降特征。

# Seepage of non-Newtonian liquid

# 非牛顿液体的渗流

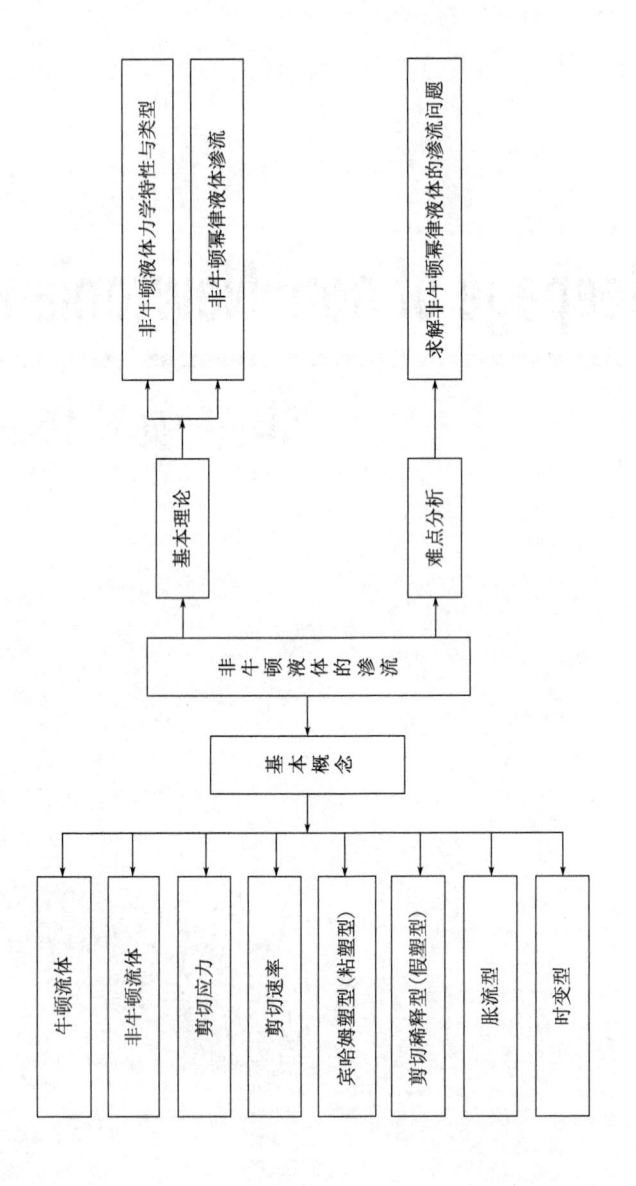

```
                    ┌─────────────────────┐
                    │ 非牛顿液体力学特性与类型 │
                    └─────────────────────┘
                              ↑
                    ┌─────────────────────┐
                    │   非牛顿幂律液体渗流   │
                    └─────────────────────┘
                              ↑
                        ┌──────────┐
                        │ 基本理论  │
                        └──────────┘
                              ↑
        ┌───────────────┐
        │ 非牛顿液体的渗流 │
        └───────────────┘
                              ↓
                        ┌──────────┐
                        │ 难点分析  │
                        └──────────┘
                              ↓
                 ┌─────────────────────────┐
                 │ 求解非牛顿幂律液体的渗流问题 │
                 └─────────────────────────┘

                        ↓
                   ┌──────────┐
                   │ 基本概念  │
                   └──────────┘
    ↓      ↓      ↓      ↓           ↓              ↓         ↓      ↓
 牛顿流体 非牛顿流体 剪切应力 剪切速率 宾哈姆塑型(粘塑型) 剪切稀释型(假塑型) 胀流型 时变型
```

第九章知识图谱

— 306 —

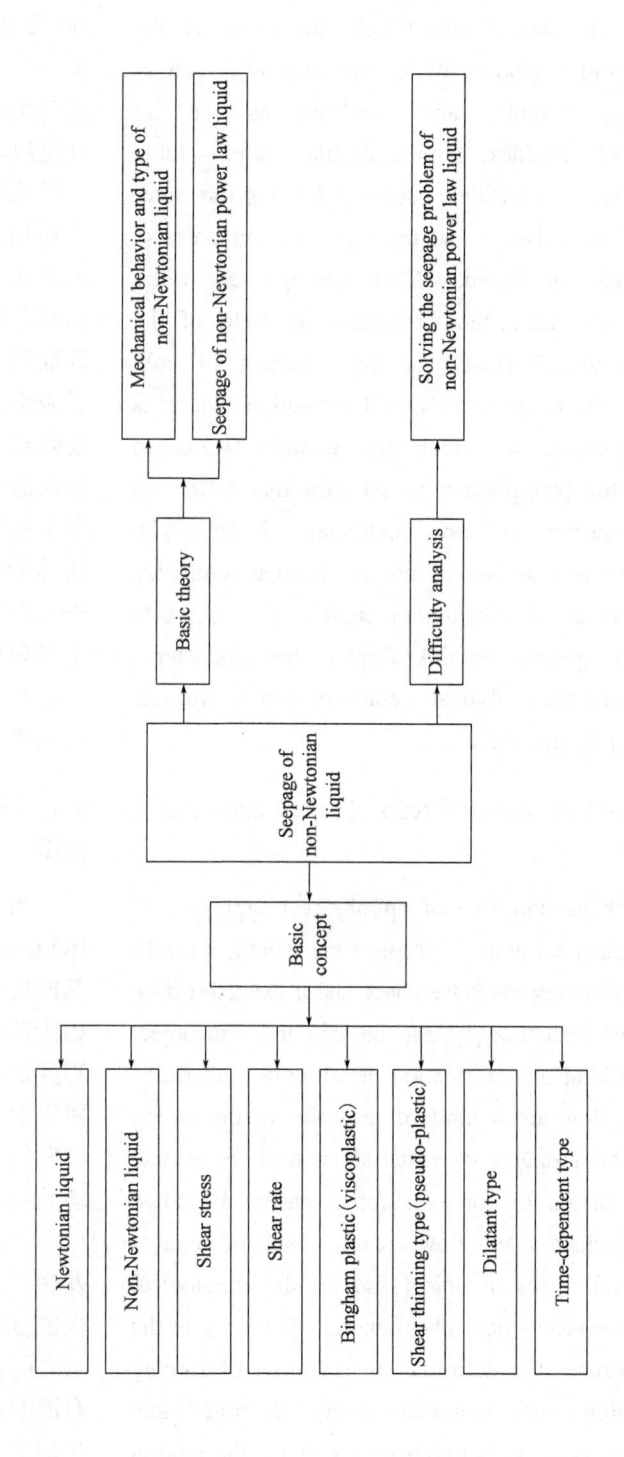

```
                ┌─────────────────────────┐
                │ Mechanical behavior and │
                │  type of non-Newtonian  │
                │         liquid          │
                ├─────────────────────────┤
                │  Seepage of non-Newto-  │
                │   nian power law liquid │
                └─────────────────────────┘
                            ▲
                            │
                     ┌──────────────┐           ┌──────────────────────┐
                     │ Basic theory │           │ Solving the seepage  │
                     └──────────────┘           │  problem of non-New- │
                            ▲                   │ tonian power law liq.│
                            │                   └──────────────────────┘
                            │                            ▲
                            │                            │
                            │                   ┌──────────────────┐
                            │                   │ Difficulty analy.│
                            │                   └──────────────────┘
                            │                            ▲
                  ┌─────────────────────┐                │
                  │ Seepage of non-New- │────────────────┘
                  │   tonian liquid     │
                  └─────────────────────┘
                            │
                            ▼
                    ┌──────────────┐
                    │Basic concept │
                    └──────────────┘
         ┌────┬────┬────┬───┼───┬────┬────┐
         ▼    ▼    ▼    ▼   ▼   ▼    ▼    ▼
```

Chapter 9  Knowledge Graph

— 307 —

At present, non-Newtonian liquid has been widely used in the oil recovery industry, and its function in oil production can be divided into two kinds which are used as the fracturing liquid and secondary or tertiary recovery such as using the foam, polymer and emulsion as the oil displacement agent. Besides, high molecular water soluble polymers have come to be widely used in adjusting the water injection profile. Also the tar and the high wax oil are also the non - Newtonian liquid. After having had water injections for a long time, the Арланское oil field of the Soviet Union showed, because of the injection of cold water, a drop in the temperature of oil formation, which is below the temperature at which the paraffin starts to precipitate, and the precipitation of paraffin makes the oil present the characters of non-Newtonian liquid. The characters and the seepage law of non-Newtonian liquid are different from those of Newtonian liquid. The opposite phenomenon even appears. In this chapter, the characters, the seepage law and the analytical method of non-Newtonian liquid will be mainly introduced.

## 9.1 Mechanical behavior and type of non-Newtonian liquid

When the shape and size of an object (aggregate of substances) are changed under a proper force system, it can be said that a deformation happens to the object, and if the deformation of object develops with time, it can be said that the object flows. Flow is a kind of deformation developing with time. Deformation and flow are a kind of response to the acting force system. The property of materials that deforms and flows under the action of force or force system is called rheological. Deformation and flow are the result of relative movement of mass points in object, so all the rheological phenomena are the mechanical phenomena. Rheology is the science that researches the deformation and flow of forcing materials. Therefore, it's comprehensively illustrated that continuum movement includes two aspects: one is the motion equation of continuous media, the other one is the rheologic equation (or called the constitutive equation)—that is, the relationship between the stress and strain. The difference of non-Newtonian liquid mechanics and the Newtonian liquid mechanics is the constitutional equation. The complexity of constitutional equation is the feature of non-Newtonian liquid, and it is also a difficult point. The experimental and theoretical method of setting up the constitutional equation will not be introduced here.

The so-called Newtonian liquid means that the shear stress $\tau$ and the shear rate $\dfrac{dv}{dr}$ present linear relationship, that is:

$$\tau = \mu \frac{dv}{dr} = \mu \dot{\gamma} \qquad (9-1)$$

Where: $v$ —Fluid velocity, cm/s;
$r$ —Tube radius, cm;
$\mu$ —The viscosity of liquid, which is a constant as to Newtonian liquid, mPa · s;
$\dot{\gamma}$ —The shearing rate, s$^{-1}$.

That is to say, the ratio of stress to the shear rate is linear constitutive relationship when the Newtonian liquid flow, and the relationship between $\tau$ and $\dfrac{dv}{dr}$ is a straight line passing through the origin as shown in Figure 9 – 1.

The mechanic's property of non-Newtonian liquid is complicated, which will be introduced from the following respects.

### 9.1.1 The viscosity

The relationship between stress and shear rate of non-Newtonian liquid is nonlinear and its viscosity is not a constant, which is relevant with the shear rate and sometimes is relevant with the shear time.

(1) The Bingham plastic, which is also called viscoplastic. It is a simpler kind in the non-Newtonian liquid, such as the drilling fluid and some oil with high viscosity. Its flowing features are: when fluid is under force, at the beginning it will not flow, like solid, and when the shear stress gradually increases and finally attains a given

critical value, it starts to flow as shown in Figure 9 – 1. This critical shear stress is called yield stress, which is called gel strength in drilling engineering. The flow regime of the Bingham plastic liquid is the same as that of the Newtonian liquid after it starts to flow and the constitutive relationship is:

$$\tau - \tau_0 = \mu \frac{dv}{dr} \qquad (9-2)$$

Where: $\tau_0$—The yield stress.

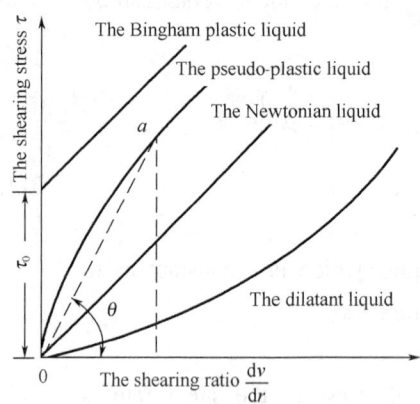

Figure 9 – 1   The rheological curve

图 9 – 1   流变曲线

The developing practice in many heavy oil and plastic-viscosity-oil fields in Azerbaijan shows the oil contained much asphaltine or tar belongs to Bingham plastic liquid, meaning there is a trigger pressure gradient, which generates by overcoming the yield stress $\tau_0$. Apparently, when the pressure gradient in formation is lower than the trigger pressure gradient, fluid will not flow, and when the pressure gradient in formation attains the trigger pressure gradient, fluid starts to flow. If the fluid flow obeys the Darcy's law (the influences of capillary pressure and gravity are neglected), the motion equation of seepage is:

$$\begin{cases} v = -\frac{K}{\mu}\left(\frac{\Delta p}{\Delta L} - G\right) & \left(\frac{\Delta p}{\Delta L} > G\right) \\ v = 0 & \left(\frac{\Delta p}{\Delta L} \leq G\right) \end{cases} \qquad (9-3)$$

Where: $G$—The trigger pressure gradient, MPa/m;

$\frac{\Delta p}{\Delta L}$—The pressure gradient in formation, MPa/m;

$K$—The permeability, $\mu m^2$.

In such reservoirs, the relation curve of oil production with pressure difference is not a straight line passing through the original point, at the full line *BAC* shown in Figure 9 – 2. Under this condition, the seepage field in reservoir will also change apparently. Even in ideal homogeneous reservoir, the stagnant region will also appear, and the reservoir heterogeneity will make the retention region and dead oil area bigger, which will make the recovery drop considerably. So when developing such reservoirs, the influence of viscoplasticity of oil must be noticed.

（2）The pseudo-plastic, which is also called shear thinning type. The flowing character of such kind of liquid is: under extra low and extra high shearing rate, the flow regime is close to the Newtonian liquid. Under general middle shear rate, with the increase of shearing rate, the apparent viscosity $\mu_a$ will decrease. The so – called apparent viscosity is the slope of the connection of a given point on the flow regime curve and the original point, as shown in the Figure 9 – 1. The slope of the connection of a point on the flow regime curve of the pseudo plastic liquid with the original point $\tan\theta$ is just the apparent viscosity $\mu_a$ under the corresponding shearing rate of point *a*. Therefore, the slower the pseudo-plastic liquid flows, the smaller the shearing rate and the higher the apparent viscosity are. When the shearing rate is much higher, the apparent viscosity will decrease instead. So it is also called shear thinning.

（3）The dilatant type. With the increase of shearing rate, the apparent viscosity will increase, so it is also called shearing thickening type as shown in Figure 9 – 1. Such type of liquid is very few, only seldom macromolecule solution belongs to dilatant type.

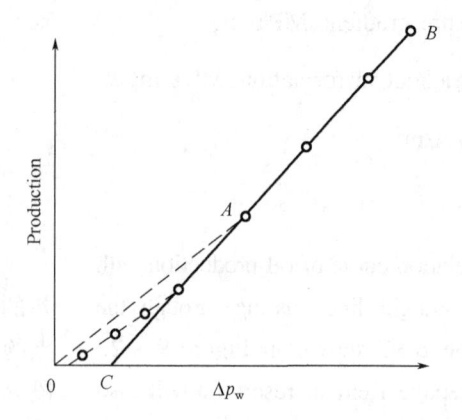

Figure 9 – 2  The relation curve of production with pressure difference
图 9 - 2  产量与压差的关系曲线

In the logarithmic coordinate of shearing stress and shearing rate of pseudo – plastic and dilatant type, under the extra low and extra high shear rate, the slope of flow regime curve is close to 0 as shown in Figure 9 – 3. Generally, within a small shearing rate region, the flow regime curve approximately is a straight line in the logarithmic coordinate. The empirical relationship between shearing stress and shearing rate can be described as power function, that is:

对于假塑型和胀流型液体，在剪切应力和剪切速率的对数坐标中，在极低及非常高的剪切速率下，流态曲线的斜率逼近为0，如图9 - 3所示，通常在不大的剪切速率范围内，其流态曲线在对数坐标系中近似为一直线，剪切应力与剪切速率的经验关系可描述为幂函数关系，即：

$$\tau = H(\gamma)^n \tag{9-4}$$
$$\gamma = \frac{dv}{dr}$$

Where: $H$—The consistency coefficient;
$\gamma$—The shearing rate;
$n$—The flow regime index.

式中: $H$ = 稠度系数；
$\gamma$ = 剪切速率；
$n$ = 流态指数。

The liquid characterized by the form of the formula (9 – 4) is called power law liquid. Although this empirical formula cannot describe the rheology of pseudo-plastic liquid precisely and its range of application is very limited, it is very simple. So it is of wide application. According to the definition of apparent viscosity $\mu_a$:

用式(9 - 4)的幂函数形式表征的液体称为幂律液体。此经验公式虽然不能很准确地描述假塑型液体的流变性，适用范围有限，然而却比较简单，因此应用十分广泛。根据视黏度 $\mu_a$ 定义为：

$$\mu_a = \frac{\tau}{\gamma}$$

So the following expression can be obtained:  从而求得：

$$\mu_a = H(\gamma)^{n-1} \tag{9-5}$$

Apparently, as to the Newtonian liquid, $n = 1$ and $\mu_a =$ constant. As to the pseudo-plastic liquid, $0 < n < 1$. As to the dilatant type liquid, $n > 1$.

(4) The time-dependent type. There is also some non-Newtonian liquid whose viscosity is not only relevant with the shear rate, but also with the sheared time. Under definite shear rate, the fluid whose viscosity decreases with time is called thixotropic fluid. On the contrary, the fluid whose viscosity increases with time is called rheopectic fluid. The relationship between the apparent viscosity and the shear lasting time are shown in Figure 9-4. For example, some oil is a kind of thixotropic fluid, and gypsum slurry is a kind of rheopectic fluid. The thixotropy is also called thixotropy and the rheopecticity is also called rheopexy.

显然，对牛顿液体，$n = 1$，$\mu_a =$ 常数；对假塑型液体，$0 < n < 1$；对胀流型液体，$n > 1$。

(4) 时变型。还有一些非牛顿液体的黏度，不仅与剪切率的大小有关，而且与受剪切的时间有关，在一定剪切率下，黏度随时间减少的流体称为触变型流体；反之，黏度随时间增加的称为震凝型液体，其视黏度与剪切持续时间的关系如图9-4所示。例如某些原油就是一种触变型液体，石膏浆是一种震凝型液体。触变性又称为摇溶性，震凝性又称为流凝性。

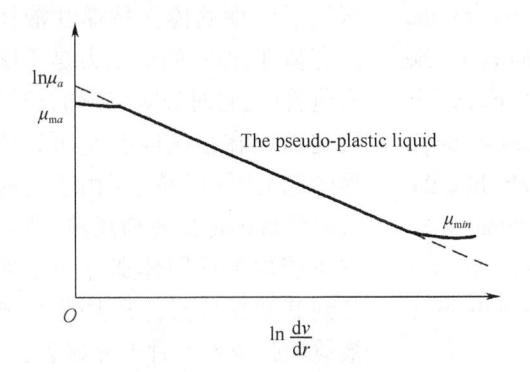

Figure 9-3 The relation curve of shear stress against shear rate

图9-3 剪切应力和剪切速率关系曲线

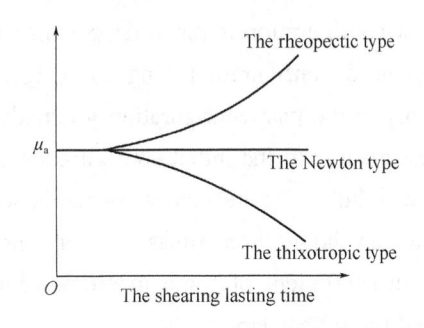

Figure 9-4 The relation curve of apparent viscosity against shear lasting time

图9-4 视黏度与剪切持续时间的关系曲线

The several kinds of non-Newtonian liquid above are all called pure viscosity non-Newtonian liquid, and its main characters are presented on the change of viscosity or are relevant with shear rate and the shear lasting time.

## 9.1.2 The viscoelasticity

Actually, the pure viscosity liquid and absolute elastic solid are all idealized. The rheological behavior of many materials is partial solid character and partial liquid character; that is, the strain neither recover completely like the elastic solid nor all the strain work can be dissipated as the form of heat, and this kind of liquid is called viscoelastic liquid, such as the high polymer liquid.

上述几种非牛顿液体统称为纯黏性非牛顿液体，其主要特点表征在黏度的变化上，或与剪切速率有关，或与剪切持续时间有关。

## 9.1.2 黏弹性

实际上纯黏性液体和完全弹性固体都是理想化的，许多现实材料的流变行为都是部分带有固体特点，又部分带有液体特点，即应变既非像弹性固体那样全部回复，亦非像液体那样全部应变功以热的形式耗散，这种液体称为黏弹性液体，如高聚物液体。

Release the prolonged spring, and it will recover to its initial length at once. This is the resilience phenomenon of elastomer. The Newtonian liquid is only of viscosity and without elasticity, it is not of the resilience phenomenon. While the non-Newtonian liquid is not like this, Eapoor. N. N made the resilience experiment of viscoelastic fluid in 1963 and he found the viscoelastic fluid is of elasticity. Under impressed force it will flow, and after the force is removed, it is of the ability to recover to the previous state, but it cannot recover to its initial state completely. This kind of character is called degenerative memory. The viscoelastic fluid has. The former effect makes it different from the Newtonian liquid, and the latter makes it different from the elastic solid. Although the viscoelastic fluid is of elasticity. Its memory is degenerative constantly, Its memory to the present configuration is much deeper than the memory to the experienced configuration long time ago. Just because its memory to the past configuration gradually becomes weak, it cannot recover to the initial configuration completely like the elastic solid. The resilience phenomenon of some non-Newtonian liquid sometimes is very noticeable, and the resilient coefficient of some macromolecular melt will even exceed the rubber tape.

Besides the characters above, the non-Newtonian liquid following also has the follwing phenomena, such as the tubeless siphon, pole climbing phenomenon (the Weissenberg effect) and the Uebler effect of oil-gas two-phase flow and so on, which are different from the Newtonian liquid. Definitely many phenomena are not found until today and the mechanisms of some phenomena are still not clarified.

When temperature is constant, the flowing character of Newtonian liquid can be ascertained only with two constants (the density $\rho$ and viscosity $\mu$). Once the two constants are ascertained, the distribution of velocity and the pressure in any flow can be both obtained. However, as to the non-Newtonian liquid, the problem is much more complicated. Whether the viscosity is a constant or not, the liquid is of other complicated mechanical properties, which are not only

将拉长的弹簧放松,它就立刻缩回到原始长度,这就是弹性体的回弹现象。牛顿液体只有黏性而无弹性,因此没有回弹现象。非牛顿液体则不然,Eapoor. N. N于1963年做了黏弹性流体的回弹实验,发现黏弹性流体具有弹性,在外力推动下可使其流动,外力消除后,它有立即回复原状的能力,但又不能完全回复原始状态,这种性质称为衰退记忆。黏弹性液体有弹性和衰退记忆两种效应,前者使它区别于牛顿液体,后者又使它区别于弹性固体。黏弹性液体虽有弹性,但它的记忆力是不断衰退着的,它对现时位形的记忆要比对它在很久以前经历的位形的记忆深刻得多,正由于它对过去位形的记忆逐渐淡薄,所以它不能像弹性固体那样完全回弹到其初始位形。某些非牛顿液体回弹现象有时十分显著,高分子熔体回弹的恢复系数甚至会超过橡皮带。

除上述特性外,非牛顿液体还具有无管虹吸、爬杆现象(Weissenberg效应)、油气两相流的Uebler效应等不同于牛顿液体的流动现象。另外必定还有不少现象尚未被发现,不少现象的机理至今尚不清楚。

在温度不变时,牛顿液体的流动特性只要用两个常数(密度$\rho$和黏度$\mu$)就可以确定。一旦这两个常数被确定,任何流动中流体的速度分布和压力分布都可求得。但对非牛顿液体来说,问题要复杂得多,黏度将不是常数,还存在其他复杂的力学性质,

relevant with the physical properties of liquid, but also connected with the flow form of non-Newtonian liquid. More importance has been attached to the seepage problem of non-Newtonian liquid with the development of secondary and tertiary recovery, such as injecting viscous water and micellar solution and so on. However, the research on the seepage of non-Newtonian liquid is not very mature so far, and some obtained results will be introduced next.

## 9.2 Seepage of non-Newtonian power law liquid

The non-Newtonian liquid is of complicated mechanical properties, so its seepage problem passing through the porous media is much more complicated. At present, many researches have been conducted on the seepage of pseudo plastic fluid and the constitutive equation is expressed in the form of the power function (9 – 4), so it is also called power law liquid. Some researching results of the power law liquid will be introduced next.

### 9.2.1 The method of Van Poollen and Jargon

1. The linear steady state seepage of powerlaw liquid

Assume the power law liquid passing through the core obeys the Darcy's law, and then the following expression can be obtained:

$$\Delta p = \frac{Q\mu_x L}{A_x K} \qquad (9-6)$$

As to the power law liquid, its viscosity changes with velocity, assume:

$$\mu_x = a\left(\frac{Q}{F_x}\right)^m \qquad (9-7)$$

Where: $a$—constant;
  $L$—The length of core, cm;
  $Q$—The flow rate, cm$^3$/s;
  $\mu_x$—The apparent viscosity of liquid in the position of $x$, mPa·s;
  $A_x$—The area of seepage cross section in the position of $x$, cm$^2$;
  $m$—The index, when $m = 0$ it is Newtonian liquid;
  $K$—The permeability, μm$^2$.

这些特殊的力学性质不只与液体的物理属性有关,而且还与非牛顿流体的流动形态有联系。非牛顿液体的渗流问题随着二次和三次采油的发展而被重视起来,例如注入增黏水、胶束液等,但是到目前为止,对非牛顿流体的渗流研究得很不成熟,下面将介绍已经取得的一些成果。

## 9.2 非牛顿幂律液体的渗流

非牛顿液体具有复杂的力学性质,因而它通过多孔介质的渗流问题是一个更加复杂的课题。目前对假塑型流体的渗流研究较多,其本构方程表示为幂函数(9-4)形式,故而又称为幂律液体。下面将介绍几种对幂律液体研究的结果。

### 9.2.1 Van Poollen 和 Jargon 法

1. 幂律液体线性稳定渗流

设通过线性岩心的幂律液体服从达西定律,则:

对幂律液体其黏度随速度而变,设:

式中: $a$ = 常数;
  $L$ = 岩心长度,cm;
  $Q$ = 流量,cm$^3$/s;
  $\mu_x$ = $x$ 处液体的视黏度, mPa·s;
  $A_x$ = $x$ 处渗流断面的面积,cm$^2$;
  $m$ = 指数, $m = 0$ 时则为牛顿液体;
  $K$ = 渗透率,μm$^2$。

And the following expression can be obtained:

$$\Delta p = \frac{a}{K}\frac{Q^{m+1}}{A_x^{m+1}}L \quad (9-8)$$

$$\lg\frac{\Delta p}{(Q/A_x)} = \lg C + m\lg\frac{Q}{A_x} \quad (9-9)$$

Or

$$C = \frac{aL}{K}$$

If the relation curve of $\Delta p/(Q/A_x)$ against $Q/A_x$ is drawn on the log-log coordinate, a straight line with slope $m$ will be obtained. Then make $Q/A_x = 1$, the value of $C = \Delta p/(Q/A_x)$ will be finally obtained.

2. The radial steady state seepage of power law liquid

Assume the motion equation of liquid flow between the $r$ and $r + \mathrm{d}r$ is:

$$\mathrm{d}p = \frac{Q\mu_r}{2\pi Kh}\frac{\mathrm{d}r}{r} \quad (9-10)$$

Where: $\mu_r$—The apparent viscosity of liquid at the position $r$, mPa·s.

As to the power law liquid:

$$\mu_r = a\left(\frac{Q}{A_r}\right)^m \quad (9-11)$$

$$A_r = 2\pi rh$$

Where: $A_r$— The area of seepage cross section in the position $r$, cm$^2$.

The apparent viscosity can be expressed as:

$$\mu_r = a\left(\frac{Q}{2\pi h}\right)^m\left(\frac{1}{r}\right)^m \quad (9-12)$$

And

$$\mathrm{d}p = \frac{aQ^{m+1}}{(2\pi h)^{m+1}K}\frac{\mathrm{d}r}{r^{m+1}} \quad (9-13)$$

Integrate the formula (9-13) from $r_w$ to $r_e$, and the following expression can be obtained:

$$\Delta p = \frac{aQ^{m+1}}{(2\pi h)^{m+1}K}\int_{r_w}^{r_e}\frac{1}{r^{m+1}}dr = \frac{-aQ^{m+1}}{(2\pi h)^{m+1}Km}(r_e^{-m} - r_w^{-m}) \qquad (9-14)$$

Assume the total volume of non-Newtonian liquid within the region of $r_e$ is $Q_T$, and the following expression can be obtained:

设 $r_e$ 范围内非牛顿液体总量为 $Q_T$,则:

$$Q_T = \pi r_e^2 h\phi \text{ 或 } r_e = \left(\frac{Q_T}{\pi h\phi}\right)^{\frac{1}{2}}$$

Change the formula (9-14) to be the following form:

式(9-14)改写为:

$$\Delta p = \frac{-aQ^{m+1}}{(2\pi h)^{m+1}Km}\left[\left(\frac{Q_T}{\pi h\phi}\right)^{-\frac{m}{2}} - r_w^{-m}\right] \qquad (9-15)$$

Because of $r_e \gg r_w$, within the region of $-1 < m < 0$, the last item $r_w^{-m}$ in the parenthesis on the right of the formula (9-15) can be neglected. Assume:

由于 $r_e \gg r_w$,在 $-1 < m < 0$ 范围内,式(9-15)等号右边的括号中后一项 $r_w^{-m}$ 可略去不计。设:

$$C = -\frac{a\phi^{\frac{m}{2}}}{2^{m+1}mK(\pi h)^{\frac{m}{2}+1}} \qquad (9-16)$$

Then the following expressions can be obtained:

从而得到:

$$\Delta p = CQ_T^{-\frac{m}{2}}Q^{m+1} \qquad (9-17)$$

$$\lg\Delta p = \lg C - \frac{m}{2}\lg Q_T + (m+1)\lg Q \qquad (9-18)$$

Or

或

$$\lg\frac{\Delta p}{Q} = \lg C + \frac{m}{2}\lg\frac{Q^2}{Q_T} \qquad (9-19)$$

When $Q_T$ = constant, draw the relation curve between $Q$ and $\Delta p$ on the log-log coordinate, and a straight line with slope $m+1$ will be obtained. If the relation curve of $\lg\frac{\Delta p}{Q}$ with $\lg\frac{Q^2}{Q_T}$ is drawn, the value of $\lg C$ and $m$ will be obtained.

在 $Q_T$ 为常数时,在双对数坐标上绘制 $Q$ 与 $\Delta p$ 的关系曲线,得到斜率为 $m+1$ 的直线。若绘制 $\lg\frac{\Delta p}{Q}$ 与 $\lg\frac{Q^2}{Q_T}$ 的关系曲线,可求得 $\lg C$ 及 $m$ 值。

### 9.2.2 The method of Ikoku and Ramey

The method of Van Poollen and Jargon cannot be used to analyze the data of unsteady well testing. The analysis and explanation of the unsteady state test data have become a problem which needs to be solved urgently since the non-Newtonian liquid has been used to displace oil.

### 9.2.2 Ikoku 和 Ramey 法

Van Poollen 和 Jargon 法不能用于分析不稳定试井资料。自从采用注入非牛顿液体驱油以来,对这类井的不稳定测试资料的分析、解释成为急需解决的问题。

It is known that the flow regime of pseudo-plastics liquid, such as the dilute polymer solution, microemulsion and macroemulsion, are shown in figure 9 – 3. When the shear rate is extremely low, the apparent viscosity is close to the maximum. When the shear rate is extremely high, the apparent viscosity is close to the minimum. Within the general region, with the increase of shear rate, the apparent viscosity will decrease. Within this interval, the relationship of apparent viscosity $\mu$ and shear rate $\gamma$ is:

$$\mu_a = \frac{\tau}{\gamma} = H\gamma^n \cdot \frac{1}{\gamma} = H\gamma^{n-1} \quad (0 < n < 1) \quad (9-20)$$

It is known that when the Newtonian liquid passes through the artificial core and it is laminar flow, the semi-empirical relational expression (the Blake-Kozeny equation) of seepage velocity with the parameters is:

$$v = \frac{D_p^2 \phi^3 \Delta p}{150\mu(1-\phi)^2 L} \quad (9-21)$$

Where: $v$ —The seepage velocity that is the volume speed passing through the unit cross-section, m/s;
$L$ —The length of core, m;
$D_p$ —The diameter of core particles, m;
$\frac{\Delta p}{L}$ —The pressure drop per unit length, Pa/m;
$\mu$ —The Newton viscosity, Pa · s;
$\phi$ —The porosity, %.

Therefore, the permeability of core is:

$$K = \frac{D_p^2 \phi^3}{150(1-\phi)^2} \quad (9-22)$$

Compared with the Blake-Kozeny equation, the one-dimensional motion equation of power law liquid passing through the porous media is:

$$v = \frac{n\phi}{3n+1} \left[ \frac{D_p \phi}{3(1-\phi)} \right]^{\frac{n+1}{n}} \left( \frac{6\Delta p}{25HL} \right)^{\frac{1}{n}} \quad (9-23)$$

Where: $H$ —The consistency coefficient, which is a viscosity characteristic parameter of non-Newtonian liquid;
$n$ —The flow regime index.

When $n = 1$, $H = \mu =$ constant, formula (9 – 23) is the same as formula (9 – 21). Put formula (9 – 23) into Darcy's formula, and we get:

$$v = \left(\frac{K}{\mu_{\text{eff}}} \frac{\Delta p}{L}\right)^{\frac{1}{n}} \quad (9-24)$$

$$\mu_{\text{eff}} = \frac{H}{12}\left(9 + \frac{3}{n}\right)^n (150K\phi)^{\frac{1-n}{2}} \quad (9-25)$$

When $n = 1$, $\mu_{\text{eff}} = H =$ constant, $\mu_{\text{eff}}$ is called the effective viscosity, which is a parameter relevant with the non-Newtonian liquid rheological parameter $(H, n)$ and the properties $(K, \phi)$ of rock. Therefore, compared with the Darcy's law, the motion equation of radial (injection) flow of non – Newtonian power-law liquid is:

$$v_r^n = \frac{-K_r}{\mu_{\text{eff}}} \frac{\partial p}{\partial r} \quad (9-26)$$

Where: $K_r$ — The radial permeability;
$v_r$ — The radial seepage velocity.

Under the following conditions, the seepage equation of power law liquid can be derived:

(1) The reservoir is homogeneous and isopachous, and the thickness of formation and the permeability are all constants.

(2) The compressibility coefficient of system (matrix and liquid) is very small and is a constant.

(3) The influence of gravity is neglected.

(4) Liquid is pseudo plastics, that is, $0 < n < 1$.

The continuity equation of liquid flow passing through the porous media is:

$$\frac{1}{r}\frac{\partial}{\partial r}(r\rho v_r) = -\frac{\partial}{\partial t}(\rho\phi) \quad (9-27)$$

Where: $\rho$ — The density of liquid, kg/m³.

The following expression can be obtained after replacing the formula (9 – 26) into the formula (9 – 27):

$$\frac{1}{r}\frac{\partial}{\partial r}\left[r\rho\left(-\frac{K_r}{\mu_{\text{eff}}}\frac{\partial p}{\partial r}\right)^{\frac{1}{n}}\right] = -\frac{\partial}{\partial t}(\rho\phi) \quad (9-28)$$

As to liquid under the isothermal condition, it is known:

$$C_L = -\frac{1}{\overline{V}}\left(\frac{\partial \overline{V}}{\partial p}\right) = \frac{1}{\rho}\left(\frac{\partial \rho}{\partial p}\right) \quad (9-29)$$

$$\rho = \rho_0 e^{C_L(p-p_0)} \qquad (9-30)$$

Where: $C_L$—The compressibility coefficient of non-Newtonian liquid, $MPa^{-1}$;

$\overline{V}$—The volume of non-Newtonian liquid, $m^3$;

$\rho_0$—The density of liquid when pressure is $p_0$, $kg/m^3$.

Thus, the following expressions can be derived:

$$\begin{cases} \dfrac{\partial \rho}{\partial t} = C_L \rho \dfrac{\partial p}{\partial t} \\ \dfrac{\partial \rho}{\partial r} = C_L \rho \dfrac{\partial p}{\partial r} \end{cases} \qquad (9-31)$$

The following expression can be obtained after expending the formula $(9-28)$:

$$\frac{\rho}{r}\frac{\partial}{\partial r}\left[r\left(-\frac{K_r}{\mu_{eff}}\frac{\partial p}{\partial r}\right)^{\frac{1}{n}}\right] + r\left(-\frac{K_r}{\mu_{eff}}\frac{\partial p}{\partial r}\right)^{\frac{1}{n}}\frac{\partial \rho}{\partial r} = -\phi\frac{\partial \rho}{\partial t} - \rho\frac{\partial \phi}{\partial t} \qquad (9-32)$$

And the following expressions can be obtained after substituting the formula $(9-31)$ into the formula $(9-32)$:

$$\frac{\partial^2 p}{\partial r^2} + \frac{n}{r}\frac{\partial p}{\partial r} + C_L n\left(-\frac{\partial p}{\partial r}\right)^2 = C_t \phi n\left(\frac{\mu_{eff}}{K_r}\right)^{\frac{1}{n}}\left(-\frac{\partial p}{\partial r}\right)^{\frac{n-1}{n}}\frac{\partial p}{\partial t} \qquad (9-33)$$

$$C_t = C_L + C_f; \quad C_f = \frac{1}{\phi}\frac{\partial \phi}{\partial p}$$

When the compressibility coefficient $C_L$ of liquid is very small and the radial pressure gradient $\dfrac{\partial p}{\partial r}$ is also very small, the third item on the left side of the formula $(9-33)$ can be neglected. Then the formula can be simplified as:

$$\frac{\partial^2 p}{\partial r^2} + \frac{n}{r}\frac{\partial p}{\partial r} = C_t \phi n\left(\frac{\mu_{eff}}{K_r}\right)^{\frac{1}{n}}\left(-\frac{\partial p}{\partial r}\right)^{\frac{n-1}{n}}\frac{\partial p}{\partial t} \qquad (9-34)$$

Let $Q$ = constant, and the following expression can be obtained:

$$\left(-\frac{\partial p}{\partial r}\right)^{\frac{1}{n}} = \left(\frac{\mu_{eff}}{K_r}\right)^{\frac{1}{n}} v_r \approx \left(\frac{\mu_{eff}}{K_r}\right)^{\frac{1}{n}}\frac{Q}{2\pi h r} \qquad (9-35)$$

The formula $(9-34)$ can be simplified as:

$$\frac{\partial^2 p}{\partial r^2} + \frac{n}{r}\frac{\partial p}{\partial r} = G r^{1-n} \frac{\partial p}{\partial t} \qquad (9-36)$$

$$G = \frac{n\phi C_t \mu_{eff}}{K_r}\left(\frac{2\pi h}{Q}\right)^{1-n}$$

1. The radial steady state seepage of power law liquid

When power law liquid flow is steady state, the seepage

equation can be written as:

$$\frac{d^2 p}{dr^2} + \frac{n}{r}\frac{dp}{dr} = 0 \tag{9-37}$$

And the definite conditions are:

$$\left(-\frac{dp}{dr}\right)_{r_w} = \left(\frac{Q}{2\pi h r_w}\right)^n \frac{\mu_{eff}}{K_r} \tag{9-38}$$

$$p\bigg|_{r=r_e} = p_i$$

In the position of fluid-supply radius $r_e$, the pressure maintains the initial formation pressure $p_i$.

When the boundary conditions are the formula (9-38), the following expression can be obtained after solving the formula (9-37):

$$p - p_i = \left(\frac{Q}{2\pi h}\right)^n \frac{\mu_{eff}}{(1-n)K_r}(r_e^{1-n} - r^{1-n}) \tag{9-39}$$

In the formula above: $p$ is the pressure at any point in the formation after the power law liquid is injected from the injection well. Thus the bottom hole flowing pressure in injection well $p_j$ is:

$$p_j - p_i = \left(\frac{Q}{2\pi h}\right)^n \frac{\mu_{eff}}{(1-n)K_r}(r_e^{1-n} - r_w^{1-n}) \tag{9-40}$$

In the formula above: $\dfrac{\mu_{eff}}{K_r}$ will be determined by the unsteady state well testing.

2. The radial unsteady state seepage of power law liquid

Firstly, change the seepage equation (9-36) to be dimensionless quantity and assume:

$$p_D = \frac{p - p_i}{\left(\dfrac{Q}{2\pi h}\right)^n \dfrac{\mu_{eff} r_w^{1-n}}{K_r}} \tag{9-41}$$

$$r_D = \frac{r}{r_w}; \quad t_D = \frac{t}{G r_w^{2-n}}$$

And the following expression can be obtained:

$$\frac{\partial^2 p_D}{\partial r_D^2} + \frac{n}{r_D}\frac{\partial p_D}{\partial r_D} = r_D^{1-n}\frac{\partial p_D}{\partial t_D} \tag{9-42}$$

The initial and boundary conditions are:

$$p_D(r_D, 0) = 0; \quad p_D(\infty, t_D) = 0$$

$$\left(\frac{\partial p_D}{\partial r_D}\right)_{r_D=1} = -1, \quad t_D > 0 \quad (\text{the injection volume } Q \text{ is a constant}) \tag{9-43}$$

Doing Laplace transformation, we obtain dimensionless pressure at any point in formation at any time:

$$p_D \approx \frac{(3-n)^{\frac{2(1-n)}{3-n}} t_D^{\frac{1-n}{3-n}}}{(1-n)\Gamma\left(\frac{2}{3-n}\right)} - \frac{1}{1-n} \qquad (9-44)$$

In the formula above, $\Gamma(x)$ is the gamma function.

After simplification, the obtained bottom pressure of injection well after long time ($t_D \rightarrow \infty$) is:

$$p_{wf} - p_i = \frac{\left(\frac{Q}{2\pi h}\right)^{\frac{n+1}{3-n}} \left(\frac{\mu_{eff}}{K_r}\right)^{\frac{2}{3-n}} \left[\frac{(3-n)^2 t}{n\phi C_t}\right]^{\frac{1-n}{3-n}}}{(1-n)\Gamma\left(\frac{2}{3-n}\right)} \qquad (9-45)$$

Take the logarithm of both sides of the formula (9-45), and the following expression can be obtained:

$$\lg(p_{wf} - p_i) = \left(\frac{1-n}{3-n}\right)\lg t + \lg \frac{\left(\frac{Q}{2\pi h}\right)^{\frac{n+1}{3-n}} \left(\frac{\mu_{eff}}{K_r}\right)^{\frac{2}{3-n}} \left[\frac{(3-n)^2}{n\phi C_t}\right]^{\frac{1-n}{3-n}}}{(1-n)\Gamma\left(\frac{2}{3-n}\right)} \qquad (9-46)$$

Draw the relation curve of $\lg(p_{wf} - p_i)$ against $\lg t$, and the slope of straight line section is $\frac{1-n}{3-n}$. Thus the value of $n$ can be obtained.

When $t = 1$, the corresponding Y-coordinate is $\Delta p_1$, if $\phi C_t$ is known, the effective mobility $\lambda_{eff}$ is as follows:

$$\lambda_{eff} = \frac{K_r}{\mu_{eff}} = \frac{\left(\frac{Q}{2\pi h}\right)^{\frac{n+1}{2}} \left[\frac{(3-n)^2}{n\phi C_t}\right]^{\frac{1-n}{2}}}{\left[\Delta p_1 (1-n)\Gamma\left(\frac{2}{3-n}\right)\right]^{\frac{3-n}{2}}} \qquad (9-47)$$

### 9.2.3 The method of Odeh and Yang

The processing method proposed by Ikoku etc. concerning the relationship between seepage velocity and the pressure gradient is exponential when the power law liquid flows. While Odeh and others proposed the seepage equation of power law liquid according to the Darcy's law.

It is known that the seepage equation of single-phase compressible liquid, when it makes radial fluid seepage, is:

$$\frac{1}{r}\frac{\partial}{\partial r}\left(\frac{r}{\mu}\frac{\partial p}{\partial r}\right) = \frac{\phi C_t}{K}\frac{\partial p}{\partial t} \qquad (9-48)$$

As to the non-Newtonian liquid, $\mu$ is a function of shearing rate. When the shearing rate is very high or very low, the viscosity is a constant. Generally, the relationship between viscosity and shear rate can be written as:

$$\mu(\gamma)^{1/N} = \mu_a \qquad (1 < N < \infty) \qquad (9-49)$$

In the formula above: $\mu_a$ is the apparent viscosity of non-Newtonian liquid. After comparing the formula (9-49) and the formula (9-20), $\frac{1}{N} = 1 - n$ can be obtained.

The formula (9-48) does not suit for the seepage of formation around the injection wells and the production wells, because around the well bottom exists very high shear rate. The following expression can be obtained after substituting the formula (9-49) into the formula (9-48):

$$\frac{1}{r}\frac{\partial}{\partial r}\left[\frac{r(\gamma)^{1/N}}{\mu_a}\frac{\partial p}{\partial r}\right] = \frac{\phi C_t}{K}\frac{\partial p}{\partial t} \qquad (9-50)$$

The shear rate $\gamma$ was derived by Savins on the basis of theoretical analysis and experience and the expression is as follows:

$$\gamma = \frac{1500Q}{hr(K\phi)^{1/2}} \qquad (9-51)$$

Where: $Q$—The flow rate, cm$^3$/s;
$K$—The permeability, $\mu$m$^2$;
$h$—The thickness of formation, cm;
$r$—The radius, cm.

Change the formula (9-50) to be the following form:

$$\frac{\partial^2 p}{\partial r^2} + \left(1 - \frac{1}{N}\right)\frac{1}{r}\frac{\partial p}{\partial r} = r^{1/N}B\frac{\partial p}{\partial t} \qquad (9-52)$$

$$B = \frac{\mu_a \phi^{1+\frac{1}{2N}} C_t h^{2/N}}{K^{1-\frac{1}{2N}}(1500Q)^{1/N}} \qquad (9-53)$$

As to the Newtonian liquid, $N \to \infty$, and $\mu = \mu_a$ = constant. The formula (9-52) can be changed into the following form:

$$\frac{\partial^2 p}{\partial r^2} + \frac{1}{r}\frac{\partial p}{\partial r} = \frac{\mu \phi C_t}{K}\frac{\partial p}{\partial t}$$

The law of pressure distribution when the injection well is injected with constant $Q$ will be analyzed next, the initial and boundary conditions are:

界条件为:

$$p(r,0) = p_i; \quad p(\infty,t) = p_i$$

$$\left.\frac{\partial p}{\partial r}\right|_{r=r_w} = -\frac{Q\mu}{2\pi K h r_w} \quad (9-54)$$

In infinite formation, in the first phase of pressure transmission, the bottom pressure of injection well $p_j$ will be obtained by the Laplace transformation:

在无限大地层中,在压力传导第一阶段,应用拉普拉斯变换,得到注入井井底压力 $p_j$ 为:

$$p_j - p_i = \frac{Q\mu_{rw}}{2\pi K h}\left[\frac{N^{\frac{2N-1}{2N+1}}(2N+1)^{\frac{2}{2N+1}}t^{\frac{1}{2N+1}}}{\Gamma\left(\frac{2N}{2N+1}\right)r_w^{\frac{1}{N}}B^{\frac{1}{2N+1}}} - N\right] \quad (9-55)$$

Where: $\mu_{rw}$—The viscosity of liquid on the position of well radius $r_w$.

式中: $\mu_{rw}$ = 井半径 $r_w$ 处的液体黏度, mPa·s。

Draw the relation curve between viscosity and shear rate in the logarithmic coordinate, from which the value of $N$ can be determined. Draw the relation curve between $p_w - p_i$ and $t^{\frac{1}{2N+1}}$. If the relation curve is a straight line, the value of $N$ obtained is right; if it is not a straight line, the value of $N$ needs to be solved again with trial method, until the relation curve is a straight line. The permeability $K$ can be obtained with the slope of the straight line.

在对数坐标中绘制黏度与剪切速率的关系曲线,确定 $N$ 值。在直角坐标系中绘制 $p_w - p_i$ 与 $t^{\frac{1}{2N+1}}$ 的关系曲线,如为直线说明所求 $N$ 值是正确的,如不为直线,需用试算法重求 $N$ 值,直到 $p_w - p_i$ 与 $t^{\frac{1}{2N+1}}$ 成直线时为止。由直线斜率可算出渗透率 $K$。

## Exercises

## 习 题

9-1  Assume the constitutive equation of non-Newtonian fluid can be written as:

9-1  设非牛顿流体的本构方程可写成:

$$\tau = \frac{1}{B}\text{sh}^{-1}\left(-\frac{1}{A}\frac{dv}{dr}\right)$$

Please try to derive its seepage differential equation with the non-uniform capillary model.

试用不均匀毛管组模型导出其渗流微分方程。

9-2  If the constitutive equation of non-Newtonian fluid can be written as:

9-2  设非牛顿流体的本构方程可写成:

$$\tau = \frac{1}{\varphi_0 + \varphi_1 \tau^{\alpha-1}}\gamma$$

Please try to derive its seepage differential equation.

试导出其渗流微分方程。

9-3  Please try to derive the seepage differential equation of Reiner-Philipoff non-Newtonian fluid.

9-3  试对 Reiner-Philipoff 型非牛顿流体导出其渗流微分方程。

9-4 Assume the constitutive equation of a kind of non-Newtonian fluid can be written as:

$$\gamma = \tau \left\{ \varphi_\infty - (\varphi_\infty - \varphi_0) \exp\left[ \frac{\tau^2 (d\varphi/d\tau)^2}{\varphi_\infty - \varphi_0} \right] \right\}$$

Please try to derive its seepage differential equation.

# References/参考文献

[1] 贝尔. 多孔介质流体动力学. 李竞生,陈崇希,译. 北京:中国建筑工业出版社,1983.
[2] 恰尔内. 地下水—气动力学. 陈钟祥,郎兆新,译. 北京:石油工业出版社,1982.
[3] 科林斯. 流体通过多孔材料的流动. 陈钟祥,译. 北京:石油工业出版社,1984.
[4] 刘尉宁. 渗流力学基础. 北京:石油工业出版社,1985.
[5] 葛家理. 油气层渗流力学. 北京:石油工业出版社,1982.
[6] 李璗,陈军斌. 油气渗流力学. 北京:石油工业出版社,2009.
[7] 李璗,陈军斌. 油气渗流力学(英文版). 北京:石油工业出版社,2012.
[8] 陈军斌,黄海. 油气渗流力学学习指南. 北京:石油工业出版社,2010.

# Appendix

附录

# Appendix A  Seepage equation in cylindrical coordinate/柱坐标形式的渗流方程

In section 6 of chapter 2, the continuity equation of single-phase fluid flow and the basic differential equation of seepage of single-phase liquid in rectangular coordinate have been derived. However, in this course, the continuity equation and the basic differential equation in cylindrical coordinate is used more widely and conveniently. Therefore, it is of great necessity to master the equations in cylindrical coordinate.

1. The continuity equation

At the arbitrary point $M$ in oil formation, take an element with the point $M$ as the center, at which the radial length is $dr$, the vertical height is $dz$ and the circumferential angle is $d\theta$ as shown in figure A-1. Assume at the point $M$ the density of fluid is $\rho$ and the seepage velocity is $v = (v_r, v_\theta, v_z)$. There is no harm to assume the direction of velocity component is consistent with coordinates, and then the mass difference of fluid flowing in and flowing out of the element along the direction of $r$ in time $dt$ is:

在第二章第六节中已经导出了直角坐标形式的单相流体渗流的连续性方程以及单相液体渗流的基本微分方程。然而，在渗流力学这门课程中柱坐标形式的连续性方程和基本微分方程的使用更为方便和普遍，因此，有必要掌握柱坐标形式的连续性方程和基本微分方程。

1. 连续性方程

在油层中任一点 $M$ 处，以点 $M$ 为中心取一个径向长度为 $dr$、垂向高度为 $dz$、周向角度为 $d\theta$ 的单元体，如图 A-1 所示。设 $M$ 点流体的密度为 $\rho$，渗流速度为 $v = (v_r, v_\theta, v_z)$。不妨设渗流分速度的方向与坐标方向一致，则 $dt$ 时间内沿 $r$ 方向流入和流出单元体的流体质量之差为：

$$\left(\rho v_r - \frac{1}{2}\frac{\partial(\rho v_r)}{\partial r}dr\right)\left(r - \frac{1}{2}dr\right)d\theta dzdt - \left(\rho v_r + \frac{1}{2}\frac{\partial(\rho v_r)}{\partial r}dr\right)\left(r + \frac{1}{2}dr\right)d\theta dzdt$$
$$= -\left(\rho v_r + r\frac{\partial(\rho v_r)}{\partial r}\right)drd\theta dzdt$$

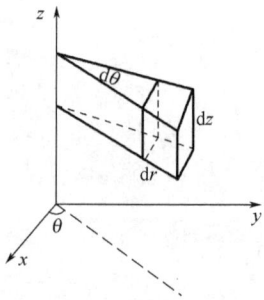

Figure A-1  The infinitesimal element in cylindrical coordinate
图 A-1  微元示意图

The mass difference of fluid flowing in and flowing out of the element along the direction of $\theta$ in time $dt$ is:

$dt$ 时间内沿 $\theta$ 方向流入和流出单元体的流体质量之差为：

$$\left[\rho v_\theta - \frac{1}{2}\frac{\partial(\rho v_\theta)}{\partial \theta}d\theta\right]drdzdt - \left[\rho v_\theta + \frac{1}{2}\frac{\partial(\rho v_\theta)}{\partial \theta}d\theta\right]drdzdt = -\frac{\partial(\rho v_\theta)}{\partial \theta}drd\theta dzdt$$

While the mass difference of fluid flowing in and flowing out of the element along the direction of $z$ in time $dt$ is:

$$\left[\rho v_z - \frac{1}{2}\frac{\partial(\rho v_z)}{\partial z}dz\right]rd\theta drdt - \left[\rho v_z + \frac{1}{2}\frac{\partial(\rho v_z)}{\partial z}dz\right]rd\theta drdt = -\frac{\partial(\rho v_z)}{\partial z}rd\theta drdzdt$$

So the mass difference of fluid flowing in and flowing out of the element along each direction in time $dt$ is:

$$-\left[\rho v_r + r\frac{\partial(\rho v_r)}{\partial r} + \frac{\partial(\rho v_\theta)}{\partial \theta} + r\frac{\partial(\rho v_z)}{\partial z}\right]drd\theta dzdt$$

While the mass of fluid in the element is:

$$\rho\phi\left[\left(r+\frac{dr}{2}\right)^2\frac{d\theta}{2} - \left(r-\frac{dr}{2}\right)^2\frac{d\theta}{2}\right]dz = \rho\phi rdrd\theta dz$$

So in time $dt$ the increasing mass of fluid in the element is:

$$\frac{\partial(\rho\phi)}{\partial t}rdrd\theta dzdt$$

According to the law of conservation of mass, the mass difference of fluid flowing in and flowing out of the element in time $dt$ should be equal to the increasing mass of fluid in the element at the same time, so the following expression can be obtained:

$$\frac{1}{r}\rho v_r + \frac{\partial(\rho v_r)}{\partial r} + \frac{1}{r}\frac{\partial(\rho v_\theta)}{\partial \theta} + \frac{\partial(\rho v_z)}{\partial z} = -\frac{\partial(\rho\phi)}{\partial t} \quad (A-1)$$

The expression above is the continuity equation of the single-phase fluid flow in cylindrical coordinate. Supposing the formation is homogeneous, isopachous and isotropic, fluid makes radial fluid seepage. Then the expression (A – 1) can be simplified as:

$$\frac{1}{r}\rho v_r + \frac{\partial(\rho v_r)}{\partial r} = -\frac{\partial(\rho\phi)}{\partial t} \quad (A-2)$$

When the compressibility of matrix and fluid is not taken into account, the expression above can be further simplified as:

$$\frac{1}{r}\rho v_r + \frac{\partial(\rho v_r)}{\partial r} = 0$$

2. The basic differential equation of single-phase liquid flow

For the seepage of single – phase liquid, the Darcy's law is:

$$\begin{cases} v_r = -\dfrac{K}{\mu}\dfrac{\partial p}{\partial r} \\ v_\theta = -\dfrac{K}{\mu}\dfrac{\partial p}{\partial s} = -\dfrac{K}{\mu r}\dfrac{\partial p}{\partial \theta} \\ v_z = -\dfrac{K}{\mu}\dfrac{\partial p}{\partial z} \end{cases} \quad (A-3)$$

The equation of state of fluid and matrix respectively are:

$$\rho = \rho_o e^{C_1(p-p_o)} \quad (A-4)$$

$$\phi = \phi_o e^{C_f(p-p_o)} \quad (A-5)$$

After substituting equations (A-3) through (A-5) are substituted into the equation (A-1), some items on the left side of the equation (A-1) are changed to be:

$$-\frac{\partial(\rho v_r)}{\partial r} = \frac{\partial}{\partial r}\left[\rho_o \frac{K}{\mu} e^{C_1(p-p_o)} \frac{\partial p}{\partial r}\right] = \rho_o e^{C_1(p-p_o)} \frac{K}{\mu}\left(\frac{\partial p}{\partial r}\right)^2 + \rho \frac{\partial}{\partial r}\left(\frac{K}{\mu}\frac{\partial p}{\partial r}\right)$$

$$= \rho C_1 \frac{K}{\mu}\left(\frac{\partial p}{\partial r}\right)^2 + \rho \frac{\partial}{\partial r}\left(\frac{K}{\mu}\frac{\partial p}{\partial r}\right) \quad (A-6)$$

Likewise, the following expressions can be obtained:

$$-\frac{\partial(\rho v_\theta)}{\partial \theta} = \rho C_1 \frac{K}{\mu}\left(\frac{\partial p}{\partial \theta}\right)^2 + \frac{\rho}{r}\frac{\partial}{\partial \theta}\left(\frac{K}{\mu}\frac{\partial p}{\partial \theta}\right) \quad (A-7)$$

$$-\frac{\partial(\rho v_z)}{\partial z} = \rho C_1 \frac{K}{\mu}\left(\frac{\partial p}{\partial z}\right)^2 + \rho \frac{\partial}{\partial z}\left(\frac{K}{\mu}\frac{\partial p}{\partial z}\right) \quad (A-8)$$

While the item on the right side of the equation (A-1) is:

$$\frac{\partial(\rho\phi)}{\partial t} = \rho\phi C_1 \frac{\partial p}{\partial t} \quad (A-9)$$

After substitute the equation (A-5) to the equation (A-8) into the equation (A-1) and make $C_t = C_f + C_L$, then the following expression can be obtained:

$$\frac{K}{\mu r}\frac{\partial p}{\partial r} + \frac{KC_1}{\mu}\left[\left(\frac{\partial p}{\partial r}\right)^2 + \left(\frac{\partial p}{\partial \theta}\right)^2 + \left(\frac{\partial p}{\partial z}\right)^2\right]$$
$$+ \left[\frac{\partial}{\partial r}\left(\frac{K}{\mu}\frac{\partial p}{\partial r}\right) + \frac{1}{r^2}\frac{\partial}{\partial \theta}\left(\frac{K}{\mu}\frac{\partial p}{\partial \theta}\right) + \frac{\partial}{\partial z}\left(\frac{K}{\mu}\frac{\partial p}{\partial z}\right)\right] = \phi C_t \frac{\partial p}{\partial t} \quad (A-10)$$

The equation above is the basic differential equation of single-phase liquid flow in cylindrical coordinate. If the pressure gradient is very small in the seepage course, then the result will be much smaller after its square multiplies $C_t$.

Thus the equation (A-10) can be simplified as:

$$\frac{K}{\mu}\frac{\partial p}{\partial r} + \frac{\partial}{\partial r}\left(\frac{K}{\mu}\frac{\partial p}{\partial r}\right) + \frac{1}{r^2}\frac{\partial}{\partial \theta}\left(\frac{K}{\mu}\frac{\partial p}{\partial \theta}\right) + \frac{\partial}{\partial z}\left(\frac{K}{\mu}\frac{\partial p}{\partial z}\right) = \phi C_t \frac{\partial p}{\partial t} \quad (A-11)$$

If the formation and fluid is homogeneous, then the expression above can be simplified as:

若地层和流体是均质的,则上式可以简化成:

$$\frac{1}{r}\frac{\partial p}{\partial r} + \frac{\partial^2 p}{\partial r^2} + \frac{1}{r^2}\frac{\partial^2 p}{\partial \theta^2} + \frac{\partial^2 p}{\partial z^2} = \frac{1}{\eta}\frac{\partial p}{\partial t} \quad (A-12)$$

In the formula above, $\eta = \frac{K}{\phi\mu C_t}$ is called diffusivity coefficient. When fluid makes radial seepage, the following expression can be obtained:

式中,$\eta = \frac{K}{\phi\mu C_t}$ 称为导压系数。当流体作平面径向渗流时则有:

$$\frac{1}{r}\frac{\partial p}{\partial r} + \frac{\partial^2 p}{\partial r^2} = \frac{1}{\eta}\frac{\partial p}{\partial t} \quad (A-13)$$

If fluid makes radial steady flow, the following expression can be obtained:

如果液体作平面径向稳定渗流,则:

$$\frac{1}{r}\frac{dp}{dr} + \frac{d^2 p}{dr^2} = 0 \quad (A-14)$$

These basic differential equations are the foundation of researching the law of liquid seepage.

这些基本微分方程是研究液体渗流规律的基础。

## Appendix B  To obtain the solution of one-dimensional unsteady seepage by the Laplace transformation/用拉氏变换求液体作平面一维不稳定渗流的解

As introduced in the section 2 of chapter 4, assume liquid makes planar one-dimensional unsteady state flow in the semi-infinite reservoir (Figure 4-5), and the bottom pressure is constant, and then the seepage problem can be described with the following mathematical definite problem:

如第四章第二节所述,设液体在半无限大带状油气藏(图4-5)中作平面一维不稳定渗流,井底压力恒定,则该渗流问题可用以下数学定解问题来描述:

$$\begin{cases} \dfrac{\partial^2 p}{\partial^2 x} = \dfrac{1}{\eta}\dfrac{\partial p}{\partial t} \\ p(x,0) = p_i \\ p(\infty,t) = p_i \\ p(0,t) = p_w \end{cases} \quad (B-1)$$

In order to solve this problem by the Laplace transformation, make:

为了便于用拉氏变换求解这一问题,令:

$$U = p - p_i$$

Then the definite conditions (B-1) can be written as:

$$\begin{cases} \dfrac{\partial^2 U}{\partial x^2} = \dfrac{1}{\eta}\dfrac{\partial U}{\partial t} \\ U(x,0) = 0 \\ U(\infty,t) = 0 \\ p(0,t) = p_w - p_i \end{cases} \quad (B-2)$$

Make the Laplace transformation to the definite problem (B-2) and assume $L[U] = \overline{U}$, and then the basic differential equation is changed to be:

$$\dfrac{d^2 \overline{U}}{dx^2} - \dfrac{S}{\eta}\overline{U} = 0 \quad (B-3)$$

In the course of transformation of the expression above, initial condition is applied. Apparently, the solution of this ordinary differential equation is:

$$\overline{U} = C_1 e^{\sqrt{\frac{S}{\eta}}x} + C_2 e^{-\sqrt{\frac{S}{\eta}}x} \quad (B-4)$$

And the corresponding inner and outer boundary conditions are changed to be the following expressions after the Laplace transformation:

$$\overline{U}(0,t) = (p_w - p_i)/S \quad (B-5)$$
$$\overline{U}(\infty,t) = 0 \quad (B-6)$$

After the formula (B-5)、the formula (B-6) are substituted into the formula (B-4), the following expressions can be obtained:

$$C_1 = 0; \; C_2 = \dfrac{p_w - p_i}{S}$$

So:

$$\overline{U} = \dfrac{p_w - p_i}{S} e^{-\sqrt{\frac{S}{\eta}}x} \quad (B-7)$$

After making the reverse transformation of Laplace transformation, the following expressions can be obtained:

$$U = L^{-1}[\overline{U}] = (p_w - p_i)\,\mathrm{erfc}\left(\dfrac{x}{2\sqrt{\eta t}}\right)$$

That is:

$$p(x,t) - p_i = (p_w - p_i)\left[1 - \mathrm{erfc}\left(\dfrac{x}{2\sqrt{\eta t}}\right)\right] \quad (B-8)$$

# Appendix C  Calculation procedure of error function erf(x)/误差函数 erf(x)计算程序

$$\mathrm{erf}(x) = \frac{2}{\sqrt{\pi}} \int_0^x \mathrm{e}^{-t^2} \mathrm{d}t$$

## The program of FUNCTION Language /FUNCTION 语言程序

```
      C     ERF. F77
              IMPLICIT   REAL*8(A—H,O—Z)
              OPEN(6, FILE = 'ERFX')
              WRITE(6, *)'     X       ERF(X)'
              C = 2.00/SQRT(4.00*ATAN(1.00))
              DO  30   R = 1.0 , 45.0
              X = R/1. D1
              EF = X
              EFS = X
              DO  10   Q = 0.0 ,100.0
              EF = -(2.*Q+1)*X*X*EF/(Q+1)/(2.*Q+3)
              EFS = EFS + EF
              IF(DABS(EF). LT. 1. D - 8) GO TO 20
      10      CONTINUE
      20      ERF = C*EFS
              WRITE(6,25)   X, ERF
      25      FORMAT(2X,F7.2,5X,F10.6)
      30      CONTINUE
              CLOSE(6)
              END
```

## The program of C++ language /C 语言程序

```c
#include <iostream. h>
#include <math. h>
#include <iomanip. h>
#include <stdlib. h>
#include <fstream. h>
#include <string>
#include <process. h>

double gammln(double xx)
{
    int j;
```

```
    float temp;
    double cof[6],stp,half,one,fpf,x,tmp,ser;
    cof[1] = 76.18009173;
    cof[2] = -86.50532033;
    cof[3] = 24.01409822;
    cof[4] = -1.231739516;
    cof[5] = 0.00120858003;
    cof[6] = -0.00000536382;
    stp = 2.50662827465;
    half = 0.5;
    one = 1.0;
    fpf = 5.5;
    x = xx - one;
    tmp = x + fpf;
    tmp = (x + half) * log(tmp) - tmp;
    ser = one;
    for (j = 1;j< =6;j++)
    {
        x = x + one;
        ser = ser + cof[j] / x;
    }
    temp = tmp + log(stp * ser);
    return temp;
}

void gser(double& gamser, double& a, double& x, double& gln)
{
    int itmax,n;
    double ap,sum,del,eps;
    itmax = 100;
    eps = 0.0000003;
    gln = gammln(a);
    if (x < = 0.0)
    {
        if (x < 0.0)
        {
            cout < <" pause";
            _c_exit( );
        }
        gamser = 0.0;
```

```
        _c_exit( );
    }
    ap = a;
    sum = 1.0 / a;
    del = sum;
    for( n = 1;n < = itmax;n + + )
    {
        ap = ap + 1.0;
        del = del * x / ap;
        sum = sum + del;
        if (fabs(del) < (fabs(sum) * eps))
        {//goto loop;
        gamser = sum * exp( -x + a * log(x) - gln);
        break;
        }
    }
}

void gcf( double& gammcf, double& a, double& x, double& gln)
{
    int itmax,n;
    double eps,a0,a1,b0,b1,fac,an,ana,anf,gold,g;
    itmax = 100;
    eps = 0.0000003;
    gln = gammln(a);
    gold = 0.0;
    a0 = 1.0;
    a1 = x;
    b0 = 0.0;
    b1 = 1.0;
    fac = 1.0;
    for (n = 1;n < = itmax;n + + )
    {
        an = n;
        ana = an - a;
        a0 = (a1 + a0 * ana) * fac;
        b0 = (b1 + b0 * ana) * fac;
        anf = an * fac;
        a1 = x * a0 + anf * a1;
        b1 = x * b0 + anf * b1;
```

```cpp
            if (a1 != 0.0)
            {
                fac = 1.0 / a1;
                g = b1 * fac;
                if (fabs((g - gold) / g) < eps)
                {
                    gammcf = exp(-x + a * log(x) - gln) * g;
                    break;
                }
                //goto yi;
                gold = g;
            }
        }
}

double gammq(double a, double x)
{
    double temp, gamser, gammcf, gln;
    if (x < 0.0 || a <= 0.0)
    {
        cout << "pause";
        exit(1);
    }
    if (x < a + 1.0)
    {
        gser(gamser, a, x, gln);
        temp = 1.0 - gamser;
    }
    else
    {
        gcf(gammcf, a, x, gln);
        temp = gammcf;
    }
    return temp;
}

double gammp(double a, double x)
{
    double temp, gamser, gln, gammcf;
    if (x < 0.0 || a <= 0.0)
```

```cpp
    {
        cout << "pause";
        exit(1);
    }
    if (x < a + 1.0)
    {
        gser(gamser, a, x, gln);
        temp = gamser;
    }
    else
    {
        gcf(gammcf, a, x, gln);
        temp = 1.0 - gammcf;
    }
    return temp;
}

double erfc(double x)
{
    double temp;
    if (x < 0.0)
    {
        temp = 1.0 + gammp(0.5, x*x);
    }
    else
    {
        temp = gammq(0.5, x*x);
    }
    return temp;
}

double erf(double x)
{
    double temp;
    if (x < 0.0)
    {
        temp = -gammp(0.5, x*x);
    }
    else
    {
```

```
            temp = gammp(0.5, x * x);
    }
    return temp;
}

void main( )
{
    double nval, actual, x;
    const double pi = 3.1415926;
    cout << "Error Function " << endl;
    cout << endl;
    cout << " x          erf(x)" << endl;
    cin >> x;
    cout << erf(x) << endl;
}
```

## Appendix D  Calculation procedure of exponential integral function -Ei(-x)/指数积分函数 –Ei( – x)计算程序

$$\mathrm{Ei}(x) = -\mathrm{Ei}(-x) = \int_{x}^{\infty} \frac{e^{-u}}{u} du \quad (x > 0)$$

### The program of C++ language/ C 语言程序

```
#include <iostream.h>
#include <math.h>
#include <iomanip.h>
#include <stdlib.h>
#include <fstream.h>
#include <string>
#include <process.h>
double ex(double x)
{
    double r1, r2, r3, r4, r5, r6, t1, t2, t3, t4, s1, s2, s3, s4, temp, aaa, y, w;
    r1 = 0.00107857;
    r2 = -0.00976004;
    r3 = 0.05519968;
    r4 = -0.24991055;
    r5 = 0.99999193;
    r6 = -0.57721566;
    t1 = 8.5733287401;
```

```
    t2 = 18.059016973;
    t3 = 8.6347608925;
    t4 = 0.2677737343;
    s1 = 9.5733223454;
    s2 = 25.6329561486;
    s3 = 21.0996530827;
    s4 = 3.9584969228;
    if (x < 1.0)
    {
        aaa = (r1 * x + r2) * x + r3;
        temp = ((aaa * x + r4) * x + r5) * x + r6 - log(x);
    }
    else
    {
        y = (((x + t1) * x + t2) * x + t3) * x + t4;
        w = (((x + s1) * x + s2) * x + s3) * x + s4;
        temp = (y / w) / (exp(x) * x);
    }
    return temp;
}

void main()
{
    double nval, value, x;
    const double pi = 3.1415926;
    cout << "Exponential integral Ex " << endl;
    cout << endl;
    cout << " x         Ex(x)" << endl;
    cin >> x;
    cout << ex(x) << endl;
```

## Appendix E  Calculation procedure of function Γ(x)/Γ(x)函数计算程序

$$\Gamma(x) = \int_0^\infty t^{x-1} e^{-t} dt$$

### The program of C++ language/C 语言程序

ln Γ(x)

```
#include <iostream.h>
#include <math.h>
#include <iomanip.h>
```

```
#include <stdlib.h>
#include <fstream.h>
#include <string>

    double gammln(double xx)
    {
    int j;
    float temp;
    double cof[6],stp,half,one,fpf,x,tmp,ser;
    cof[1] = 76.18009173;
    cof[2] = -86.50532033;
    cof[3] = 24.01409822;
    cof[4] = -1.231739516;
    cof[5] = 0.00120858003;
    cof[6] = -0.00000536382;
    stp = 2.50662827465;
    half = 0.5;
    one = 1.0;
    fpf = 5.5;
    x = xx - one;
    tmp = x + fpf;
    tmp = (x + half) * log(tmp) - tmp;
    ser = one;
    for (j = 1;j<=6;j++)
    {
        x = x + one;
        ser = ser + cof[j] / x;
    }
    temp = tmp + log(stp * ser);
    return temp;
}
void main()
{
    double calc,x;
    const double pi = 3.1415926;
    cout<<"Log of Gamma function"<<endl;
    cout<<endl;
    cout<<"    x        gamma(x)"<<endl;
        cin>>x;
    if (x > 0.0)
```

```
            {
                if ( x > = 1.0)
            {
                calc = gammln(x);
            }
            else
            {
                calc = gammln(x + 1) - log(x);
            }
            cout < < calc < < endl;
            }
}
```

## Appendix F  Conversion relationship between the commonly used units in reservoir engineering/油气藏工程常用单位之间的换算关系

1. Length/长度

$1m = 100cm = 1000mm = 3.281ft = 39.37in$

$1ft = 0.305m = 30.5cm = 3050mm = 12in$

$1km = 0.621mile; 1mile = 1.61km$

$1Å(Angstrom) = 10^{-8}cm = 10^{-4}\mu m = 10^{-10}m$

2. Area/面积

$1m^2 = 10000cm^2 = 1000000mm^2 = 10.76ft^2 = 1549in^2$

$1km^2 = 100hm^2 = 100ha = 247acres$

$1hm^2(ha) = 10000m^2 = 2.47acres$

$1sq\ mile = 1\ section = 2.59km^2 = 259hm^2 = 640acres$

$1acre = 43560ft^2 = 0.405hm^2(ha) = 4050m^2$

3. Volume/体积

$1m^3 = 1000L = 1000dm^3 = 35.32ft^3 = 6.29bbl = 264gal$

$1L = 1\ dm^3 = 0.001m^3 = 1000cm^3 = 0.035ft^3 = 61in^3 = 0.264gal$

$1ft^3 = 0.0283m^3 = 28.3L$

$1bbl = 5.615ft^3 = 0.159m^3 = 159L = 42U.S.gal = 35U.K.gal$

$1acre.ft = 1233.5m^3 = 43560ft^3 = 7758.4bbl$

$1bbl/(acre.ft) = 0.1289m^3/m^3 = 1.289m^3/(hm^2 \cdot m) = 128.9m^3/(km^2 \cdot m)$

4. Mass/质量

$1kg = 2.205\ lbm = 1000g$

$1lbm = 0.454kg = 454g$

$1t = 1000kg = 2205lbm$

$1c(Carat) = 0.2g = 200mg$

$1oz = 31.104g$

5. Density/密度

$1kg/m^3 = 0.001g/cm^3 = 0.001t/m^3 = 0.0624lb/ft^3$

$1lb/ft^3 = 16.02kg/m^3 = 0.01602g/cm^3 = 0.1334lb/gal$

$1g/cm^3 = 1000kg/m^3 = 1t/m^3 = 1kg/L = 62.4bm/ft^3 = 8.33lb/gal$

6. Force/力

$1N = 10^5 dyn = 0.102kgf = 0.225lbf$

$1kgf = 9.82N = 9.81 \times 10^5 dyn = 2.205lbf$

$1lbf = 4.45N = 0.454kgf$

7. Pressure/压力

$1MPa = 10^6 Pa = 9.8692atm = 10.2at = 145.04psi$

$1atm = 0.1013MPa = 1.033at = 14.7psi; 1at = 1kgf/cm^2$

$1psi = 0.00689MPa = 6.89kPa = 0.068atm = 0.070at$

8. Temperature/温度

$1℃ = \dfrac{5}{9}(℉ - 32); 1K = 1℃ + 273$

$1℉ = 1.8℃ + 32; 1R = 1℉ + 460$

$1K = R/1.8$

9. Viscosity/黏度

$1mPa \cdot s = 1cP (\text{The dynamic viscosity})$

$1mm^2/s = 1cSt = 1.08 \times 10^{-5} ft^2/s (\text{The kinematic viscosity})$

10. Permeability/渗透率

$1\mu m^2 = 10^{-12} m^2 = 1.01325D = 1.01325 \times 10^3 mD \approx 1D$

$1D(darcy) = 0.9869 \times 10^{-8} cm^2 = 0.9869 \times 10^{-12} m^2 = 0.9869 \mu m^2 \approx 1\mu m^2$

$1mD = 10^{-3}D = 0.9869 \times 10^{-15} m^2 = 0.9869 \times 10^{-3} \mu m^2 \approx 10^{-3} \mu m^2$

$1\mu m^2 = 1D = 1000mD = 10^{-12} m^2 = 10^{-8} cm^2$

11. The interfacial (surface) tension/表面(界面)张力

$1mN/m = 1dyn/cm$

12. Work and heat/功与热

$1kJ = 0.948Btu = 1000N \cdot m = 0.239kcal$

$1kcal = 4.19kJ = 3.97Btu$

$1Btu = 1.055kJ = 0.252kcal$

13. The electrical energy/电力

$1kW = 3600kJ/h = 860kcal/h = 3415Bu/h = 1.341HP$

$1HP = 0.746kW = 641kcal/h = 2690kJ/h = 2545Btu/h$

14. Thermal energy of oil and gas/油与气的热能当量换算关系

$1t(oil) = 1111m^3 (\text{natural gas}) = 39218ft^3 (\text{natural gas}) = 4 \times 10^7 Btu$
$= 422 \times 10^7 Joules = 6173kWh = 0.405t(LNG)$

$$= 3.86 \text{bbl(oil)} = 0.526 \text{t(oil)}$$

$1\text{m}^3\text{LNG(liquified natural gas)} = 584\text{m}^3\text{(natural gas)} = 20631\text{ft}^3\text{(natural gas)} = 2.104 \times 10^7 \text{Btu}$
$$= 221.9 \times 10^8 \text{Joules} = 6173 \text{kWh} = 0.405 \text{t(LNG)} = 3.86 \text{bbl(oil)}$$
$$= 0.526 \text{t(oil)}$$

$1\text{m}^3\text{Gas(natural gas)} = 35.31\text{ft}^3\text{(natural gas)} = 3.6 \times 10^4 \text{Btu} = 3.8 \times 10^7 \text{Joules}$
$$= 10.54 \text{kWh} = 0.00171 \text{m}^3\text{(LNG)} = 0.000725 \text{t(LNG)} = 0.0066 \text{bbl(oil)}$$
$$= 0.0009 \text{t(oil)}$$

15. Particular units/特定单位

$1\text{scf/STB} = 0.178 \text{m}^3/\text{m}^3$

$1\text{m}^3/\text{m}^3 = 5.615 \text{scf/STB}$

$1\text{bbl/lb} \cdot \text{mol} = 0.350 \text{m}^3/\text{kmol}$

$1\text{psi/ft} = 0.0226 \text{MPa/m}$

$1\text{MPa/m} = 44.25 \text{psi/ft}$